Mathematical Engineering

For further volumes:
http://www.springer.com/series/8445

Wolfgang Koch

Tracking and Sensor Data Fusion

Methodological Framework and Selected Applications

Springer

Wolfgang Koch
Fraunhofer FKIE
Wachtberg
Germany

ISSN 2192-4732 ISSN 2192-4740 (electronic)
ISBN 978-3-642-39270-2 ISBN 978-3-642-39271-9 (eBook)
DOI 10.1007/978-3-642-39271-9
Springer Heidelberg New York Dordrecht London

Library of Congress Control Number: 2013944545

Printed on acid-free paper

Springer is part of Springer Science+Business Media (www.springer.com)

To my beloved wife Dorothea who makes everthing possible

Foreword

Tracking and sensor data fusion have a long tradition in the Fraunhofer Research Institute for Communications, Information Systems, and Ergonomics (FKIE) and its predecessor FFM (FGAN Research Institute for Radio Technology and Mathematics). Established in 1963, mainly aspects of air traffic control have been the driving factors for applied research in these pioneering years. Radar digitization, distributed radar systems, fusion with background information such as flight plans or target tracking have been keywords describing the challenges at this time. Under Günther van Keuk—a young physicist from the University of Hamburg, student of Harry Lehmann and Lothar Collatz, joining FFM in 1965—these activities were related to distributed target tracking and data fusion in multiple radar networks for the German Agency of Air Traffic Security (DFS).

Over many years, active sensor management, tracking, and data fusion for the phased-array radar system ELRA (Elektronisches Radar, a dominating project over a long time) was an important focal point. Günther van Keuk was among the first, who proposed and realized a sequential track initiation scheme based on an optimal criterion related to state estimates. In this context, he developed a performance prediction model for phased-array radar, which has been called "Van-Keuk-Equation" in the tracking literature.

In summer 1990, another young physicist, Wolfgang Koch, joined van Keuk's department. Educated at the RWTH Aachen and a student of Gert Roepstorff, he began under van Keuk's mentorship to apply his fundamental theoretical knowledge to the application oriented world of sensor data fusion. He was a member of the team which has done pioneering work in multiple emitter tracking within networks of electromagnetic and acoustic sensors under the effect of hostile measures in challenging Cold-War reconnaissance scenarios.

In the following years—since 2002 as the successor of van Keuk as the department head of "Sensor Data and Information Fusion"—he contributed remarkable results to the field of sensor data fusion. He did it successfully and with passionate enthusiasm. So he became a well-known member of the world wide sensor data fusion community and the academic scene in Germany, especially at the University of Bonn. Today, the research activities at FKIE cover a wide range of topics in the area of sensor data fusion related to localization and navigation, wide-area surveillance, resource management, self protection, and threat recognition for defence and security applications.

The reader of this book will get both, a fairly comprehensive overview of the field of tracking and sensor data fusion and deeper insight in the specific scientific results, reached in the last two decades. I am very proud to have had the opportunity to follow this development and to be able to support these activities as the former director of FFM and FKIE, respectively, for almost 25 years. Enjoy reading this book as I did.

Adendorf, September 2013 Jürgen Grosche

Preface

Sensor Data Fusion is the process of combining incomplete and imperfect pieces of mutually complementary sensor information in such a way that a better understanding of an underlying real-world phenomenon is achieved. Typically, this insight is either unobtainable otherwise or a fusion result exceeds what can be produced from a single sensor output in accuracy, reliability, or cost. Appropriate collection, registration, and alignment, stochastic filtering, logical analysis, space-time integration, exploitation of redundancies, quantitative evaluation, and appropriate display are part of Sensor Data Fusion as well as the integration of related context information. The technical term "Sensor Data Fusion" was created in George Orwell's very year 1984 in the US defence domain, but the applications and scientific topics in this area have much deeper roots. Today, Sensor Data Fusion is evolving at a rapid pace and present in countless everyday systems and civilian products.

Although a vast research literature with specialized journals and conference proceedings, several handbooks, and scientific monographs deal with Sensor Data Fusion, it often seems difficult to find access to the underlying general methodology and to apply the inventory of various fusion techniques to solving individual application problems. To facilitate the transfer of notions and algorithms of Sensor Data Fusion to problem solving in engineering and information systems design is the main objective of this book. The idea of it has grown from both the author's lecturing on Sensor Data Fusion at Bonn University since 2002 and extensive research work at Fraunhofer FKIE on improving defence- and security-related surveillance and reconnaissance systems by Sensor Data Fusion. The inner structure of the book directly follows from these considerations.

Sensor Data Fusion, as an information technology as well as a branch of engineering science and informatics, is discussed in an introductory chapter, put into a more general context, and related to information systems. Basic elements and concepts are introduced.

Part I presents a coherent methodological framework of Sensor Data Fusion, thus providing the prerequisites for discussing selected applications in Part II of the book in four chapters. The presentation reflects the author's views on the subject and emphasizes his own contributions to the development of particular aspects.

Based on a more general notion of object states, probabilistic models of their temporal evolution and the underlying sensors are discussed. Their proper combination within a Bayesian framework provides iterative update formulae for probability densities that represent the knowledge about objects of interest extracted from imperfect sensor observations and context information. Various data fusion algorithms appear as limiting cases and illustrate the more general Bayesian approach. Particular emphasis is placed on fusing data produced at different instants of times, i.e., on-time series of sensor data. The resulting multiple sensor tracking problem is a key issue in Sensor Data Fusion. A discussion of track initiation and fusion of locally preprocessed information, i.e., track-to-track fusion, concludes Part I.

Progress in fusion research is based on precise and methodical work on relevant, well-posed, but sufficiently specialized research questions. Besides answering them appropriately and evaluating the result in comparison to alternatives, the identification of such questions in itself is an essential part of scientific work and often far from trivial.

Following this observation, selected applications are discussed in Part II, where specific problems of Sensor Data Fusion are highlighted. Their solutions are based on the methods previously introduced, which are crucial for meeting challenging user requirements. At the same time, the application examples illustrate the inner structure and practical use of the underlying Bayesian formalism. The very success of Bayesian Sensor Data Fusion may serve as retrospective justification of the approach as well as a motivation to apply this formalism to an even broader field of applications.

The discussed examples are chosen from the author's own contributions to this area and are grouped around the following over-all topics:

1. *Integration of Advanced Sensor Properties*
2. *Integration of Advanced Object Properties*
3. *Integration of Topographical Information*
4. *Feedback to Acquisition: Sensor Management*,

which define the four chapters of Part II. The material discussed in the individual sections of these chapters is collected from journal publications and a handbook chapter by the author. Although the presentation of the key points with respect to specialized methodology and application aspects is self-contained on the methodological basis provided by Part I, a related publication of the author is displayed in each section, where more details and numerical results can be found.

The results of Part II are input for large ISR Systems (Intelligence, Surveillance, and Reconnaissance). Since the examples have been selected from sufficiently different, but mutually complementary areas in Sensor Data Fusion, the detailed analysis of the specialized problems involved and their individual solutions provide a fairly comprehensive overview of various aspects of Sensor Data Fusion for situation picture production. This type of "example-driven" discussion is perhaps better suited to stimulate research work and progress on analogous problems in different applications than a more abstract and generalizing presentation might do.

With some delay, Sensor Data Fusion is likely to develop along lines similar to the evolution of another modern key technology whose origin is rooted in the military domain, the Internet. It is the author's firm conviction that until now, scientists and engineers have only scratched the surface of the vast range of opportunities for research, engineering, and product development that still waits to be explored: the Internet of the Sensors.

This text book would not have been possible without two eminent scientists, who greatly formed the author's mind and apprehension over many years. Günther van Keuk, his teacher in tracking and Sensor Data Fusion and former department head, who died far too early in 2003, introduced him into the exciting field of Sensor Data Fusion and shaped his scientific habit. Jürgen Grosche generously accompanied the author's research as a Fraunhofer director with personal interest, valuable advice, and clear directions. In particular, Jürgen Grosche mediated the author's lecturing activities on Sensor Data Fusion at Bonn University and encouraged him to summarize his research results in this book.

Of course, the merits of many scientific colleagues should also be mentioned here, who contributed greatly through countless scientific discussions and joint work over the years, especially Klaus Becker, Richard Klemm, Martin Ulmke, and Ulrich Nickel. Furthermore, the author is indebted to Jane Stannus and Diana Dorau for their help in editorial and layout issues.

Since the inner strength for his professional life is given to the author by his family, his beloved wife Dorothea and his children Maria, Veronika, Theresia, Katharina, and Johannes, as well as by his parents and brothers, it might be appropriate to express his deep gratitude to them here as well.

Rolandswerth, September 2013 Johann WolfgangKoch

Reference

1. M. Wunder, J. Grosche (eds.), *Verteilte Führungsinformationssysteme* (Springer, Heidelberg, 2009).

Contents

Chapter 1
Notion and Structure of Sensor Data Fusion

Sensor data fusion is an omnipresent phenomenon that existed prior to its technological realization or the scientific reflection on it. In fact, all living creatures, including human beings, by nature or intuitively perform sensor data fusion. Each in their own way, they combine or "fuse" sensations provided by different and mutually complementary sense organs with knowledge learned from previous experiences and communications from other creatures. As a result, they produce a "mental picture" of their individual environment, the basis of behaving appropriately in their struggle to avoid harm or successfully reach a particular goal in a given situation.

1.1 Subject Matter

As a sophisticated technology with significant economic and defence implications as well as a branch of engineering science and applied informatics, modern sensor data fusion aims at automating this capability of combining complementary pieces of information. Sensor data fusion thus produces a "situation picture," a reconstruction of an underlying "real situation," which is made possible by efficiently implemented mathematical algorithms exploiting even imperfect data and enhanced by new information sources. Emphasis is not only placed on advanced sensor systems, technical equivalents of sense organs, but also on spatially distributed networks of homogeneous or heterogeneous sensors on stationary or moving platforms and on the integration of data bases storing large amounts of quantitative context knowledge. The suite of information sources to be fused is completed by the interaction with human beings, which makes their own observations and particular expertise accessible.

The information to be fused may comprise a large variety of attributes, characterized, for example, by sensor ranges from less than a meter to hundreds of kilometers, by time scales ranging from less than a second to a few days, by nearly stationary or rapidly changing scenarios, by actors behaving cooperatively, in-cooperatively, or even hostile, by high precision measurements or sensor data of poor quality.

W. Koch, *Tracking and Sensor Data Fusion*,
Mathematical Engineering, DOI: 10.1007/978-3-642-39271-9_1,

Sensor data fusion systems emerging from this branch of technology have in effect the character of "cognitive tools", which enhance the perceptive faculties of human beings in the same way conventional tools enhance their physical strength. In this type of interactive assistance system, the strengths of automated data processing (dealing with mass data, fast calculation, large memory, precision, reliability, robustness etc.) are put into service for the human beings involved. Automated sensor data fusion actually enables them to bring their characteristically "human" strengths into play, such as qualitatively correct over-all judgment, expert knowledge and experience, intuition and creativity, i.e. their "natural intelligence" that cannot be substituted by automated systems in the foreseeable future. The user requirements to be fulfilled in a particular application have a strong impact on the actual fusion system design.

1.1.1 Origins of Modern Development

Sensor data fusion systems have been developed primarily for applications, where a particular need for support systems of this type exists, for example in time-critical situations or in situations with a high decision risk, where human deficiencies must be complemented by automatically or interactively working data fusion techniques. Examples are fusion tools for compensating decreasing attention in routine and mass situations, for focusing attention on anomalous or rare events, or complementing limited memory, reaction, and combination capabilities of human beings. In addition to the advantages of reducing the human workload in routine or mass tasks by exploiting large data streams quickly, precisely, and comprehensively, fusion of mutually complementary information sources typically produces qualitatively new and important knowledge that otherwise would remain unrevealed.

The demands for developing such support systems are particularly pressing in defence and security applications, such as surveillance, reconnaissance, threat evaluation, and even weapon control. The earliest examples of large sensor data fusion projects were designed for air defence against missiles and low-flying bombers and influenced the development of civilian air traffic control systems. The development of modern sensor data fusion technology and the underlying branch of applied science was stimulated by the advent of sufficiently powerful and compact computers and high frequency devices, programmable digital signal processors, and last but not least by the "Strategic Defence Initiative (SDI)" announced by US President RONALD REAGAN on March 23, 1983.

After a certain level of maturity has been reached, the Joint Directors of Laboratories (JDL), an advisory board to the US Department of Defense, coined the technical term "Sensor Data and Information Fusion" in George Orwell's very year 1984 and undertook the first attempt of a scientific systematization of the new technology and the research areas related to it [1, Chap. 2, p. 24]. To the present day, the scientific fusion community speaks of the "JDL Model of Information Fusion" and its subsequent generalizations and adaptations [1, Chap. 3], [2]. The JDL model provides a structured and integrated view on the complete functional chain from dis-

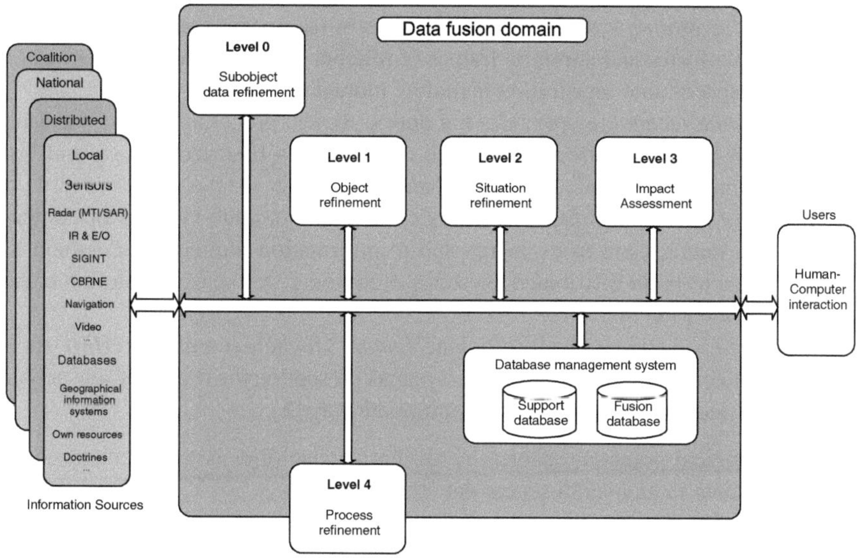

Fig. 1.1 Overview of the JDL-Model of Sensor Data and Information Fusion [1, Chap. 3], which provides a structured and integrated view on the complete functional chain from distributed sensors, data bases, and human reports to the users and their options to act including various feed-back loops at different levels

tributed sensors, data bases, and human reports to the users and their options to act including various feed-back loops at different levels (Fig. 1.1). It seems to be valid even in the upcoming large fields of civilian applications of sensor data fusion and cyber security [3]. Obviously, the fundamental concepts of sensor data fusion have been developed long before their full technical feasibility and robust realizability in practical applications.

1.1.2 General Technological Prerequisites

The modern development of sensor data fusion systems was made possible by substantial progress in the following areas over the recent decades:

1. Advanced and robust *sensor systems*, technical equivalents of sense organs with high sensitivity or coverage are made available that may open dimensions of perception usually unaccessible to most living creatures.
2. *Communication links* with sufficient bandwidths, small latencies, stable connectivity, and robustness against interference are the backbones of spatially distributed networks of homogeneous or heterogeneous sensors.

3. Mature *navigation systems* are prerequisites of (semi-)autonomously operating sensor platforms and common frames of reference for the sensor data based on precise space–time registration including mutual alignment.
4. *Information technology* provides not only sufficient processing power for dealing with large data streams, but also efficient data base technology and fast algorithmic realizations of data exploitation methods.
5. *Technical interoperability*, the ability of two or more sub-systems or components to interact and to exchange and to information mutually understood, is inevitable to build distributed "systems of systems" for sensor exploration and data exploitation [4].
6. Advanced and ergonomically efficient *Human–Machine Interaction (HMI)* tools are an integral part of man-machine-systems presenting the results of sensor data fusion systems to the users in an appropriate way [5].

The technological potential enabled by all these capabilities is much enhanced by integrating them in an overall sensor data fusion system.

1.1.3 Relation to Information Systems

According to this technological infrastructure, human decision makers on all levels of hierarchy, as well as automated decision making systems, have access to vast amounts of data. In order to optimize use of this high degree of data availability in various decision tasks, however, the data continuously streaming in must not overwhelm the human beings, decision making machines, or actuators involved. On the contrary, the data must be fused in such a way that at the right instant of time the right piece of high-quality information relevant to a given situation is transmitted to the right user or component and appropriately presented. Only if this is the case, the data streams can support goal-oriented decisions and coordinated action planing in practical situations and on all levels of decision hierarchy.

In civilian applications, management information or data warehouse systems are designed in order to handle large information streams. Their equivalents in the defence and security domain are called C^4ISTAR Systems [4]. This acronym denotes computer-assisted functions for C^4 (Command, Control, Communications, Computers), I (Intelligence), and STAR (Surveillance, Target Acquisition and Reconnaissance) in order to enable the coordination of defence-related operations. While management information or data warehouse systems are primarily used to obtain competitive advantages in economic environments, C^4ISTAR systems aim at information dominance over potential military opponents. The observation that more or less the same terminology is used in both areas for characterizing the struggle to avoid harm or successfully reach goals, is an indication of far-reaching fundamental commonalities of decision processes in defence command & control as well as in product development and planing, in spite of different accentuations in particular aspects.

A basic component of C^4ISTAR information systems, modular and flexibly designed as "systems of systems," is the combination of sensor systems and data bases with appropriate sensor data and information fusion sub-systems. The objective at this level is the production of timely, consistent and, above all, sufficiently complete and detailed "situation pictures," which electronically represent a complex and dynamically evolving overall scenario in the air, on the ground, at sea, or in an urban environment. The concrete operational requirements and restrictions in a given application define the particular information sources to be considered and data fusion techniques to be used.

A Characteristic Example

A particularly mature example of an information system, where advanced sensor data fusion technology is among its central pillars, is given by a distributed, coalition-wide C^4ISTAR system of systems for wide-area ground surveillance. It mirrors many of the aspects previously addressed and has been carried out within the framework of a multinational technology program called MAJIIC (Multi-Sensor Aerospace-Ground Joint ISR Interoperability Coalition) [4, Chap. 20]. By collaboratively using interoperable sensor and data exploitation systems in coalition operations, MAJIIC has been designed to improve situational awareness of military commanders over the various levels of the decision making hierarchy.

Based on appropriate concepts of deployment and the corresponding tactical procedures, technological tools for Collection, Coordination and Intelligence Requirements Management (CCIRM) are initiated by individual sensor service requests of deployed action forces. The CCIRM tools produce mission plans according to superordinate priorities, task sensor systems with appropriate data acquisition missions, initiate data exploitation and fusion of the produced sensor data streams in order to obtain high-quality reconnaissance information, and, last but not least, guarantee the feedback of the right information to the requesting forces at the right instant of time.

Under the constraint of leaving existing C^4ISTAR system components of the nations participating in MAJIIC unchanged as far as possible, the following aspects are addressed with particular emphasis:

1. The integration of advanced sensor technology for airborne and ground-based wide-area surveillance is mainly based on Ground Moving Target Indicator Radar (GMTI), Synthetic Aperture Radar (SAR), electro-optical and infrared sensors (E/O, IR) producing freeze and motion imagery, Electronic Support Measures (ESM), and artillery localization sensors (radar- or acoustics-based).
2. Another basic issue is the identification and implementation of common standards for distributing sensor data from heterogeneous sources including appropriate data and meta-data formats, agreements on system architectures as well as the design and implementation of advanced information security concepts.
3. In addition to sensor data fusion technology itself, tools and procedures have been developed and are continuously enhanced for co-registration of hetero-

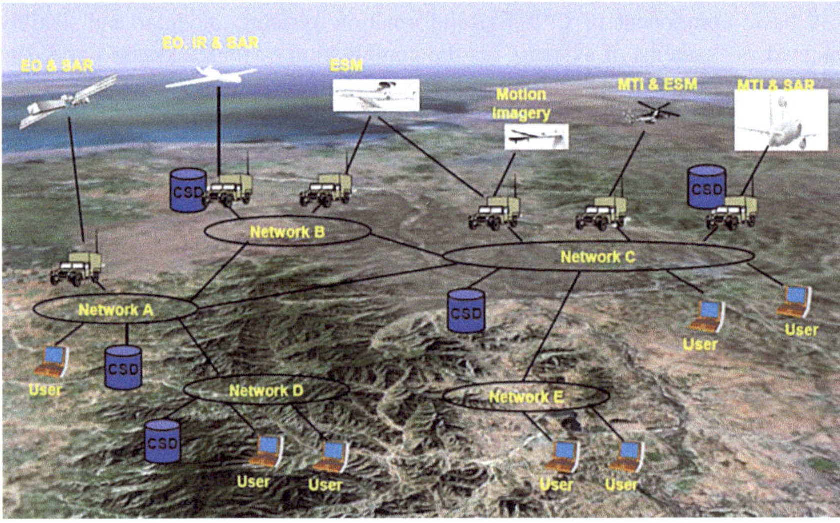

Fig. 1.2 MAJIIC system architecture emphasizing the deployed sensors, databases, and distributed sensor data fusion systems (Interoperable ISR Exploitation Stations)

geneous sensors, cross-cueing between the individual sensors of a surveillance system, the sensors of different systems, and between sensors and actuators, as well as for exploitation product management, representation of the "Coalition Ground Picture," for coordinated mission planning, tasking, management, and monitoring of the MAJIIC sub-systems.

4. MAJIIC-specific communications have been designed to be independent of network-types and communication bandwidths, making it adaptable to varying requirements. Commercially available and standardized internet- and crypto-technology has been used in both the network design and the implementation of interfaces and operational features. Important functionalities are provided by collaboration tools enabling ad-hoc communication between operators and exchange of structured information.
5. The central information distribution nodes of the MAJIIC C^{4}ISTAR system of systems are so-called Coalition Shared Data servers (CSD) making use of modern database technology. Advanced Data Mining and Data Retrieval tools are part of all MAJIIC data exploitation and fusion systems.
6. From an operational point of view, a continuous interaction between Concept Development and Experimentation (CD&E process, [6]) by planning, running, and analyzing simulated and live C^{4}ISTAR experiments is an essential part of the MAJIIC program, fostering the transfer of MAJIIC capabilities into national and coalition systems.

Figure 1.2 provides an overview of the MAJIIC system architecture and the deployed sensor systems.

1.2 Characterization as a Branch of Applied Science

The object of knowledge in sensor data fusion as a branch of applied science is sensor data fusion technology discussed previously. In other words, it aims at the acquisition of knowledge required to build automated sensor data fusion systems, often being part of larger information systems, by using appropriately developed scientific methodologies. This includes the elicitation, collection, analysis, modeling, and validation of this knowledge.

In order to reach this goal, scientific research in sensor data fusion is performed in an interdisciplinary way by applying fundamental results gathered from other sciences, such as natural sciences dealing with physical object properties perceptible by sensors and the underlying sensing principles, engineering sciences, mainly sensor engineering, metrology, automation, communications, and control theory, but also applied mathematics and statistics, and, last but not least, applied informatics. Two characteristic features of sensor data fusion can be identified.

1. The available sensor data and context knowledge to be fused typically provide incomplete and imperfect pieces of information. These deficiencies have manifold reasons and are unavoidable in real-world applications. For dealing with imperfect sensor and context data, sophisticated mathematical methodologies and reasoning formalisms are applied. Certain aspects of them are developed by extending the underlying methodology, thus providing contributions to fundamental research. Reasoning with uncertain information by using probabilistic or other formalisms is therefore a major scientific feature characterizing sensor data fusion.
2. As a branch of applied science, sensor data fusion is closely related to the practical design of surveillance and reconnaissance components for information systems. In implementing fundamental theoretical concepts, a systematic way of finding reasonable compromises between mathematical exactness and pragmatic realization issues as well as suitable approximation methodologies are therefore inevitable. System aspects such as robustness and reliability even in case of unforeseeable nuisance phenomena, priority management, and graceful degradation are of particular importance in view of practicability. This is equally true for comprehensive evaluation and prediction of fusion system performance and identification of relevant factors for system control and operation, based, for example, on extensive Monte-Carlo-simulations and the analysis of theoretical bounds [7].

1.2.1 Pioneers of Sensor Data Fusion

Since sensor data fusion can be considered as a branch of automation with respect to imperfect sensor data and non-sensor information, a historical reflection on its roots could identify numerous predecessors in automation engineering, cybernetics, and Bayesian statistics, who developed fundamental notions and concepts relevant to sensor data fusion. Among many other pioneers, CARL FRIEDRICH GAUSS, THOMAS BAYES and the Bayesian statisticians, as well as RUDOLF E. KALMAN have created the methodological and mathematical prerequisites of sensor data fusion that made the modern development possible.

Carl Friedrich Gauß

Many achievements in science and technology that have altered today's world can be traced back to the great mathematician, astronomer, geodesist, and physicist CARL FRIEDRICH GAUSS (1777–1855). This general tendency seems also to be true in the case of sensor data fusion. After finishing his opus magnum on number theory, GAUSS re-oriented his scientific interests to astronomy. His motive was the discovery of the planetoid Ceres by the Theatine monk GIUSEPPE PIAZZI (1746–1826) on Jan 1, 1801, whose position was lost shortly after the first astronomical orbit measurements. GAUSS succeeded in estimating the orbit parameters of Ceres from a few noisy measurements by using a recursively defined least-squares error compensation algorithm [8], a methodology, which can be interpreted as a limiting case of Kalman filtering, one of the most important backbone algorithms of modern target tracking and sensor data fusion. Based on his results, HEINRICH OLBERS (1758–1840) was able to rediscover Ceres on Jan 1, 1802. The discovery of three other planetoids followed (Pallas 1802, Juno 1804, Vesta 1807). Although until then, GAUSS was well-known to mathematical experts only, this success made his name popular, leading to his appointment at Göttingen University in 1807 as a Professor of Astronomy and Director of the Observatory. GAUSS' personal involvement in this new scientific branch of reasoning with imprecise observation data is indicated by the fact that he called his first borne child Joseph, after Father GUISEPPE PIAZZI [9, p. 15]. Three others of his children were named after the discoverers of Pallas, Juno, and Vesta.

Bayesian Statisticians

In sensor data fusion, the notion of "Bayesian probability" is of fundamental importance. It interprets the concept of probability as "a measure of a state of knowledge" (see [10], e.g.) and not as a relative frequency as in classical statistics. According to this interpretation, the probability of a hypothesis given the sensor data is proportional to the product of the likelihood function multiplied by the prior probability. The likelihood function represents the incomplete and imperfect information provided by the sensor data themselves as well as context information on the sensor performance and the sensing environment, while the prior probability the belief in the hypothesis before the sensor data were available (see Chap. 3 *Bayesian Knowledge Propagation* of this book).

The term 'Bayesian' refers to THOMAS BAYES (1702–1761), a British mathematician and Presbyterian minister, who proved a special case of this proposition, which is now called Bayes' theorem (published posthumously by his friend RICHARD PRICE (1723–1791) in 1763, [11]). The roots of 'subjective probability' can even be traced back to the great Jewish philosopher MOSES MAIMONIDES (1135/38-1204) and the medieval rabbinic literature [12, Chap. 10]. It was PIERRE- SIMON LAPLACE (1749–1827), however, who introduced a more general version of Bayes' theorem, apparently unaware of Bayes' work, and used it to approach problems in celestial mechanics, medical statistics, reliability, and jurisprudence [13, Chap. 3]. In the sequel, the foundations of Bayesian statistics were laid by many eminent statisticians.

Of particular importance is ABRAHAM WALD (1902–1950, [14]), an Austro-Hungarian mathematician, who immigrated to the USA in 1938, where he created *Sequential Analysis*, a branch of applied statistical decision making, which is of enormous importance for sensor data fusion, especially in track management and consistency testing (see Chap. 4 *Sequential Track Extraction* of this book). In his influential work on *Statistical Decision Functions* [15], he recognized the fundamental role of Bayesian methods and called his optimal decision methods 'Bayes strategies'.

Rudolf E. Kalman and his Predecessors

The beginning of modern sensor data fusion is inextricably bound up with the name of RUDOLF E. KALMAN (*1930), a Hungarian-American system theorist, though he had many predecessors. The Kalman filter is a particularly influential example of a processing algorithm for inferring a time variable object state from uncertain data assuming an uncertain object evolution, which can elegantly be derived from Bayesian statistics. Among Kalman's predecessors, THORVALD NICOLAI THIELE (1838–1910), a Danish astronomer, actuary and mathematician, derived a geometric construction of a fully developed Kalman filter in 1889 [16, Chap. 4]. Also RUSLAN L. STRATONOVICH (1930–1997), a Russian physicist, engineer, probabilist, and PETER SWERLING (1929–2000), one of the most influential RADAR theoreticians in the second half of the twentieth century [17, Appendix], developed Kalman-type filtering algorithms earlier using different approaches.

STANLEY F. SCHMIDT (*1926) is generally credited with developing the first application of a Kalman filter to the problem of trajectory estimation for the NASA Apollo Spaceflight Program in 1960, leading to its incorporation in the Apollo navigation computer. The state-of-the-art until 1974 is summarized in the influential book *Applied Optimal Estimation*, edited by ARTHUR GELB [18].

Contemporary Researchers

Independently of each other, GÜNTHER VAN KEUK (1940–2003) and SINGER first applied Kalman filtering techniques to single air target tracking problems in multiple radar data processing [19, 20]. The foundations of multiple hypothesis tracking methods for dealing with data of uncertain origin related to multiple objects were laid by ROBERT W. SITTLER, who first posed the problem [21], while DONALD B. REID published a method for solving it [22]. VAN KEUK, SAM S. BLACKMAN, and YAAKOV BAR- SHALOM were among the first, who transformed Reid's method into practical algorithms (see [23, 24] for an overview of the development until 2004).

In the vast research literature published since then, however, it is impossible to identify all important scientists and engineers. The following discussion of significant contributions is therefore by no means complete, reflects the author's personal point of view, and is related to methodological framework presented in Part 1 of this book.

In particular due to their monographs on target tracking and sensor data fusion issues, YAAKOV BAR- SHALOM [25], SAM S. BLACKMAN [26], and ALFONSO FARINA [27] are highly influential researchers and have inspired many developments. HENK A. P. BLOM introduced stochastic hybrid processes into data fusion [28], which under the name of "Interacting Multiple Models" still define the state-of-the-art in target dynamics modeling. He in particular applied Bayesian data fusion to large air traffic control systems under severe reliability constraints. Countless realization aspects in fusion systems design are covered by OLIVER DRUMMOND's contributions. Already in his PhD thesis [29], where he has addressed many important issues in multiple object tracking at a very early time. LARRY STONE is a pioneer in Bayesian sonar tracking and data fusion in complex propagation environments [30]. NEIL GORDON was among the first, who applied sequential random Monte-Carlo-techniques to non-linear tracking problems, known under the name of "Particle Filtering", and inspired a rapid development in this area [31]. Numerous contributions to problems at the borderline between advanced signal processing, distributed detection theory, and target tracking were made by PETER K. WILLETT. XIAO- RONG LI provided important solutions to radar data fusion. The integration of modern mathematical non-linear filtering to practical radar implementation is among the merits of FRED DAUM. Numerous achievements in non-linear filtering, distributed sensing, and resources management were provided by UWE D. HANEBECK. HUGH FRANCIS DURRANT-WHYTE is generally credited with creating decentralized data fusion algorithms as well as with simultaneous localization and navigation. The stormy development of efficient multitarget tracking based on random set theory with Probabilistic Hypothesis Density Filtering (PHD) as an efficient realization has been developed by RONALD MAHLER [32]. Finally, ROY STREIT first introduced Expectation Maximization techniques to solve efficiently the various data association problems in target tracking and sensor data fusion and exploited the use of Poisson-point precesses in this area [33].

A well readable introduction to sensor data fusion was published by H. B. MITCHELL [34]. The handbook "Advanced Signal Processing: Theory and Implementation for Sonar, Radar, and Non-Invasive Medical Diagnostic Systems" [35] deals with many advanced sensor data fusion applications. MARTIN E. LIGGINS, JAMES LLINAS, AND DAVID L. HALL edited the compendium "Handbook of Multisensor Data Fusion: Theory and Practice" [1]. An excellent introduction to more advanced techniques with emphasis on particle filtering is provided by FREDRIK GUSTAFSSON [36].

1.2.2 Organization of the Research Community

The interdisciplinary significance of sensor data fusion is illustrated by the fact that numerous institutions with different profiles are working world-wide on particular aspects of it. For this reason, the "International Society of Information Fusion (ISIF)" was founded in 1998 as a scientific framework organization. According to its constitution, it is "an independent, non-profit organization dedicated to advancing the knowledge, theory and applications of information fusion" [37]. Since that year, ISIF has been organizing the annual *International Conferences on Information Fusion*, the main scientific conference of the international scientific information fusion community.

1.2.3 Important Publication Platforms

To publish high-quality scientific papers on sensor data and information fusion, several well-established scientific journals are available, such as the *IEEE Transactions on Aerospace and Electronic Systems* and *on Signal Processing*, the most visible publication platforms, the *ISAF Journal of Advances in Information Fusion*, or the *Elsevier Journal on Information Fusion*. Besides the proceedings of the FUSION conferences, the annual SPIE Conference Series *Signal and Data Fusion of Small Targets (SPIE SMT)* organized by OLIVER E. DRUMMOND since 1989 in the USA, numerous special sessions at radar and automated control conferences as well as several national fusion workshops, such as the German IEEE ISIF Workshop Series *Sensor Data Fusion: Trends, Solutions, Applications (SDF)* [41], provide forums, where the latest advances and research results are presented and discussed among researchers and application engineers.

1.3 From Imperfect Data to Situation Pictures

Sensor data fusion typically provides answers to questions related to objects of interest such as: Do objects exist at all and how many of them are moving in the sensors' fields of view? Where are they located at what time? Where will they be in the future with what probability? How can their overall behavior be characterized? Are anomalies or hints to their possible intentions recognizable? What can be inferred about the classes the objects belong to or even their identities? Are there clues for characteristic interrelations between individual objects? In which regions do they have their origin? What can be said about their possible destinations? Are there observable over-all object flows? Where are sources or sinks of traffic? and many other questions.

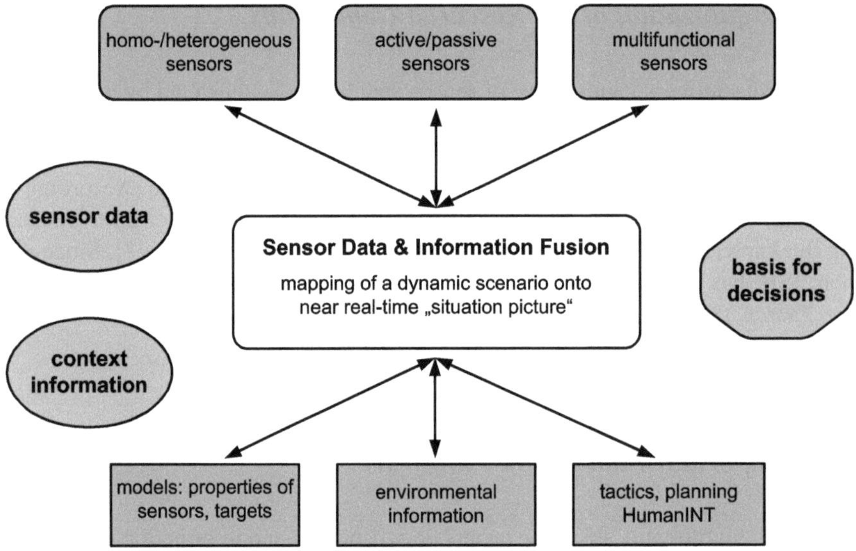

Fig. 1.3 Sensor data and information fusion for situation pictures: overview of characteristic aspects and their mutual interrelation

The answers to those questions are the constitutive elements, from which near real-time situation pictures can be produced that electronically represent a complex and dynamically evolving overall scenario in the air, on the ground, at sea, under water, as well as in out- or in-door urban environments, and even more abstract spaces. According to the previous discussion, these "situation elements" must be gained from the currently received sensor data streams while taking into account all the available context knowledge and pre-history. Since situation pictures are fundamental to any type of computer-aided decision support, the requirements of a given application define which particular information sources are to be fused.

The sensor data to be fused are usually inaccurate, incomplete, or ambiguous. Closely spaced moving objects are often totally or partially irresolvable. The measured object parameters may be false or corrupted by hostile measures. The context information is in many cases hard to formalize and even contradictory in certain aspects. These deficiencies of the information to be fused are unavoidable in any real-world application. Therefore, the extraction of 'information elements' for situation pictures is by no means trivial and requires a sophisticated mathematical methodology for dealing with imperfect information. Besides a precise requirement analysis, this is one of the major scientific features that characterizes and shapes sensor data fusion as branch of applied science.

1.3.1 Discussion of Characteristic Aspects

Figure 1.3 provides an overview of different aspects within this context and their mutual interrelation, which should be emphasized:

1. The underlying sensor systems can be located in different ways (collocated, distributed, mobile) producing measurements of the same or of different type. A multisensor system potentially increases the coverage or data rate of the total system and may help to resolve ambiguities.
2. Even by fusing homogeneous sensors, information can be obtained that is unaccessible to each sensor individually, such as in stereoscopic vision, where range information is provided by fusing two camera images taken from different viewpoints.
3. Fusion of heterogeneous sensor data is of particular importance, such as the combination of kinematic measurements with measured attributes providing information on the classes to which objects belongs to. Examples for measured attributes are Signal Intelligence (SIGINT), Jet Engine Modulation (JEM), radial or lateral object extension, chemical signatures, etc.
4. Especially for defense and security applications, the distinction between active and passive sensing is important as passive sensors enable covert surveillance, which does not reveal itself by actively emitting radiation.
5. Multi-functional sensor systems, such as phased-array radar, offer additional operational modes, thus requiring more intelligent strategies of sensor management that provide feedback to the process of information acquisition via appropriate control or correction commands. By this, the surveillance objectives can often be reached much more efficiently.
6. Context information is given, for example, by available knowledge on sensor and object properties, which is often quantitatively described by statistical models. Context knowledge is also given by environmental information on roads or topographical occlusions and provided by Geographical Information Systems (GIS). Seen from a different perspective, context information, such as road-maps, can also be extracted from real-time sensor data directly.
7. Relevant context knowledge (e.g. doctrines, planning data, tactics) and human observer reports (HUMINT: Human Intelligence) is also important information in the fusion process. The exploitation of context information of this kind can significantly improve the fusion system performance.

1.3.2 Remarks on the Methods Used

Situation elements for producing timely situation pictures are provided by integratively and spatio-temporally processing various pieces of information that in themselves often may have only limited value for understanding the situation. Essentially, logical cross-references, inherent complementarity, and redundancy are exploited.

More concretely speaking, the methods used are characterized by a stochastic approach (estimating relevant state quantities) and a more heuristically defined knowledge-based approach (modeling actual human behavior when exploiting information).

Among the data exploitation products of data fusion systems, object 'tracks' are of particular importance. Tracking faces an omnipresent aspect in every real-world application insofar as it is dealing with fusion of data produced at *different instants of time*; i.e. tracking is important in all applications where particular emphasis is placed on the fact that the sensor data to be exploited have the character of a time series.

Tracks thus represent currently available knowledge on relevant, time-varying quantities characterizing the instantaneous "state" of individual targets or target groups of interest, such as aircraft, ships, submarines, vehicles, or moving persons. Quantitative measures that reliably describe the quality of this knowledge are an integral part of a track. The information obtained by 'tracking' algorithms [25, 26, 42] also includes the history of the targets. If possible, a one-to-one association between the target trajectories in the sensors' field of view and the produced tracks is to be established and has to be preserved as long as possible (track continuity). The achievable track quality does not only depend on the performance of the sensors used, but also on target properties and the operational conditions within the scenario to be observed. If tracks 'match' with the underlying real situation within the bounds defined by inherent quality measures being part of them, we speak of 'track consistency."

Tracking algorithms, including Bayesian multiple hypothesis trackers as particularly well-understood examples, are iterative updating schemes for conditional probability density functions representing all available knowledge on the kinematic state of the objects to be tracked at discrete instants of time t_l. The probability densities are conditioned on both, the sensor data accumulated up to some time t_k, typically the current data acquisition time, as well as on available context information, such as on sensor characteristics, the object dynamics, the environment, topographical maps, or on certain rules governing the object behavior. Depending on the time instant t_l at which estimates for the state $\mathbf{x}_l$ are required, the related estimation process is referred to as prediction ($t_l > t_k$), filtering ($t_l = t_k$), or retrodiction ($t_l < t_k$) [43, 44].

1.3.3 A Generic Sensor Data Fusion System

Figure 1.4 shows a generic scheme of functional building blocks within a multiple sensor tracking and data fusion system along with its relation to the underlying sensors. In the case of multi-functional sensors, there is feedback from the tracking system to the process of sensor data acquisition (sensor management). The following aspects should be emphasized:

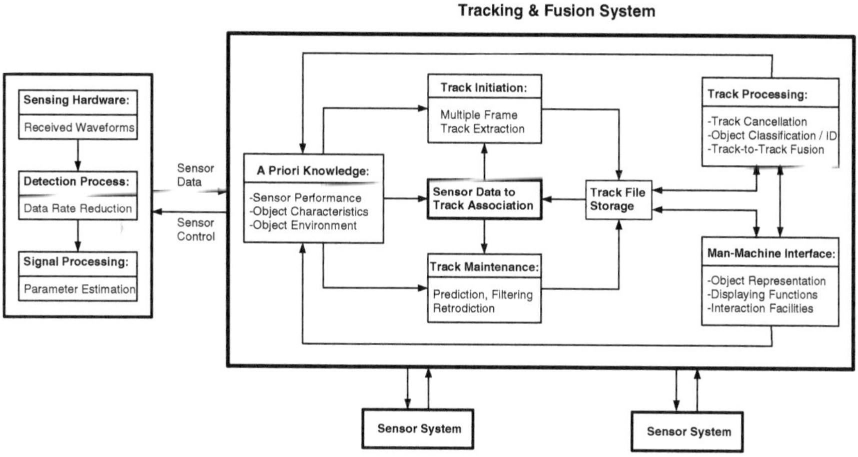

Fig. 1.4 Generic scheme of functional building blocks within a tracking/fusion system along with its relation to the sensors (centralized configuration, type IV according to O. Drummond)

Sensor Systems

After passing a detection process, essentially working as a means of data rate reduction, the signal processing provides estimates of parameters characterizing the waveforms received at the sensors' front ends (e.g. radar antennas). From these estimates sensor reports are created, i.e. measured quantities possibly related to objects of interest, which are the input for the tracking and sensor data fusion system. By using multiple sensors instead of one single sensor, among other benefits, the reliability and robustness of the entire system is usually increased, since malfunctions are recognized easier and earlier and often can be compensated without risking a total system breakdown.

Interoperability

A prerequisite of all further processing steps, which at first sight seems to be trivial, is technical interoperability. It guarantees that all relevant sensor data are transmitted properly, in a timely way, and completely including all necessary meta-data describing the sensor performance, the platform parameters, and environmental characteristics. This type of meta-data is necessary to transform the sensor data into common frames of reference, to identify identical pieces of data, and to merge similar pieces of data into one single augmented piece of information. The process of combining data from different sources and providing the user with a unified view of these data is sometimes also referred to as data integration. Often interoperability acts as a bottleneck in designing real-world data fusion systems of systems [4, Chap. 20].

Fusion Process

All sensor data that can be associated to existing tracks are used for track maintenance (using, e.g., prediction, filtering, and retrodiction). The remaining data are processed for initiating new tentative tracks (multiple frame track extraction). Association techniques thus play a key role in tracking/fusion applications. Context information in terms of statistical models (sensor performance, object characteristics, object environment) is a prerequisite for track maintenance and initiation. Track confirmation/termination, classification/identification, and fusion of tracks related to the same objects or object groups are part of the track management functionalities.

Human–Machine Interface

The scheme is completed by a human–machine interface with display and interaction functions. Context information can be updated or modified by direct human interaction or by the track processor itself, for example as a consequence of object classification or road-map extraction. For an introduction to the vast literature on the related problems in human factors engineering and on practical systems solutions see Ref. [5].

1.3.4 On Measuring Fusion Performance

In sensor data fusion, the underlying 'real' situation is typically unknown. Only in expensive and time-consuming experiments certain aspects of a dynamically evolving situation are monitored, sometimes even with questionable accuracy. For this reason, experiments are valuable for demonstrating the "proof of concept" as well as to understand the underlying physical phenomena and operational problems, for example. They are of limited use, however, in performance evaluation and prediction. This underlines the role of comprehensive Monte-Carlo-simulations in fusion system performance evaluation.

According to the previous discussion, sensor data fusion systems try to establish one-to-one relations between objects in the sensors' fields of view and identified object tracks in the situation picture. Strictly speaking, this is only possible under ideal conditions regarding the sensor performance and the underlying target scenario. It seems thus reasonable to measure the performance of a given tracking/fusion system by its characteristic deficiencies when compared to this ideal goal. In general, two categories of deficiencies can be distinguished that are either caused by mismatch regarding the input data or by non-optimal processing and unfavorable application constraints.

Selected Performance Measures

Selected performance measures or 'measures of deficiency' in the sense of the previous discussion, which have practical relevance in fusion systems design should be emphasized in the following.

1. Usually a time delay is involved until a track has been extracted from the sensor data. A corresponding performance measure is thus given by the 'extraction delay' between the first detection of a target by a sensor and a confirmed track.
2. False tracks, i.e. tracks related to unreal or unwanted targets, are unavoidable in the case of a high density of false or unwanted data (e.g. by clutter, jamming/deception). Corresponding 'deficiencies' are: mean number of falsely extracted targets per time and mean life time of a false track before its deletion.
3. Targets should be represented by one and the same track until leaving the field of view. Related performance measures are: mean life time of true target tracks, probability of an 'identity switch', and probability of a target not being represented by a track.
4. The track inaccuracy (given by the error covariance matrix of a state estimate, e.g.) should be as small as possible. Furthermore, the deviations between the estimated and actual target characteristics should correspond with the error covariance matrices produced (consistency). If this is not the case, 'track loss' usually occurs.

In a given application it is by no means simple to achieve a reasonable compromise between the various, competing performance measures and the user requirements. Optimization with respect to one measure may easily degrade other performance measures, finally deteriorating the entire system performance. This is especially true under more challenging conditions.

1.3.5 Tracking-Derived Situation Elements

The primary objective of multiple sensor target tracking is to explore the underlying target kinematics such as position, velocity, or acceleration. In other words, standard target tracking applications gain information related to 'Level 1 Fusion' according to the well-established terminology of the JDL model of information fusion (see e.g. [1, Chap. 2] and the literature cited therein). Kinematic data of this type, however, are by no means the only information to be derived from target tracks. In many cases, reliable and quantitative higher level information according to the JDL terminology can be obtained. To be more concrete, wide-area air and ground surveillance is considered here as an important real-world example serving as a paradigm for other challenging tracking and fusion applications.

Inferences based on Retrodicted Tracks

The first type of higher JDL level information to be inferred from tracking data is based on a closer analysis of the histories of the kinematic object states provided by retrodiction techniques. The statements derived typically refer to object characteristics that are either time invariant or change with time on a much larger scale than kinematics quantities usually tend to do. This is the main reason why the gain in accuracy achievable by retrodiction techniques can be exploited.

- *Velocity History.* The analysis of precisely retrodicted velocity histories enables the distinction of objects belonging to different classes such as moving persons, boats, vehicles, vessels, helicopters, or jet aircraft. If the object speed estimated with sufficiently high accuracy has exceeded a certain threshold, certain object classes can be reliably be excluded. As an example, uncertainty whether an object is a helicopter or a wing aircraft can be resolved if in the track history a velocity vector 'Zero' exists. Depending on the context of the underlying application, classifications of this type can be essential to generate an alert report.
- *Acceleration History.* Similar considerations are valid if acceleration histories are taken into account. High normal accelerations, e.g., are a clear indication of a fighter aircraft. Moreover, one can safely conclude that a fighter aircraft observed with a normal acceleration > 6 g, for example, is not carrying a certain type of weaponry (any more). In other words, conclusions on the threat level connected with the objects observed can be drawn by analyzing kinematic tracks.
- *Heading, Aspect Angle.* Precise reconstructions of the targets' heading vectors are not only important input information for threat evaluation and weapon assignment in themselves, but also enable estimates of the aspect angle of an object at a given instant of time with respect to other sensors, such as those producing high range or Doppler resolution spectra. Track-derived information of this type is basic for fusing spectra distributed in time and can greatly improve object classification thus providing higher-JDL-level information.
- *Rare Event Detection.* Analysis of JDL-level-1 tracks can be the key to detecting rare or anomalous events by fusing kinematic tracks with other context information such as annotated digital road-maps and general rules of behavior. A simple example in the area of continuous-time, wide-area ground surveillance can be the production of an alert message if a large freight vehicle is observed at an unusual time on a dirt road in a forest region. There are analogous examples in the maritime or air domain.

Inferences based on Multiple Target Tracking

A second type of higher JDL level information related to mutual object interrelations can be inferred from JDL level 1 tracking data if emphasis is placed on the results of *multiple target* tracking.

- *Common History.* Multiple target tracking methods can identify whether a set of targets belongs to the same collectively moving group, such as an aircraft formation or a vehicle convoy, whose spatial extension may be estimated and tracked. If an aircraft formation has split off after a phase of penetration, e.g., the interrelation between the individual objects is to be preserved and provides valuable higher-JDL-level information that is important, e.g., when a former group target is classified as 'hostile' since this implies that all other targets originally belonging to the same group are likely to be hostile as well.
- *Object Sources and Sinks.* The analysis of large amounts of target tracks furthermore enables the recognition of sources and sinks of moving targets. By this type of reasoning, certain areas can be identified as air fields, for example, or an area of concentration of military forces. In combination with available context information, the analysis of multiple object tracks can also be used for target classification by origin or destination. A classification as hostile or suspect directly leads to an alert report.
- *Split-off Events.* By exploiting multiple target tracking techniques, certain split-off events can be identified as launches of air-to-air or air-to-surface missiles. The recognition of such an event from JDL-level-1 tracking information not only has implications on classifying the original target as a fighter aircraft, but can also establish a certain type of 'book-keeping', such as counting the number of missile launches. This enables estimates of the residual combat strength of the object, which has direct implications on countermeasures, e.g.
- *Stopping Events.* In the case of MTI radar (Moving Target Indicator), Doppler blindness can be used to detect the event 'A target under track has stopped', provided this phenomenon is described by appropriate sensor models. If there is previous evidence for a missile launcher, e.g., missing data due to Doppler blindness may indicate preparation for launch with implications on potential countermeasures. In combination with other tracks, a stopping event may also establish new object interrelations, for example, when a target is waiting for another and then moving with it.

1.3.6 Selected Issues in Anomaly Detection

Anomaly detection can be regarded as a process of information fusion that combines incomplete and imperfect pieces of mutually complementary sensor data and context information in such a way that the attention of human decision makers or decision making systems is focused on particular events that are "irregular" or may cause harm and thus require special actions, such as exploiting more specialized sensors or initiating appropriate activities by military or security personnel [45]. Fusion-based anomaly detection thus improves situational awareness. What is actually meant by "regular" or "irregular" events is higher-level information itself that depends on the context of the underlying application. Here, it is either assumed to be a priori known or to be learned from statistical long-time analysis of typical situations.

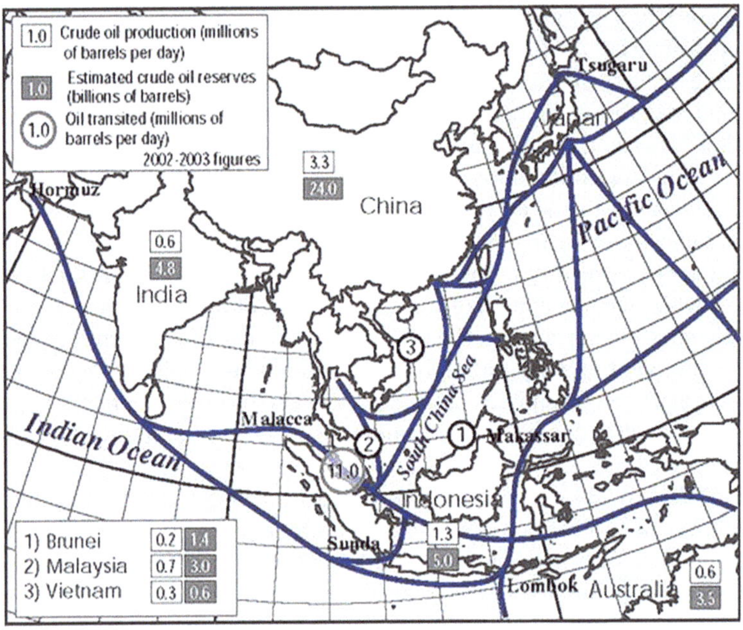

Fig. 1.5 Illustration of sea lanes and strategic passages in Pacific Asia

In complex surveillance applications, we can often take advantage of context information on the sensing environment insofar as it is the stationary or slowly changing "stage" where a dynamic scenario evolves. Typical examples of such environmental information are digital road or sea-/air-lane maps and related information, which can essentially be regarded as spatial motion constraints (see Fig. 1.5 as an illustration). In principle, this information is available by Geographical Information Systems (GIS). Another category of context information is provided by visibility models and littoral or weather maps indicating regions, where a high clutter background is to be taken into account, for example. Moreover, rather detailed planning information is often available. This category of information is not only important in mission planning or in the deployment and management of sensor systems, but can be used to decide whether an object is moving on a lane or leaving it, for example. In addition, ground-, sea- or air-lane information information can be used to improve the track accuracy of lane-moving vehicles and enhance track continuity. See Sect. 9.1 for a more detailed discussion.

Integration of Planning Information

In certain applications, rather detailed planning information is available, which provides valuable context knowledge on the temporal evolution of the objects involved and can in principle be incorporated into the tracking formalism. Planning information is often approximately described by space–time waypoints that have to be passed by the individual objects during a preplanned operation, i.e. by a set of position vectors to be reached at given instants of time and possibly via particular routes (roads, lanes) between the waypoints. In addition, we assume that the acceptable tolerances related to the arrival of the objects at the waypoints are characterized by known error covariance matrices, possibly individually chosen for each waypoint and object, and that the association between the waypoints and the objects is predefined.

The impact of waypoints on the trajectory to be estimated from future sensor data (under the assumption that the plan is actually kept) can simply be obtained by processing the waypoints as additional artificial 'measurements' via the standard Bayesian tracking paradigm, where the tolerance covariance matrices are taken into account as the corresponding 'measurement error covariances'. If this is done, the processing of sensor measurements with a younger time stamp are to be treated as "out-of sequence" measurements with respect to the artificial waypoint measurements processed earlier. For dealing with out-of-sequence measurements see Sect. 5.1. According to these considerations, planning information can well improve both track accuracy and continuity as well as facilitate the sensor-data-to-track association problems involved, provided the plan is actually kept.

Detecting Regularity Pattern Violation

A practically important class of anomalies results from a violation of regularity patterns such as those previously discussed (motion on ground-, sea-, or air-lanes or following preplanned waypoints and routes). An anomaly detector thus has to decide between two alternatives:

- The observed objects obey an underlying pattern.
- The pattern is not obeyed (e.g. off-lane, unplanned).

Decisions of this type are characterized by decision errors of first and second kind. In most cases, it is desirable to make the decisions between both alternatives for given decision errors to be accepted. A "sequential likelihood ratio" test fulfills this requirement and has enormous practical importance. For a more detailed discussion see Chap. 9.2. As soon as the test decided that the pattern is obeyed, the calculation of the likelihood ratio can be restarted since it is more or less a by-product of track maintenance. The output of subsequent sequential ratio tests can serve to re-confirm "normality" or to detect a violation of the pattern at last. The most important theoretical result on sequential likelihood ratio tests is the fact that the test has a *minimum decision length on average* given predefined statistical decision errors of first and second kind.

Tracking-derived Regularity Patterns

We have discussed moving targets that obey certain space–time constraints that are a priori known (roads/lanes, planned waypoints). A violation of these constraints was quite naturally interpreted as an anomaly. Seen from a different perspective, however, moving targets that are assumed to obey a priori *unknown* space–time constraints and to be observed by wide-area sensors, such as vehicles on an unknown road network, produce large data streams that can also be used for extracting the underlying space–time constraint, e.g. a road-map. After a suitable post-processing, the produced tracks of motion-constrained targets simply define the corresponding constraints and can thus be extracted from tracking-based results. See Sect. 9.2 for a more detailed discussion. Extracted road-maps can be highly up-to-date and precise. A discussion where such ideas are used in wide-area maritime surveillance using AIS data can be found in [46] (AIS: Automatic Identification System).

1.4 Future Perspectives of Sensor Data Fusion

Due to the increasing availability of inexpensive, but powerful sensor, communication, and information technology, its technical prerequisites, sensor data fusion, or more general, information fusion, increasingly emancipates from its roots in defense related applications. A commonplace example of this trend is the advent of navigation systems, which have developed a mass market by fusing military global navigation satellite system data with digital road-maps in combination with an appealing graphical interface. We can therefore expect that information fusion will become a key technology driver for developing numerous innovative products penetrating everyone's daily life and changing it profoundly. In this context, many new research questions are expected to emerge that will foster the further evolution of information fusion as an also economically eminent branch of applied informatics.

1.4.1 New Everyday Life Applications

Even now, intelligent filtering, analysis, evaluation, and graphical presentation of multiple sensor information enable numerous products that make everyday life safer or more secure. For example, in intelligent car-driver assistance systems, image and video data from cameras and miniaturized automotive radar sensors are automatically fused in order to perceive road obstacles and pedestrians or to exclude "ghost objects." At airport security checks, assistance systems can be used, which directly take advantage of military surveillance technology. By fusing signatures of stand-off chemical sensors and miniaturized gamma-spectrometers, for example, with person trajectories, carry-on items contaminated with hazardous materials or explosives can be detected. This may be a contribution to avert threats or avoid terrorist attacks.

Other areas where information fusion based assistance systems will increasingly be important are medical and health care, process control, logistics, industrial production, precision agriculture, and traffic monitoring. A particularly stormy evolution can currently be observed for assistance systems, where physical activities and the health status of elderly or handicapped human beings can be monitored, allowing them to live in their usual everyday environment much longer than now. In the vast fields of fire, disaster, and pollution control, quick exploitation and fusion of complex data streams can be essential for safety analysis and designing corresponding concepts as well as for developing sophisticated emergency information and management systems.

Since sensor data fusion has actually evolved into a mature technology in major fields and provides a coherent and powerful inventory of methodologies and algorithms already proven in ambitious applications, the further realization of its inherent application potential is much alleviated by the very fact that research and development for new products can be done on a sound technology base that does not need to be created in a time-consuming and expensive way. For this reason, the expected development cycles for innovative products are short, while the development risks involved are calculable. Due to its traditional strengths in high-tech industries, such as system technology or software engineering, sensor or RFID technology, highly industrialized and research-intensive countries like Germany can use their potential especially in those branches where they are traditionally well-positioned—for example in automotive technology, automation and aerospace industries, in security, safety and medical technology, and last but not least, in information system technology in general.

1.4.2 Discussion of Large-Scale Trends

More generally speaking, information fusion technology already provides mature results with profitable market opportunities, especially in those areas where physical or technical sensor data are to be fused with quantitative context information on the basis of well-understood mathematical algorithms, often making use of Bayesian reasoning.

Human Assistance Systems

Typically "human" fusion processes, however, characterized by associative reasoning, negotiating of reasonable compromises, or extrapolating incomplete information creatively and in an intuitive way, seem to be still unfit for automation, perhaps fundamentally unfit. Nevertheless, technical data fusion systems can offer assistance functionalities also here, by which specifically human competencies of judgment are freed from routine or mass tasks, quite in the sense of a "cognitive tool" as discussed earlier. Moreover, highly promising research areas are and will increasingly be those that aim at modeling and formalizing this specific human expert knowledge and expertise of situation assessment and incorporate it into the process of automated multiple sensor data.

Context Data Integration

Furthermore, a large-scale technology tend to be highlighted is given by the large potential of quantitative non-sensor information available in comprehensive databases, such as Geographical Information Systems (GIS), which is still waiting to be integrated into multiple sensor data fusion systems. This is especially true in the vast area of ground, air, sea, and underwater robotics, but has also strong implications in guaranteeing high levels of air transportation security, even in the case of high traffic densities, and in advanced logistics support systems, such as container monitoring and tracking, topics with direct implications for global economy.

Network-centric Operations

A predominant trend in defence applications is given by the demand of supporting "Network-centric Operations", which will still be in effect for the next decade. Sensor data and information fusion technology is one of the major forces shaping this process of transformation from more standard operational doctrines. Especially for out-of-area operations and operations in an urban terrain, as well as for dealing with "asymmetric" opponents, distributed high-performance reconnaissance is inevitable. In particular, wide-area ground, sea, and underwater surveillance, belong to this field, specially by making use of unmanned reconnaissance robots (unmanned ground, aerial, or underwater vehicles). Moreover, intelligent security systems for harbors, critical infrastructure, or camp protection are likely to raise many research intensive data fusion problem.

Pervasive Passive Surveillance

A particularly exciting topic of recent research is advanced distributed signal and data fusion for passive radar systems, where radio, TV, or mobile phone base stations are used as sources for illuminating targets of interest. Even in remote regions of the world, each transmitter of electromagnetic radiation becomes a potential radar transmitter station, which enables air surveillance by passively receiving reflections of non-cooperatively emitted signals of opportunity. In this way, the reconnaissance process remains covert and is not revealed by actively transmitting radiation. Analogous considerations are valid for sub-sea surveillance.

Fusion-driven Communications

The communications sub-systems within a large sensor network are typically characterized by many internal degrees of freedom, which can be controlled and adapted. This opens the vast area of fusion-driven communications, where communications and the distributed data fusion system architectures are closely tied and optimized with respect to the particular surveillance goals to be reached [48]. In the focus are multi-component system consisting of sensors, data bases, and communication infrastructures that collectively behave as a single dynamically adaptive system. Important aspects are network scalability given a limited communication bandwidth, adaptive and optimal spectrum sharing protocols, sensor data against network objectives, and in-network information. In addition, the growing use and ubiquitous nature of sensor networks pose issues when networks deployed for multiple applications need to be combined or need to exchange information at the network level.

'Add-on' Research Efforts

Since a stormy evolution of civilian information fusion applications is to be expected in the near future, defence-related research and development on information fusion technology will increasingly show the character of "add-on" research, which adapts existing civilian problem solutions to specifically military requirements. This trend is analogous to the evolution in advanced communication systems, a technology that also had its roots in the military domain, before the civilian market opportunities became the predominant force driving its technological and scientific progress.

References

1. M.E. Liggins, D.L. Hall, J. Llinas (eds.), *Handbook of Multisensor Data Fusion—Theory and Practice*, 2nd edn. (CRC Press, Boca Raton, 2008)
2. E. Bosse, J. Roy, S. Wark, *Concepts, Models, and Tool for Information Fusion* (Artech House, Norwood, 2007)
3. B.V. Dasarathy (ed.), A special Issue on information fusion in computer security, in *Information Fusion*, vol 10, No. 4 (Elsevier, New York, 2009)
4. M. Wunder, J. Grosche (eds.), *Verteilte Führungsinformationssysteme* (Springer, New York, 2009)
5. L. Schmidt, C.M. Schlick, J. Grosche (eds.), *Ergonomie und Mensch-Maschine-Systeme* (Springer, Berlin, 2008)
6. D.S. Alberts, R.E. Hayes, *Code of Best Practice for Experimentation* (CCRP Publication series, Washington, DC, 2002)
7. H.L.V. Trees, K.L. Bell (eds.), *Bayesian Bounds for Parameter Estimation and Nonlinear Filtering/Tracking* (Wiley-IEEE Press, New York, 2007)
8. H.W. Sorenson, Least-squares estimation: from Gauss to Kalman. IEEE Spectr. **7**, 63–68 (1970)
9. K.-R. Biermann (ed.), *Carl Friedrich Gauß in Briefen und Gesprächen* (C. H. Beck, München, 1990)

10. E.T. Jaynes, *Probability Theory: The Logic of Science* (Cambridge University Press, Cambridge, 2003)
11. T. Bayes, An essay towards solving a problem in the doctrine of chances. Philos. Trans. Roy. Soc. London **53**, 370418 (1763) (Reprinted in Biometrika, **45**, 1958. http://www.stat.ucla.edu/history/essay.pdf)
12. N.L. Rabinovitch, *Probability and Statistical Inference in Ancient and Medieval Jewish Literature* (University of Toronto Press, Toronto, 1973)
13. S.M. Stigler, *The History of Statistics* (Harvard University press, Cambridge, 1986)
14. J. Wolfowitz, Abraham Wald, 1902–1950. Ann. Math. Stat. **23**(1), 1–13 (1952)
15. A. Wald, *Statistical Decision Functions* (Wiley, New York, 1950)
16. S.L. Lauritzen, *Thiele—Pioneer in Statistics* (Oxford University Press, Oxford, 2002)
17. E. Brookner, *Tracking and Kalman Filtering Made Easy* (Wiley, Sudbury, 1998)
18. A. Gelb (ed.), *Applied Optimal Estimation* (M.I.T Press, Cambridge, 1974)
19. G. van Keuk, Description of a tracking strategy for phased array radar, in *Proceedings of Advanced Radar Systems, NATO AGARD Advisory Group for Aerospace Research and Development Conference*, No. 66, p. 36-1–36-10, Istanbul, Turkey, May 1970
20. R.A. Singer, Estimating optimal tracking filter performance for manned maneuvering targets. IEEE Trans. Aerosp. Electron. Syst. TAES **5**, 473–483 (1970)
21. R.W. Sittler, An optimal data association problem in surveillance theory. IEEE Trans. Mil. Electron. **MIL–8**, 2 (1964)
22. D.B. Reid, An algorithm for tracking multiple targets. IEEE Trans. Autom. Control **AC–24**, 6 (1979)
23. G. van Keuk, PDAB—Ein Verfahren zur Einzelzielvervolgung in stark gestörter Umgebung. FFF Research Report 363, Wachtberg (1987)
24. S.S. Blackman, Multiple hypothesis methods. IEEE AES Mag. **19**(1), 41–52 (2004)
25. Y. Bar-Shalom, P.K. Willett, X. Tian, *Tracking and Data Fusion. A Handbook of Algorithms* (YBS Publishing, Storrs, 2011)
26. S.S. Blackman, R. Popoli, *Design and Analysis of Modern Tracking Systems* (Artech House, Norwood, 1999)
27. A. Farina, & F.A. Studer, *Radar Data Processing*, Vol. 1 Introduction and Tracking (Wiley, New York, 1985) (Vol. 2 Advanced Topics and Applications, 1986)
28. H.A.P. Blom, J. Jygeros (eds.), *Stochastic Hybrid Systems: Theory and Safety Critical Applications* (Springer, Berlin, 2006)
29. O.E. Drummond, Multiple-Object Tracking, PhD Thesis, University of California (1975)
30. L.D. Stone, T.L. Corwin, C.A. Barlow, *Bayesian Multiple Target Tracking* (Artech House, Norwood, 1999)
31. B. Ristic, S. Arulampalam, N. Gordon, *Beyond the Kalman Filter: Particle Filters for Tracking Applications* (Artech House Radar Library, Norwood, 2004)
32. R.P.S. Mahler, *Statistical Multisource-Multitarget Information Fusion* (Artech House, Norwood, 2007)
33. R.L. Streit, *Poisson Point Processes. Imaging, Tracking, and Sensing* (Springer, New York, 2010)
34. H.B. Mitchell, *Multi-Sensor Data Fusion* (Springer, Berlin, 2007)
35. S. Stergiopoulos (ed.), *Advanced Signal Processing: Theory and Implementation for Sonar, Radar, and Non-Invasive Medical Diagnostic Systems* (CRC Press, Boca Raton, 2001)
36. F. Gustafsson, *Statistical Sensor Fusion* (Gazelle, Lancaster, 2010)
37. http://www.isif.org
38. W.-D. Wirth, *Radar Techniques Using Array Antennas* (IEE Press, London, 2001)
39. G. van Keuk, S. Blackman, On phased-array tracking and parameter control. IEEE Trans. Aerosp. Electron. Syst. **AES–29**(1), 186 (1993)
40. R. Baltes, G. van Keuk, Tracking multiple maneuvering targets in a network of passive radars, in *Proceedings of the IEEE International Radar Conference*, RADAR'95, Washington DC (1995)

41. W. Koch (ed.), IEEE ISIF Sensor Data Fusion: Trends, Solutions, Applications. Annual Workshop published in IEEE Xplore. 2005: Bonn, 2006: Dresden, 2007: Bremen, 2009: Lübeck, 2010: Leipzig, 2011: Leipzig, since 2012: Bonn
42. W. Koch, Target tracking, Chap. 8, in *Advanced Signal Processing: Theory and Implementation for Sonar, Radar, and Non-Invasive Medical Diagnostic Systems*, ed. by S. Stergiopoulos (CRC Press, Boca Raton, 2001)
43. O.E. Drummond, Target tracking with retrodicted discrete probabilities. Proc. of SPIE Sig. Data Process Small Targets **3163**, 249 (1997)
44. W. Koch, Fixed-interval retrodiction approach to Bayesian IMM-MHT for maneuvering multiple targets. IEEE Trans. Aerosp. Electron. Syst. **AES–36**(1), 2–14 (2000)
45. W. Koch, J. Biermann, Anomaly detection from tracking-derived situation elements, in *Processdings of the NATO Workshop on Data Fusion and Anomaly Detection for Maritime Situational Awareness—MSA 2009*, La Spezia, Italy, September 2009
46. B. Ristic, B. La Scala, M. Morelande, & N. Gordon, Statistical analysis of motion patterns in AIS data: Anomaly detection and motion prediction, in Proceedings of the 11th International Conference on Information Fusion, FUSION 2008, Cologne, July 2008
47. D. Fox, Activity recognition from wearable sensors. Keynote Lecture at 11th International Conference on Information Fusion, Cologne (2008)
48. S.C. Mukhopadhyay, H. Leung, P.K. Varshney, & R. Schober (eds.), Guest editorial special issue. IEEE Sens. J. Spec. Issue Cogn. Sens. Netw. **11**(3), 519–521 (2010)

Part I
Sensor Data Fusion: Methodological Framework

Chapter 2
Characterizing Objects and Sensors

In most cases, not all properties characterizing observed objects in a certain application have the same importance for producing a situation picture or can be inferred by the sensor systems involved. At the very beginning, we have to identify suitable object properties relevant to the underlying requirements, which are called *state quantities*. In the context discussed here, state quantities are completely described by numbers or appropriate collections of numbers and may be time-dependent. All relevant properties characterizing an object of interest at a certain instant of time t_k, $k \in \mathbb{N}$, are gathered in a collection X_k of state quantities, which is called *object state* at time t_k. Object states can also be composed of the individual object states of an object group.

2.1 Examples of State Quantities

1. As a first example, consider a vehicle moving on a road approximately modeled by a curve. If the vehicle's position or speed on the road at a time t_k only has interest, the corresponding object state is composed by two real numbers: the arc-length x_k of a point on the curve, representing its position, and its temporal derivative $\dot{x}_k$, representing its speed. The corresponding object state is thus given by a two-dimensional vector: $X_k = \mathbf{x}_k$ with $\mathbf{x}_k = (x_k, \dot{x}_k)^\top \in \mathbb{R}^2$.
2. Another practically important example is the kinematic state X_k of an object moving in the three-dimensional space at a given instant of time t_k, which is typically given by its position $\mathbf{r}_k$, velocity $\dot{\mathbf{r}}_k$, and acceleration $\ddot{\mathbf{r}}_k$ at this time. X_k is thus represented by a 9-dimensional vector $X_k = \mathbf{x}_k$ with $\mathbf{x}_k = (\mathbf{r}_k^\top, \dot{\mathbf{r}}_k^\top, \ddot{\mathbf{r}}_k^\top)^\top \in \mathbb{R}^9$.
3. A natural generalization of this concept is the notion of the joint state of two or more objects of interest that form an object group. If kinematic object characteristics are of interest, the corresponding object state X_k is given by a possibly high-dimensional vector $X_k = \mathbf{x}_k$ with $\mathbf{x}_k = (\mathbf{x}_k^{1\top}, \mathbf{x}_k^{2\top}, \ldots)^\top$.

W. Koch, *Tracking and Sensor Data Fusion*,
Mathematical Engineering, DOI: 10.1007/978-3-642-39271-9_2,

4. The notion of object states, however, is broader and includes other characteristic state quantities. In certain applications, object attributes can be described by positive real numbers $x_k \in \mathbb{R}^+$, related to the object's backscattering properties, for example, such as its characteristic mean radar cross section. In this case, a relevant object state may be given by $X_k = (\mathbf{x}_k, x_k)$, where the individual state quantities $\mathbf{x}_k$ (e.g. kinematics) and x_k (e.g. cross section) are taken from different sets of numbers.
5. Stationary or moving objects may belong to distinct classes. Let the object property "object belongs at time t_k to class i_k" be denoted by $i_k \in \mathbb{N}$. Moving objects, for example, can be classified according to the dynamics mode currently in effect, or according to certain characteristic features indicating, e.g., their chemical signatures. Examples of object classes relevant to air surveillance are: *bird, glider, helicopter, sporting airplane, passenger jet, fighter aircraft, missile*. In this case, a characteristic object state is given by $X_k = (\mathbf{x}_k, i_k)$.
6. For describing spatially extended objects or collectively moving object clusters, the kinematic state vector $\mathbf{x}_k$ must be complemented by an additional state quantity characterizing their spatial extension. For the sake of simplicity and to deal with the extended object or cluster tracking problem as rigorously as possible, we confine the discussion to the practically important case of *ellipsoidal* object or cluster extensions. In this case, the current extension at time t_k can be described mathematically by a symmetric and positively definite matrix $\mathbf{X}_k$. According to this approach, the following object properties are covered:

   - *Size*: volume of the extension ellipsoid
   - *Shape*: ratio of the corresponding semi-axes
   - *Orientation*: direction of the semi-axes.

   The corresponding object state is thus given by $X_k = (\mathbf{x}_k, \mathbf{X}_k)$.

Since object states must be inferred from incomplete and imperfect information sources, the collection of state quantities such as

$$X_k = (\mathbf{x}_k, x_k, \mathbf{X}_k, i_k) \tag{2.1}$$

or some of them are interpreted as a *random variables*. The application of other, more general notions of uncertainty is possible (see [1], e.g.), but excluded here. According to the Bayesian interpretation of probability theory, all available knowledge on the objects of interest at time t_k is mathematically precisely represented by probability densities of their corresponding states $p(X_k)$. If only one state quantity is of interest, for example in $\mathbf{x}_k$, $p(\mathbf{x}_k)$ is given by a marginal density:

$$p(\mathbf{x}_k) = \sum_{i_k} \int \mathrm{d}x_k \mathrm{d}\mathbf{X}_k \; p(\mathbf{x}_k, x_k, \mathbf{X}_k, i_k). \tag{2.2}$$

Methods to calculate the probability density functions related to object states with at least approximate accuracy is the main goal in Bayesian sensor data fusion.

2.2 Object Evolution Models

Object states usually change in time. Their temporal evolution, however, is imperfectly known in most cases. This fundamental ignorance can often be described by a probability density function of the object state at time t_k, which is conditioned on the previous state X_{k-1}, called transition density $p(X_k|X_{k-1})$, i.e. . With an underlying Markov assumption, knowledge about future object states can be predicted from prior knowledge via the Chapman-Kolmogorov equation:

$$p(X_k) = \int \mathrm{d}X_{k-1}\; p(X_k|X_{k-1})\; p(X_{k-1}). \tag{2.3}$$

The temporal evolution described by $p(X_k|X_{k-1})$ mirrors the real object evolution insofar as it allows a Monte-Carlo-simulation of a subsequent state X_k by generating random realizations of it according to the density $p(X_k|X_{k-1})$. It is thus reasonable to call the conditional probability density $p(X_k|X_{k-1})$ the *evolution model* of an object. In the sequel, the notion of an evolution model is illustrated by examples. A wide variety of object evolution models for kinematic object states has been described in the handbook [2, Chap. 1.5] and a series of survey papers [3–7], which are adapted to the particular requirements of the underlying application.

2.2.1 Van-Keuk's Evolution Model

An early and particularly intuitive example of state evolution models in the context of tracking and sensor data fusion was proposed by Günther van Keuk in 1971 [8]. According to van Keuk, the motion of an object is described by a linear equation with additive white Gaußian noise:

$$\mathbf{x}_k = \mathbf{F}_{k|k}\mathbf{x}_{k-1} + \mathbf{G}_{k|k-1}\mathbf{v}_k, \tag{2.4}$$

referring to kinematic object states given by $\mathbf{x}_k = (\mathbf{r}_k^\top, \dot{\mathbf{r}}_k^\top, \ddot{\mathbf{r}}_k^\top)^\top$. The Gaußian random vector $\mathbf{v}_k$ is described by a zero-mean, unit-covariance Gaußian probability density $p(\mathbf{u}_k) = \mathcal{N}(\mathbf{u}_k;\, \mathbf{0},\, \mathbf{1})$. More generally, let a Gaußian be denoted by $\mathcal{N}(\mathbf{x};\, \mathbb{E}[\mathbf{x}],\, \mathbb{C}[\mathbf{x}]) = |2\pi\mathbb{C}[\mathbf{x}]|^{-\frac{1}{2}}\, \exp\{-\frac{1}{2}(\mathbf{x} - \mathbb{E}[\mathbf{x}])^\top \mathbb{C}[\mathbf{x}]^{-1}(\mathbf{x} - \mathbb{E}[\mathbf{x}])\}$ with an expectation vector $\mathbb{E}[\mathbf{x}]$ and a symmetric, positively definite covariance matrix $\mathbb{C}[\mathbf{x}]$. The matrix $\mathbf{F}_{k|k-1}$ is called *evolution matrix*,

$$\mathbf{F}_{k|k-1} = \begin{pmatrix} \mathbf{1} & (t_k - t_{k-1})\,\mathbf{1} & \frac{1}{2}(t_k - t_{k-1})^2\mathbf{1} \\ \mathbf{O} & \mathbf{1} & (t_k - t_{k-1})\,\mathbf{1} \\ \mathbf{O} & \mathbf{O} & \mathrm{e}^{-(t_k - t_{k-1})/\theta_t}\,\mathbf{1} \end{pmatrix} \tag{2.5}$$

with a modeling parameter θ_t, while the matrix $\mathbf{G}_{k|k-1}$ is given by:

$$\mathbf{G}_{k|k-1} = q_t\sqrt{1 - \mathrm{e}^{-2(t_k - t_{k-1})/\theta_t}}\,(\mathbf{O}, \mathbf{O}, \mathbf{1})^\top, \tag{2.6}$$

implying a second modeling parameter q_t. According to this evolution model, straightforward calculations show that the acceleration $\ddot{\mathbf{r}}_k$ is described by an ergodic, zero-mean Gauß-Markov process with an autocorrelation function given by:

$$\mathbb{E}[\ddot{\mathbf{r}}_k \ddot{\mathbf{r}}_l^\top] = q_t^2\,\exp[-(t_k - t_1)/\theta_t]\,\mathbf{1},\ l \le k. \tag{2.7}$$

This expression clearly defines the modeling parameters q_t (*acceleration bandwidth*) and θ_t (*maneuver correlation time*), which have characteristic values for different classes of maneuvering objects. The corresponding Gauß-Markov transition density is given by:

$$p(\mathbf{x}_k|\mathbf{x}_{k-1}) = \mathcal{N}\big(\mathbf{x}_k;\ \mathbf{F}_{k|k-1}\mathbf{x}_{k-1},\ \mathbf{D}_{k|k-1}\big) \tag{2.8}$$

where $\mathbf{D}_{k|k-1} = \mathbf{G}_{k|k-1}\mathbf{G}_{k|k-1}^\top$ is called *evolution covariance matrix*.

2.2.2 Interacting Multiple Models

In practical applications, it might be uncertain which evolution model out of a set of r possible alternatives is currently in effect. In the case of air targets, for example, we can distinguish between different flight phases (no turn, slight maneuver, high-g, turn etc.). According to the previous discussion, the maneuvering class $1 \le i_k \le r$, to which an object belongs at time t_k, can be considered as a part of its state. In general, Markovian evolution models for object states $X_k = (\mathbf{x}_k, i_k)$ are expressed by:

$$p(x_k, i_k|x_{k-1}, i_{k-1}) = p(x_k|i_k, x_{k-1}, i_{k-1})\ p(i_k|x_{k-1}, i_{k-1}). \tag{2.9}$$

A special case that implies additional assumptions is defined by:

$$p(x_k, i_k|x_{k-1}, i_{k-1}) = p(x_k|i_k, x_{k-1})\ p(i_k|i_{k-1}) \tag{2.10}$$

$$= p_{i_k i_{k-1}}\ \mathcal{N}\big(\mathbf{x}_k;\ \mathbf{F}^{i_k}_{k|k-1}\mathbf{x}_{k-1},\ \mathbf{D}^{i_k}_{k|k-1}\big) \tag{2.11}$$

and is called *IMM evolution model* (IMM: Interaction Multiple Models, see [7] and the literature cited therein) and has been introduced by Henk Blom [9]. IMM models are characterized by r purely kinematic transition densities $p(\mathbf{x}_k|\mathbf{x}_{k-1}, i_k)$, for instance of the van Keuk type, and class transition probabilities $p_{i_k i_{k-1}} = p(i_k|i_{k-1})$ that must be specified and are part of the modeling assumptions. The transition probabilities $p_{i_k i_{k-1}}$ define a stochastic matrix. According to $\sum_{i_k=1}^r p(i_k|i_{k-1}) = 1$ the columns of such matrices must add up to one.

Note that Eq. 2.11 assumes that $p(x_k|x_{k-1}, i_k, i_{k-1})$ is independent of the past maneuvering class i_{k-1} and $p(i_k|\mathbf{x}_{k-1}, i_{k-1})$ does not depend on the object's kinematic state $\mathbf{x}_{k-1}$. While the first assumption seems to be quite natural, the second may be an oversimplification in certain cases. As an example, let us consider two evolution models describing low and strong maneuvers, respectively. The probability $p(i_k = 1|i_k = 1, \mathbf{x}_{k-1})$ that an object stays in the low maneuver model increases as the object acceleration diminishes, while $p(i_k = 2|i_k = 2, \mathbf{x}_{k-1})$ increases as the acceleration increases. If q is a measure of the maximum acceleration, state-dependent transition matrices of the form

$$\begin{pmatrix} p_{11}\mathrm{e}^{-\frac{1}{2}\frac{|\ddot{\mathbf{r}}_k|^2}{q^2}} & 1 - p_{22}\big(1 - \mathrm{e}^{-\frac{1}{2}\frac{|\ddot{\mathbf{r}}_k|^2}{q^2}}\big) \\ 1 - p_{11}\mathrm{e}^{-\frac{1}{2}\frac{|\ddot{\mathbf{r}}_k|^2}{q^2}} & p_{22}\big(1 - \mathrm{e}^{-\frac{1}{2}\frac{|\ddot{\mathbf{r}}_k|^2}{q^2}}\big) \end{pmatrix} \tag{2.12}$$

can model this type of object behavior [10]. For $r = 1$, the linear-Gauß-Markov models result as a limiting case.

2.3 Sensor Likelihood Functions

Over time, one or several sensors produce sets of measurement data Z_k that potentially carry information on object states X_k characterizing one or more objects of interest at time t_k. This information is typically imprecise and corrupted by unavoidable measurement errors, e.g. In several applications, a sensor output Z_k can refer to individual properties of several neighboring objects of interest, but it is usually unknown to which particular object. In addition, some or all sensor data can be false, i.e. be originated by unwanted objects or unrelated to really existing objects. It is furthermore not necessarily true that sensors always produce measurements of objects of interest when an attempt is made. Moreover, several closely-spaced objects may produce irresolved measurements originated by two or more objects.

At discrete instants of time t_k, we consider the set $Z_k = \{Z_k^j\}_{j=1}^{m_k}$ of m_k sensor data. The accumulation of the sensor data Z_l, $1 \leq l \leq k$, up to and including the time t_k, typically the present, is an example of a time series recursively defined by $Z^k = \{Z_k, m_k, Z^{k-1}\}$. The time series produced by the measurements of individual sensors s involved are denoted by $Z_s^k = \{Z_l^s, m_l^s\}_{l=1}^k$, $1 \leq s \leq S$.

Within the framework of Bayesian reasoning, imperfect knowledge about what measured sensor data Z_k can actually say about the states of the objects involved is modeled by interpreting Z_k as a set of random variables. The statistical properties of Z_k are characterized by a probability density function $p(Z_k|X_k)$, which is conditioned on the corresponding object state X_k referring to the same time t_k. The probability densities $p(Z_k|X_k)$ are also called *likelihood functions* when considered as functions of the random variable X_k for a given sensor output Z_k. Typically, likelihood functions need to be known only up to a factor independent of X_k,

$$\ell(X_k; Z_k) \propto p(Z_k|X_k), \tag{2.13}$$

as will become clear below. The sensor data Z_k explicitly enter into the likelihood function, while all sensor properties describing the sensors' physical and technical characteristics and their measurement performance are implicitly part of it and shape its concrete functional form. In particular, all relevant sensor parameters, such as measurement accuracy, detection probability, false alarm density, sensor resolution, minimum detectable velocity, radar beam width, pulse repetition frequency etc., must be present in the likelihood function. A likelihood function thus describes what information on an object state X_k is provided by the sensor data Z_k at a given instant of time t_k. For physical reasons, often $p(Z_k|X_l, Y) = p(Z_k|X_k)$ holds for any other random variable Y that is not part of the object state.

Likelihood functions $p(Z_k|X_k)$ model the real sensor output (and thus the physics of the underlying measurement process and its interaction with the object environment). For this reason, they provide the basis for Monte-Carlo-simulations of the sensor measurements by generating random realizations of Z_k according to $p(Z_k|X_k)$. For this reason, likelihood functions are simply called "sensor models" in direct analogy to "evolution models" given by $p(X_k|X_{k-1})$. Obviously, a sensor model is more correct, the more it provides simulated measurements that correspond on average to the real sensor output.

In the sequel, the notion of a likelihood function is illustrated by selected examples.

2.3.1 *Gaußian Likelihood Functions*

For well-separated objects, perfect detection, and in absence of false sensor data, let us consider measurements $\mathbf{z}_k$ related to the kinematic state vector $\mathbf{x}_k = (\mathbf{r}_k^\top, \dot{\mathbf{r}}_k^\top, \ddot{\mathbf{r}}_k^\top)^\top$ of an object at time t_k. In constructing a sensor model $p(\mathbf{z}_k|\mathbf{x}_k)$, two questions must be answered:

1. The first question aims at what aspect of the state vector is currently in the focus of the sensor, i.e. at the identification of a *measurement function*,

$$\mathbf{h}_k : \mathbf{x}_k \mapsto \mathbf{h}_k(\mathbf{x}_k), \tag{2.14}$$

   describing what is actually measured. Sensors characterized by the same measurement function $\mathbf{h}_k$ are called *homogeneous sensors*, in contrast to *heterogeneous sensors*, where this is not the case.
2. The second question asks for the quality of such a measurement. In many applications, additive measurement errors $\mathbf{u}_k$ can be considered, given by bias-free and Gaußian distributed random variables characterized by a *measurement error covariance matrix* $\mathbf{R}_k$. The measurement errors produced at different times or by different sensors can usually be considered as independent of each other.

In this case, the measurement process can be described by a *measurement equation* $\mathbf{z}_k = \mathbf{h}_k(\mathbf{x}_k) + \mathbf{v}_k$, which is equivalent to a Gaußian likelihood function:

$$\ell(\mathbf{x}_k;\, \mathbf{z}_k) = \mathcal{N}(\mathbf{z}_k;\, \mathbf{h}_k(\mathbf{x}_k),\, \mathbf{R}_k). \tag{2.15}$$

Range-Azimuth Measurements

In a two-dimensional plane, we may, for example, consider measurements of an object's range r_k and azimuth angle φ_k with respect to the sensor position in a given Cartesian coordinate system. Let the range and azimuth measurements be independent of each other with Gaußian measurement errors described by the standard deviations σ_r, σ_φ. Hence, in polar coordinates, the measurement error covariance matrix is diagonal: $\mathrm{diag}[\sigma_r^2, \sigma_\varphi^2]$. A transformation of the original measurements into the Cartesian coordinate system, where the state vectors $\mathbf{x}_k$ are formulated, is provided by the transform $\mathbf{t}(r_k, \varphi_k) = r_k(\cos\varphi_k, \sin\varphi_k)^\top$. A well-known result on affine transforms of Gaußian random variables (see Appendix A.3) is applicable, if the non-linear function $\mathbf{t}(r_k, \varphi_k)$ is expanded in a Taylor series up to the first order. The corresponding Jacobian can be written as the product of a rotation matrix $\mathbf{D}_{\varphi_k}$ by φ_k and a dilation matrix $\mathbf{S}_{r_k}$ defined by r_k:

$$\mathbf{T}_k = \frac{\partial \mathbf{t}(r_k, \varphi_k)}{\partial (r_k, \varphi_k)} \tag{2.16}$$

$$= \underbrace{\begin{pmatrix} \cos\varphi_k & -\sin\varphi_k \\ \sin\varphi_k & \cos\varphi_k \end{pmatrix}}_{\text{rotation } \mathbf{D}_{\varphi_k}} \underbrace{\begin{pmatrix} 1 & 0 \\ 0 & r_k \end{pmatrix}}_{\text{dilation } \mathbf{S}_{r_k}}. \tag{2.17}$$

The transformed measurements $\mathbf{z}_k = \mathbf{t}(r_k, \varphi_k)$ can thus be approximately characterized as Gaußian random variables with measurement error covariance matrices $\mathbf{R}_k$ given by:

$$\mathbf{R}_k = \mathbf{D}_{\varphi_k} \begin{pmatrix} \sigma_r^2 & 0 \\ 0 & r_k^2\sigma_\varphi^2 \end{pmatrix} \mathbf{D}_{\varphi_k}^\top. \tag{2.18}$$

according to Eq. A.20. Obviously, the measurement error covariance matrix $\mathbf{R}_k$ depends on the underlying sensor-to-object geometry, i.e. differently located sensors with the same parameters σ_r, σ_φ produce measurements of the same object that are characterized by differently oriented measurement error ellipses as illustrated in Fig. 2.1. The cross-range semi-axis of the measurement error ellipses increases with increasing range, while the other semi-axis remains constant. The orientation of the measurement ellipse depends on the object's azimuth angle φ_k. With a matrix $\mathbf{H}_k = (\mathbf{1}, \mathbf{O}, \mathbf{O})$ that projects the position vector from the object state vector, $\mathbf{H}_k\mathbf{x}_k = \mathbf{r}_k$, the resulting likelihood function is thus given by:

$$\ell(\mathbf{x}_k;\, \mathbf{z}_k) \propto \mathcal{N}\big(\mathbf{z}_k;\, \mathbf{H}_k\mathbf{x}_k,\, \mathbf{R}_k\big). \tag{2.19}$$

For a discussion of problems and improvements, e.g. "Unbiased Converted Measurements", see [2, Chap. 1.7]. In many other applications, we are analogously looking for

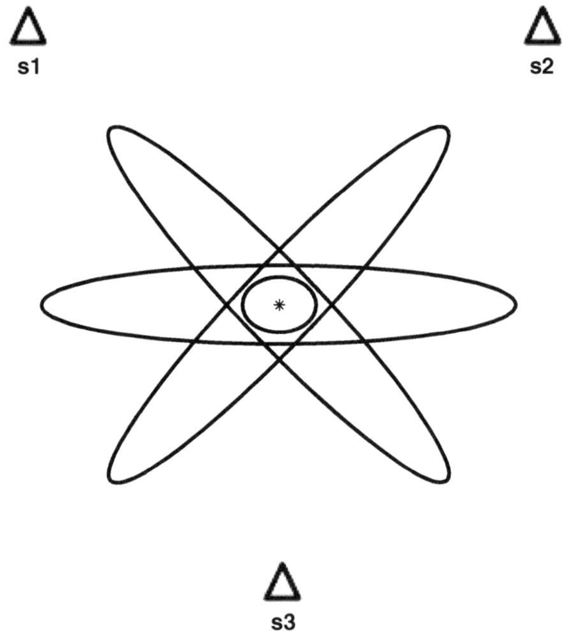

Fig. 2.1 Schematic representation of the approximate measurement error ellipses of three sensors located at s_1, s_2, s_3 measuring range and azimuth of an object at * and the impact of the sensor-to-object geometry on their mutual orientation, basic for the geometric fusion gain

formulations where a non-linear measurement function $\mathbf{h}_k$ is linearly approximated by a *measurement matrix* $\mathbf{H}_k$, possibly depending on time.

Doppler Measurements

By exploiting the Doppler effect, sensors that receive electromagnetic or acoustic wave forms reflected or emitted by objects of interest, such as radar, sonar, or ultrasonic devices, can measure the radial component $\dot{r}_k$ of an object's relative velocity $\dot{\mathbf{r}}_k - \dot{\mathbf{p}}_k$ with respect to the sensor, where $\dot{\mathbf{p}}_k$ denotes the velocity vector of the sensor platform (see Fig. 2.2). Such frequency-based measurements are often highly precise and important in certain applications such as threat evaluation. The measurement triple $(r_k, \varphi_k, \dot{r}_k)$, however, cannot be transformed into Cartesian coordinates in analogy to the previous considerations. With $(\mathbf{r}_k - \mathbf{p}_k)/|\mathbf{r}_k - \mathbf{p}_k|$, the unit vector pointing from the sensor platform at the position $\mathbf{p}_k$ to the object located at $\mathbf{r}_k$, the measurement function for range-rate measurements r_k is non-linear and given by:

$$h : \mathbf{x}_k \mapsto h(\mathbf{x}_k; \mathbf{p}_k, \dot{\mathbf{p}}_k) = \frac{(\dot{\mathbf{r}}_k - \dot{\mathbf{p}}_k)^\top (\mathbf{r}_k - \mathbf{p}_k)}{|(\dot{\mathbf{r}}_k - \dot{\mathbf{p}}_k)^\top (\mathbf{r}_k - \mathbf{p}_k)|}. \tag{2.20}$$

Note that in a practical realization sufficiently accurate navigation systems are required to estimate the platform state vector. As mentioned before, an expression following Eq. 2.19 can be obtained by a first-order Taylor expansion of the measurement function.

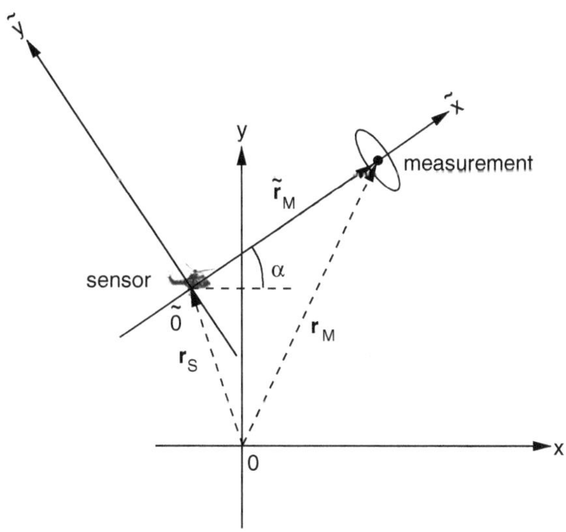

Fig. 2.2 Transformation of underlying Cartesian coordinates into a measurement-adapted system by a translation and rotation defined by the object's azimuth. Note that the $\tilde{x}$-axis points towards the object: r_k is thus a measurement of $\tilde{x}$

This type of non-linear measurement functions, however, can be handled alternatively. Consider a transformation of the underlying coordinates into a Cartesian system, where the origin is at the sensor location, while one of the axes points in a direction defined by the angular measurements (Fig. 2.2). Obviously, this transformation is simply a translation followed by a rotation. In the new coordinate system, the range-rate measurement can be interpreted as a measurement of one of the Cartesian components of the relative velocity vector of the object [11]. This means that the likelihood has a form as in Eq. 2.19 with a linear measurement equation. If a processing scheme is to be applied that requires likelihood functions of this form, a coordinate transform is therefore necessary at each processing step. In this context, Eq. A.20 is relevant, stating that a Gaußian density remains a Gaußian after this transformation. Similar considerations apply if measurements of the radial or lateral object extension is considered [12].

Evaluations with real data show that this type of dealing with range-rate measurements is significantly more robust than approaches based on Taylor expansions. The example leads to the more general observation that the appropriate formulation of sensor models requires a careful study of the individual physical quantity to be measured, quantitative performance evaluations, and comparisons with alternatives in order to achieve efficient and robust sensor models, the basic elements of sensor data fusion systems.

TDoA and FDoA Measurements

In a plane, let the kinematic state of an object emitting electromagnetic signals at time t_k be given by $\mathbf{x}_k = (\mathbf{r}_k^\top, \dot{\mathbf{r}}_k^\top)^\top$. The emitter is observed by $S = 2$ sensors on possibly moving platforms with known state vectors $(\mathbf{p}_k^{s\top}, \dot{\mathbf{p}}_k^{s\top})^\top$, $s = 1, 2$ that passively receive the emitted radiation. The *Time of Arrival (ToA)*, the time interval from transmitting a signal at the emitter position $\mathbf{r}_k$ and receiving it at a sensor position $\mathbf{p}_k^s$, is equal to the time the signal needs to travel from the emitter to the sensor. Since we know the propagation speed of the signal (speed of light c), ToA is given by $|\mathbf{r}_k - \mathbf{p}_k^s|/c$. A sensor model for *Time Difference of Arrival (TDoA)* measurements z_k^t directly follows:

$$\ell_t(\mathbf{x}_k; z_k^t) = \mathcal{N}\big(z_k^t c;\ h_t(\mathbf{x}_k),\ \sigma_t/c\big) \tag{2.21}$$

with a measurement function h_t given by:

$$h_t(\mathbf{x}_k; \mathbf{p}_k^1, \mathbf{p}_k^2) = |\mathbf{r}_k - \mathbf{p}_k^1| - |\mathbf{r}_k - \mathbf{p}_k^2|, \tag{2.22}$$

where σ_t denotes the standard deviation of the corresponding TDoA measurement errors. The locations of the sensor platforms enter as parameters. The sign of an individual measurement indicates which of the sensors is closer to the emitter. Without loss of generality, we can thus limit the discussion to one of these cases. The solid line in Fig. 2.3, a hyperbola, shows all potential emitter positions producing the same TDoA measurements, i.e. having the same distance difference from the sensors.

The Doppler shift in frequency is proportional to the radial velocity component of an emitter moving with respect to a Cartesian sensor coordinate system. The inverse wave length λ of the emitted radiation is the proportionality constant. Let $(\mathbf{r}_k - \mathbf{p}_k^s)/|\mathbf{r}_k - \mathbf{p}_k^s|$ denote the unit vector pointing from the sensor position $\mathbf{p}_k^s$ at time t_k to the emitter located at $\mathbf{r}_k$, moving with the velocity $\dot{\mathbf{r}}_k$. As before, the radial component of the relative velocity of the emitter with respect to the sensor s is given by $(\dot{\mathbf{r}}_k - \dot{\mathbf{p}}_k^s)^\top(\mathbf{r}_k - \mathbf{p}_k^s)/|\mathbf{r}_k - \mathbf{p}_k^s|$. The measurement function for *Frequency Difference of Arrival (FDoA)* measurements is thus given by:

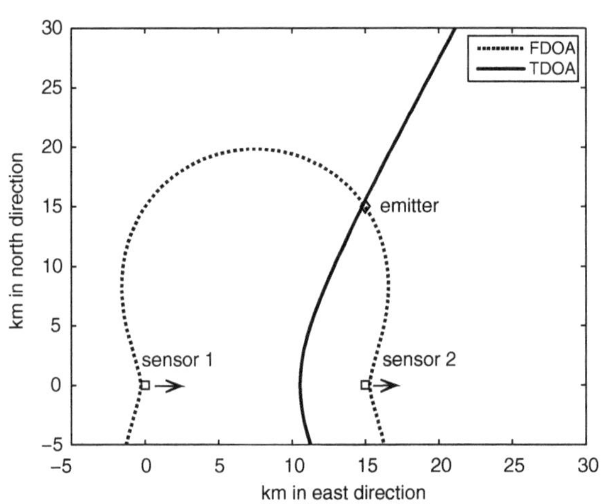

Fig. 2.3 Localization of an emitter using TDoA and FDoA measurements by two moving sensors: constant TDoA/FDoA emitter location curve

$$h_f(\mathbf{x}_k) = \frac{(\dot{\mathbf{r}}_k - \dot{\mathbf{p}}_k^1)^\top(\mathbf{r}_k - \mathbf{p}_k^1)}{|\mathbf{r}_k - \mathbf{p}_k^1|} - \frac{(\dot{\mathbf{r}}_k - \dot{\mathbf{p}}_k^2)^\top(\mathbf{r}_k - \mathbf{p}_k^2)}{|\mathbf{r}_k - \mathbf{p}_k^2|}. \tag{2.23}$$

A constant FDoA curve for a non-moving emitter is shown by the dashed curve in Fig. 2.3, where the arrows indicate the direction of the platform velocities. In this example, TDoA and FDoA are complementary in that TDoA takes the (approximate) role of bearings measurement, and FDoA, the (approximate) role of distance measurement. TDoA and FDoA measurements may be obtained simultaneously by calculating the Complex Ambiguity Function (CAF, [13]), which cross-correlates the signals received by the sensors. The likelihood functions that result from the measurement functions h_t and h_f are shown in Fig. 2.4. Techniques discussed in [14] and applied to emitter localization and tracking, make it possible to approximate the likelihood functions by sums of appropriately chosen individual Gaußians with a linear approximation of the measurement function according to Eq. 2.19.

2.3.2 Multiple Sensor Likelihood

Assume S homogeneous sensors are located at different positions that measure, at the same instant of time t_k, the same linear function $\mathbf{H}_k\mathbf{x}_k$ of an individual kinematic object state $\mathbf{x}_k$. Under conditions as considered before, let the individual likelihood functions of the sensors be given by:

$$\ell_s(\mathbf{x}_k; \mathbf{z}_k^s) \propto \mathcal{N}\big(\mathbf{z}_k^s;\ \mathbf{H}_k\mathbf{x}_k,\ \mathbf{R}_k^s\big), \quad s = 1, \ldots, S. \tag{2.24}$$

Since independently working sensors were assumed, the over-all likelihood function describing the information on an object state, which is provided by all sensors at time t_k, can be written as a product of the individual likelihood functions:

$$\ell(\mathbf{x}_k; \mathbf{z}_k^1, \ldots, \mathbf{z}_k^S) \propto \prod_{s=1}^{S} \mathcal{N}\big(\mathbf{z}_k^s;\ \mathbf{H}_k\mathbf{x}_k,\ \mathbf{R}_k^s\big). \tag{2.25}$$

In Appendix A.5, a product formula for Gaußians is proven, which is well-suited for simplification of the product representation of the over-all likelihood function. An induction argument directly yields that $\ell(\mathbf{x}_k; \mathbf{z}_k^1, \ldots, \mathbf{z}_k^S)$ can be represented by a single Gaußian,

$$\ell(\mathbf{x}_k; \mathbf{z}_k^1, \ldots, \mathbf{z}_k^S) \propto \mathcal{N}\big(\mathbf{z}_k;\ \mathbf{H}_k\mathbf{x}_k,\ \mathbf{R}_k\big) \tag{2.26}$$

with an *effective measurement* $\mathbf{z}_k$ and a corresponding *effective measurement error covariance* $\mathbf{R}_k$ defined by:

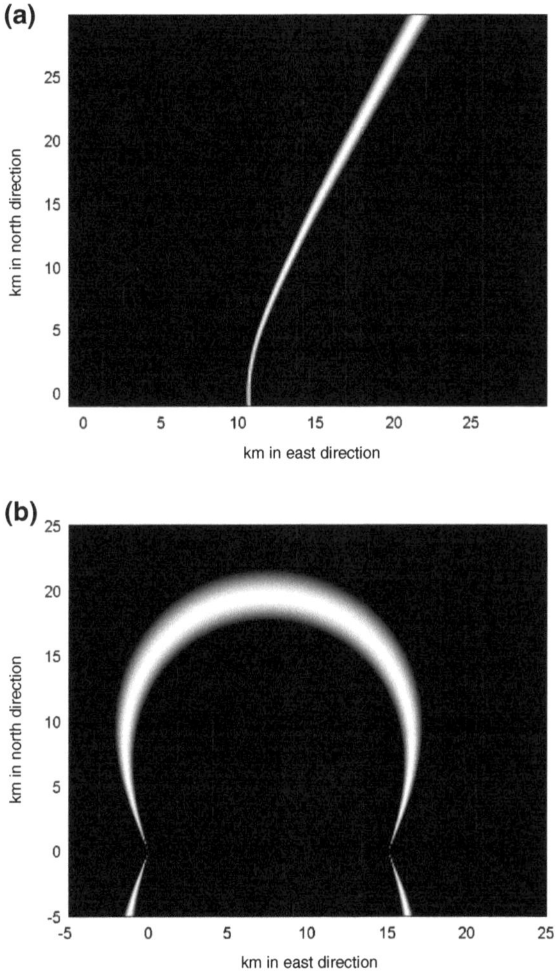

Fig. 2.4 Likelihood functions for TDoA and FDoA measurements. Idea: approximate the likelihood functions by a sum of appropriately chosen Gaußians with a linear approximation of the measurement function. **a** Likelihood of TDoA and FDoA measurements. **b** Likelihood of FDoA measurements.

$$\mathbf{R}_k = \Big(\sum_{s=1}^{S} \mathbf{R}_k^{s\;-1}\Big)^{-1} \tag{2.27}$$

$$\mathbf{z}_k = \mathbf{R}_k \sum_{s=1}^{S} \mathbf{R}_k^{s\;-1}\mathbf{z}_k^s. \tag{2.28}$$

The effective measurement is thus represented by a *weighted arithmetic mean* of the measurements $\mathbf{z}_k^s$ provided by the individual sensors involved, where the corresponding matrix-valued weighting factors are given by the inverses of the corresponding measurement error covariance matrices $\mathbf{R}_k^{s\;-1}$. Obviously, "poor" measurements, characterized by large measurement errors, provide smaller contributions to the effective measurement $\mathbf{z}_k$ than "good" ones. Dealing with data from multiple sensors in this way is an example of centralized or distributed *measurement fusion* as opposed to *track-to-track fusion* (see the discussion in Sect. 3.1.1). Figure 2.1 illustrates the *geometric fusion gain* $\mathbf{R}_k$ according to Eq. 2.27.

Geometric Fusion Gain

From these considerations several conclusions can be drawn:

1. If all individual measurement covariance matrices are identical, $\mathbf{R}_k^s = \mathbf{R}'_k$, $s = 1, \ldots, S$, the effective measurement is the simple arithmetic mean of the individual measurements: $\mathbf{z}_k = \frac{1}{S}\sum_s \mathbf{z}_k^s$. For the effective measurement error covariance, we obtain the 'square-root' law: $\mathbf{R}_k = \frac{1}{S}\mathbf{R}'_k$.
2. If all measurement error ellipses involved differ significantly in their geometrical orientation relative to each other, a much greater effect can be observed (geometrical fusion gain).
3. The 'intersection' of error ellipses is obtained by calculating the *harmonic mean* of the related error covariance matrices. The harmonic mean of error covariances quantitatively describes the gain by fusing sensor data from several sources and has been referred to as the *Fusion Equation*.
4. In the limiting case of very narrow measurement error ellipses, such as those characterized by $\sigma_r \gg r_k\sigma_\varphi$, the triangulation of an object's position from bearings only is obtained. Analogously, range-only measurements can be used (trilateration).

These considerations are also valid in three spatial dimensions as well as in more sophisticated sensor data, such as bistatic range or range-rate measurements (see for example [15, 16]).

If there is more than one object in the common field of view of bearing-only sensors, not every intersection of two bearings actually corresponds to a real object position. Figure 2.5 illustrates this situation as well as the appearance of *ghosts* that do not correspond to real objects. Of course, in the case of inaccurate, false, missing, or even irresolved bearings, the *de-ghosting* is by no means trivial. For more details and possible solutions of de-ghosting problems in certain applications, see for example [17] (bearing-only tracking) or [18] (passive radar).

Cumulative Detection

In applications with relatively large data innovation intervals between successive data collections, such as in air-to-ground wide-area surveillance, sensor data fusion is particularly important for enhancing the data rate. Assuming measurement fusion as before, we consider the *mean cumulative data innovation intervals* ΔT_c [19] resulting from the individual innovation intervals ΔT_s, $s = 1, \ldots, S$ of S sensors, which is defined by:

$$\frac{1}{\Delta T_c} = \sum_{s=1}^{S} \frac{1}{\Delta T_s}. \tag{2.29}$$

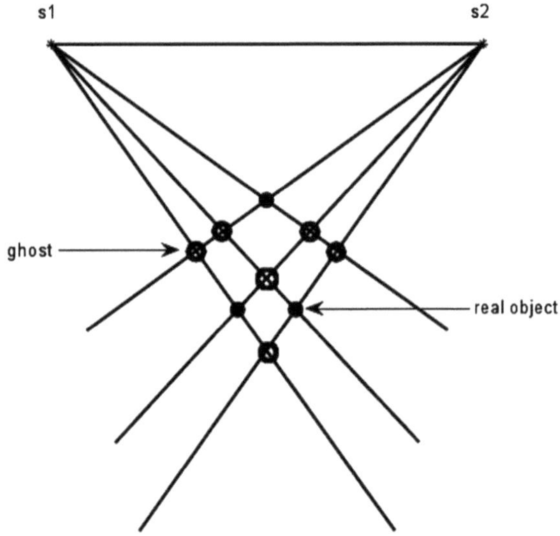

Fig. 2.5 Two objects observed by two bearings-only sensors s_1, s_2. Not all intersections of bearing correspond to real objects. Typically, the number of "ghosts" is much higher than the number of objects involved

The *cumulative detection probability* is given by $P_D^S = 1 - \prod_{s=1}^{S}(1 - P_D^s)$, where P_D^s denotes the individual detection probability of sensor s, possibly depending on the corresponding sensor-to-object geometry (see the discussion in Sect. 3.1.2). It is appropriate to introduce the notion of the *mean cumulative detection probability* P_D^c, referring to ΔT_c and defined by:

$$P_D^c = 1 - \prod_{s=1}^{S} (1 - P_D^s)^{\frac{\Delta T_c}{\Delta T_s}}. \tag{2.30}$$

The data innovation intervals ΔT_s also enter into this formula, which describes the mean improvement of the overall detection performance to be expected by sensor data fusion. The larger ΔT_s, the smaller is the effect of sensor s on the overall performance, even if the corresponding individual detection probability P_D^s is large.

2.3.3 Likelihood for Ambiguous Data

A sensor output at time t_k, consisting of m_k measurements collected in the set Z_k, can be ambiguous, i.e. the origin of the sensor data has to be explained by a set of data interpretations, which are assumed to be exhaustive and mutually exclusive. As an example, let us consider measurements $Z_k = \{\mathbf{z}_k^j\}_{j=1}^{m_k}$ possibly related to the kinematic state $\mathbf{x}_k$ of well-separated objects. 'Well-separated' here means that measurements potentially originated by one object could not have been originated by another. Even in this simplified situation, ambiguity can arise from imperfect detection, false measurements, often referred to as *clutter*, or measurements from unwanted objects.

Illustration

As a schematic illustration of a more general case, let us consider six measurements produced by two closely-spaced objects (see Fig. 2.6). Among other data interpretations, the black dots indicate two measurements assumed as real, while all other data are assumed to be false (Fig. 2.6a). The asterisks denote predicted measurements provided by the tracking system. Under assumptions discussed in (Sect. 3.2.2), object measurements are Gaußian distributed about the predicted measurements with a covariance matrix $\mathbf{S}_{k|k-1}$ determined by the ignorance on the object state as well as by the sensor and the evolution model. The difference vector $\boldsymbol{\nu}_{k|k-1}$ between an actual and a predicted measurement is called *innovation*. As will become clear below, the statistical properties of the innovation related to a particular interpretation hypothesis are essential to evaluating its statistical weight. *Gating* means that only those sensor data are considered whose innovations are smaller than a certain predefined threshold in the sense of a Mahalanobis norm: $\boldsymbol{\nu}_{k|k-1}^{\top}\mathbf{S}_{k|k-1}^{-1}\boldsymbol{\nu}_{k|k-1} < \chi^2(P_c)$. Expectation gates thus contain the measurements with a given (high) *correlation probability* P_c. Obviously, the ambiguities involved with the situation in Fig. 2.6 are not completely resolved by gating.

More feasible hypotheses, however, compete with the data interpretation previously discussed. For instance, the targets could have produced a single irresolved measurement as indicated in Fig. 2.6b, all other data being false. Alternatively, one of the objects may not have been detected or no detection may have occurred at all. The expectation gates and therefore the ambiguity of the received sensor data grow larger according to an increasing number of false measurements and missed detections as well as to large measurement errors, data innovation intervals, or expected object maneuvers.

Well-separated Objects

Let $j_k = 0$ denote the data interpretation hypothesis that the object has not been detected at all by the sensor at time t_k, i.e. all sensor data have to be considered as false measurements, while $1 \le j_k \le m_k$ represents the hypothesis that the object has been detected, $\mathbf{z}_k^{j_k} \in Z_k$ being the corresponding measurement of the object properties, the remaining sensor data being false. Obviously, $\{0, \dots, m_k\}$ denotes a set of mutually exclusive and exhaustive data interpretations.

Due to the total probability theorem and with D or $\neg$D denoting that the object has or has not been detected, the conditional probability density $p(Z_k, m_k|\mathbf{x}_k)$ can be written as a weighted sum of conditional likelihood functions:

$$p(Z_k,m_k|\mathbf{x}_k) = p(Z_k, m_k, \neg\mathrm{D}|\mathbf{x}_k) + p(Z_k, m_k, \mathrm{D}|\mathbf{x}_k) \tag{2.31}$$

$$= p(Z_k, m_k|\neg\mathrm{D}, \mathbf{x}_k)\, p(\neg\mathrm{D}|\mathbf{x}_k) + p(Z_k, m_k|\mathrm{D}, \mathbf{x}_k)\, p(\mathrm{D}|\mathbf{x}_k). \tag{2.32}$$

The sensor model $p(Z_k, m_k|\mathbf{x}_k)$ can be traced back to intuitively understandable physical/technical phenomena and related sensor parameters. As a first consequence,

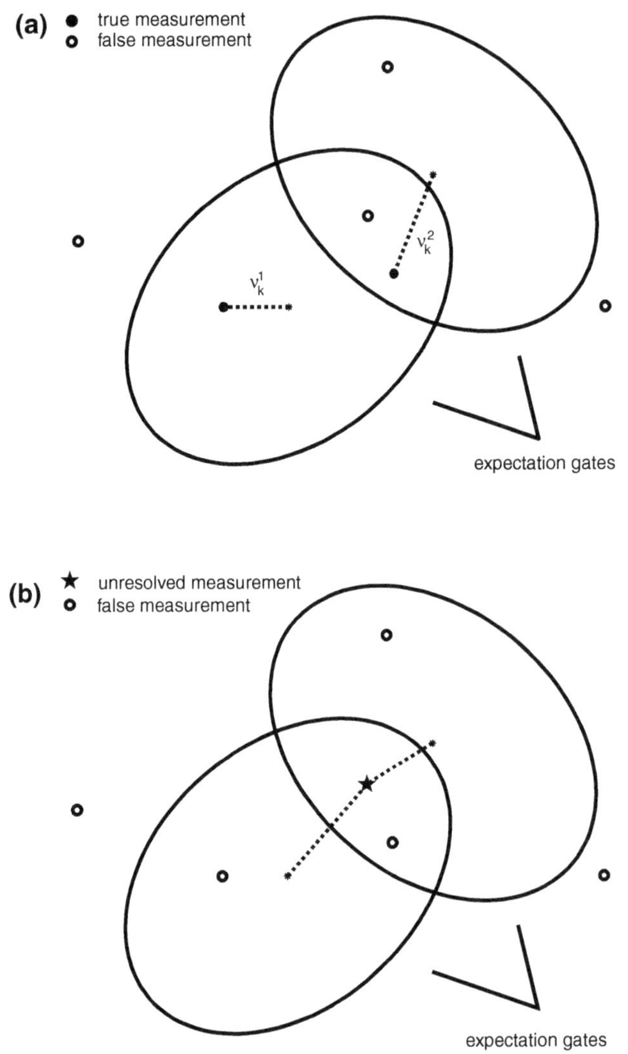

Fig. 2.6 Sensor measurements produced by two closely-spaced objects: competing data interpretations due to uncertain origin of the the sensor data including hypotheses assuming possibly irresolved measurements. **a** Two resolved targets. **b** Two irresolved targets

the *probability of detection*, $p(\mathrm{D}|\mathbf{x}_k) =: P_D$, and non-detection, $p(\neg\mathrm{D}|\mathbf{x}_k) = 1 - P_D$ enter into the likelihood as a characteristic parameter related to the detection process performed within a sensor system. For the sake of simplicity, we do not express by our notation here that detection probabilities may depend on the object state vectors $\mathbf{x}_k$. State-dependent detection probabilities, however, become relevant in several real-world applications (see the discussion in Sect. 3.1.2).

1. The conditional likelihood $p(Z_k, m_k|\neg\mathrm{D}, \mathbf{x}_k)$ in Eq. 2.32 can be rewritten as:

$$p(Z_k, m_k|\neg\mathrm{D}, \mathbf{x}_k) = p(Z_k|m_k, \neg\mathrm{D}, \mathbf{x}_k)\; p(m_k|\neg\mathrm{D}_k, \mathbf{x}_k) \tag{2.33}$$

$$= p_F(m_k)\; |\mathrm{FoV}|^{-m_k}. \tag{2.34}$$

Here, the probability of having received m_k false measurements given the object was not detected, $p(m_k|\neg\mathrm{D}_k, \mathbf{x}_k)$, is provided by a further modeling assumption, which relates the fluctuating number of false measurements to a mean *spatial clutter density* ρ_F characteristic of the sensor's detection process and the sensing environment. According to modeling assumptions, which are well-proven in many practical applications, let the probability of the number of false data involved $p(m_k|\neg\mathrm{D}_k, \mathbf{x}_k)$ be given by a Poisson distribution

$$p_F(m_k) = (\bar{m}_F^{m_k}/m_k!)\; \mathrm{e}^{-\bar{m}_F} \tag{2.35}$$

with a mean number of false measurements $\bar{m}_F$, which is related to the volume of the sensor's field of view $|\mathrm{FoV}|$ and ρ_F via $\bar{m}_F = \rho_F|\mathrm{FoV}|$. ρ_F may vary on a larger scale than the direct object neighborhood. Values for ρ_F can either be taken from so-called 'clutter maps', i.e. from related context information, or adaptively be estimated on-line [20–22]. Since false measurements are assumed to be independent from each other and equally distributed in the sensor's field of view (FoV), we have $p(Z_k|m_k, \neg\mathrm{D}, \mathbf{x}_k) = \prod_{j=1}^{m_k} p(\mathbf{z}_k^j|\neg\mathrm{D}, \mathbf{x}_k) = |\mathrm{FoV}|^{-m_k}$.
2. For the conditional likelihood $p(Z_k, m_k|\mathrm{D}, \mathbf{x}_k)$ in Eq. 2.32, we obtain analogously:

$$p(Z_k|m_k, \mathrm{D}, \mathbf{x}_k) = \sum_{j_k=1}^{m_k} p(Z_k, m_k, j_k|\mathrm{D}, \mathbf{x}_k) \tag{2.36}$$

$$= \sum_{j_k=1}^{m_k} p(Z_k|m_k, j_k, \mathrm{D}, \mathbf{x}_k)\; p(m_k|j_k, \mathrm{D}, \mathbf{x}_k)\; p(j_k|\mathrm{D}, \mathbf{x}_k) \tag{2.37}$$

$$= \frac{p_F(m_k - 1)}{m_k\; |\mathrm{FoV}|^{m_k - 1}} \sum_{j_k=1}^{m_k} \mathcal{N}\big(\mathbf{z}_k^{j_k};\; \mathbf{H}_k\mathbf{x}_k,\; \mathbf{R}_k^{j_k}\big). \tag{2.38}$$

Under the assumption j_k, we assume a Gaußian likelihood function for describing $\mathbf{z}_k^j$ according to Eq. 2.19, the other $m_k - 1$ measurements being treated as equally distributed in the sensor field of view:

$$p(Z_k|m_k, j_k, \mathrm{D}, \mathbf{x}_k) = |\mathrm{FoV}|^{-(m_k-1)}\; \mathcal{N}\big(\mathbf{z}_k^j;\; \mathbf{H}_k\mathbf{x}_k,\; \mathbf{R}_k^j\big). \tag{2.39}$$

$p(m_k|j_k, \mathrm{D}, \mathbf{x}_k)$ is given by $p_F(m_k - 1)$, while a priori the m_k data association hypotheses j_k are assumed to equally distributed, $p(j_k|\mathrm{D}, \mathbf{x}_k) = m_k^{-1}$.

By exploiting the definition of the Poisson distribution and re-arranging the terms, a likelihood function for ambiguous data is proportional to a weighted sum of Gaußians and a constant ($\rho_F > 0$):

$$\ell(\mathbf{x}_k;\, Z_k, m_k) \propto (1 - P_D)\rho_F + P_D \sum_{j_k=0}^{m_k} \mathcal{N}\big(\mathbf{z}_{j_k};\; \mathbf{H}_k\mathbf{x}_k,\, \mathbf{R}_k\big). \tag{2.40}$$

In the special case of $\rho_F = 0$ (no false measurements to be expected), Kronecker symbols can are be used to find an expression for the likelihood ($\delta_{ij} = 1$ for $i = j$, $\delta_{ij} = 0$ otherwise):

$$\ell(\mathbf{x}_k;\, Z_k, m_k) \propto (1 - P_D)\, \delta_{0m_k} + P_D\, \mathcal{N}\big(\mathbf{z}_k;\; \mathbf{H}_k\mathbf{x}_k,\, \mathbf{R}_k\big)\delta_{1m_k}. \tag{2.41}$$

Possibly Irresolved Measurements

Similar considerations can be applied to formulate appropriate likelihood functions in multiple object situations described by joint object states $\mathbf{x}_k = (\mathbf{x}_k^{1\top}, \mathbf{x}_k^{2\top}, \ldots)^\top$, where possibly irresolved measurements are to be taken into account (see Fig. 2.6b).

Among other sensor properties, in such situations the limited capability of physical sensors to resolve closely-spaced objects must be part of the sensor model. The link from from physical resolution phenomena to the likelihood functions is provided by considering the probability P_u of two objects being irresolved. P_u certainly depends on the relative distance vector $\mathbf{d}_k$ in proper coordinates between the objects at a certain time t_k: $P_u = P_u(\mathbf{d}_k)$. We qualitatively expect that P_u will be close to One for small values of $|\mathbf{d}_k|$, while $P_u = 0$ for distances significantly larger than certain resolution parameters, such as the beam-width, band-width, or coherence length of a radar. We expect a narrow transient region. In a generic model of the sensor resolution, we may describe P_u by a Gaußian-type function of $\mathbf{d}_k$ with a 'covariance matrix' serving as a quantitative measure of the sensor resolution capability, which in particular reflects the extension and spatial orientation of ellipsoidal resolution cells depending on the underlying sensor-to-object geometry.

According to these considerations, the notion of a *resolution probability* is crucial if suitable sensor models for object groups are to be designed. The underlying Gaußian structure significantly simplifies the mathematical reasoning involved and finally leads to a representation of the likelihood function by a weighted sum of Gaußians and a constant, i.e. we have to deal with the same mathematical structure as before in the case of well-separated objects. For details see the discussion in Sect. 3.1.1.

2.3.4 Incorporating Signal Strength

The strength z_k of an received object signal at time t_k carries information on the corresponding object strength x_k, which is in a radar application, for example, directly related to the object's characteristic mean radar cross section via the radar equation [23]. An individual sensor measurement related to an object state $X_k = (\mathbf{x}_k, x_k)$ is thus given by $Z_k^j = (\mathbf{z}_k^j, z_k^j)$. With this notation, the previous discussion can directly be generalized:

$$\begin{aligned} p(Z_k, m_k|X_k) =& p(Z_k, m_k|\neg \mathrm{D}, X_k)\; p(\neg \mathrm{D}|X_k) \\ &+ \sum_{j_k=1}^{m_k} p(Z_k|m_k, j_k, \mathrm{D}, X_k)\; p(m_k|j_k, \mathrm{D}, X_k)\; p(j_k|\mathrm{D}, X_k). \end{aligned} \tag{2.42}$$

We only have to consider the following conditional likelihood functions:

$$p(Z_k, m_k|\neg \mathrm{D}, X_k) = |\mathrm{FoV}|^{-m_k} \prod_{j=1}^{m_k} p(z_k^j|\neg \mathrm{D}, X_k) =: \Lambda \tag{2.43}$$

$$p(Z_k|m_k, j_k, \mathrm{D}, X_k) = \mathcal{N}\big(\mathbf{z}_k^{j_k};\, \mathbf{H}_k \mathbf{x}_k,\, \tfrac{1}{z_k}\mathbf{R}_k^{j_k}\big)\, \frac{\Lambda\; p(z_k^{j_k}|\mathrm{D}, X_k)}{|\mathrm{FoV}|^{-1}\; p(z_k^{j_k}|\neg \mathrm{D}, x_k)}. \tag{2.44}$$

We here additionally assumed a measurement error covariance matrix $\frac{1}{z_k}\mathbf{R}_k$ depending on the received signal strength z_k. This can be justified by radar signal processing theory [23] and reflects the empirical phenomenon that the weaker the signals received are the less accurate the resulting measurements.

For the sake of simplicity, we furthermore assume that $p(z_k|\neg \mathrm{D}, X_k)$ and $p(z_k|\mathrm{D}, X_k)$ do not depend on the kinematic state vector, although the received signal strength may in principle depend on the sensor-to-object geometry. The often highly complex dependency on the aspect angle is instead described by so-called Swerling models of radar cross section fluctuations [24]. According to the practically important Swerling-I-model, the received signal strengths z_k are random variables, characterized by $p(z_k|x_k) = \mathrm{e}^{-z_k/(1+x_k)}/(1 + x_k)$, i.e. simple exponential densities. Let us furthermore assume that a detector decides on "detection", denoted by 'D', if z_k exceeds a certain threshold: $z_k > \lambda$. If there is actually an object present that has been detected, P_D = p('D'|D) is thus given by:

$$P_D = \frac{1}{1 + x_k} \int_\lambda^\infty \mathrm{d}z_k\, \mathrm{e}^{-z_k/(1+x_k)} = \mathrm{e}^{-\lambda/(1+x_k)}. \tag{2.45}$$

The corresponding false alarm probability $P_F = p(\text{'D'}|\, \neg\mathrm{D})$ is given by $P_F = \int_\lambda^\infty \mathrm{d}z_k \mathrm{e}^{-z_k} = \mathrm{e}^{-\lambda}$. Here $x_k = 0$ is assumed for a noise-type target. This result directly leads to the famous *Swerling formula*, which relates the detection probability P_D to the object strength x_k and the false alarm probability P_F characterizing the detector:

$$P_D(x_k, P_F) = P_F^{\frac{1}{1+x_k}}. \tag{2.46}$$

A detected signal not belonging to a real object of interest is a clutter signal with a *clutter strength* c_k, a parameter characterizing context information on the sensing environment. After detection and according to Bayes Theorem, a received signal strength z_k is either distributed according to:

$$p(z_k|x_k, \mathrm{D}) = \begin{cases} \frac{\mathrm{e}^{(\lambda - z_k)/(1+x_k)}}{1+x_k} & \text{for } z_k > \lambda \\ 0 & \text{else} \end{cases} \tag{2.47}$$

or to:

$$p(z_k|x_k, \neg\mathrm{D}) = \begin{cases} \frac{\mathrm{e}^{(\lambda - z_k)/(1+c_k)}}{1+c_k} & \text{for } z_k > \lambda \\ 0 & \text{else.} \end{cases} \tag{2.48}$$

By inserting these densities in Eqs. 2.43 and 2.44, we directly obtain the modified likelihood function for ambiguous sensor data that include signal strength information:

$$\ell(\mathbf{x}_k, x_k; Z_k, m_k) \propto (1 - \mathrm{e}^{-\frac{\lambda}{1+x_k}})\rho_F$$
$$+ \sum_{j=1}^{m_k} \left(\frac{\mathrm{e}^{(\lambda - z_k^j)/(1+c_k)}}{1+c_k}\right)^{-1} \frac{\mathrm{e}^{-z_k^j/(1+x_k)}}{1+x_k} \mathcal{N}\left(\mathbf{z}_k^j;\ \mathbf{H}_k\mathbf{x}_k,\ \tfrac{1}{z_k^j}\mathbf{R}_k^j\right). \tag{2.49}$$

Note that this likelihood function depends on the sensor parameters $\mathbf{R}_k$ and λ, characterizing the measurement and the detection process, as well as the environmental parameters ρ_F and c_k. These parameters represent context information, which is a necessary input for the likelihood function besides the sensor data themselves.

2.3.5 Extended Object Likelihood

According to the discussion in Sect. 2.1, spatially extended objects or collectively moving object clusters, are characterized by an object state $X_k = (\mathbf{x}_k, \mathbf{X}_k)$, which consists of an kinematic state vector $\mathbf{x}_k$ and a symmetric, positively definite matrix $\mathbf{X}_k$. For the sake of simplicity, let us exclude false or unwanted measurements at present. In a first approximation, the number m_k of measurements in Z_k is assumed to be independent of the object state X_k,; i.e. $p(m_k|\mathbf{x}_k, \mathbf{X}_k)$ is assumed to be a constant.

In the case of extended or group targets, the significance of a single measurement is evidently dominated by the underlying object extension. The sensor-specific measurement error describing the precision by which a given scattering center is currently measured is the more unimportant, the larger the actual extension of the object is compared to the measurement error. The individual measurements must therefore rather be interpreted as measurements of the centroid of the extended or collective object, since it is unimportant, which of the varying scattering centers was actually responsible for the measurement.

We thus interpret each individual measurement produced by an extended object as a measurement of the object centroid with a corresponding 'measurement error' being proportional to the object extension $\mathbf{X}_k$ to be estimated. By means of this 'measurement error', however, the object extension $\mathbf{X}_k$ becomes explicitly part of the likelihood function $p(Z_k, m_k|\mathbf{x}_k, \mathbf{X}_k)$, which describes what the measured quantities Z_k, m_k can say about the state variables $\mathbf{x}_k$ and $\mathbf{X}_k$. Elementary calculations, similar to those used in Sect. 2.3.2, yield the following factorization (see Appendix A.10 for details):

$$p(Z_k, m_k|\mathbf{x}_k, \mathbf{X}_k) \propto \prod_{j=1}^{m_k} \mathcal{N}\big(\mathbf{z}_k^j;\ \mathbf{H}_k\mathbf{x}_k,\ \mathbf{X}_k\big) \tag{2.50}$$

$$\propto \mathcal{N}\big(\mathbf{z}_k;\ \mathbf{H}_k\mathbf{x}_k,\ \frac{1}{m_k}\mathbf{X}_k\big)\ \mathcal{LW}\big(\mathbf{Z}_k;\ m_k-1,\ \mathbf{X}_k\big) \tag{2.51}$$

up to a multiplicative constant independent of the state variables. In Eq. 2.51, the centroid measurement $\mathbf{z}_k$ and the corresponding scattering matrix $\mathbf{Z}_k$ are given by:

$$\mathbf{z}_k = \frac{1}{m_k}\sum_{j=1}^{m_k} \mathbf{z}_k^j \tag{2.52}$$

$$\mathbf{Z}_k = \sum_{j=1}^{m_k} (\mathbf{z}_k^j - \mathbf{z}_k)(\mathbf{z}_k^j - \mathbf{z}_k)^\top, \tag{2.53}$$

while $\mathcal{LW}\big(\mathbf{Z}_k;\ m_k-1,\ \mathbf{X}_k\big)$ is proportional to a Wishart density with m_k-1 degrees of freedom, a matrix-variate probability density function describing the properties of the random variable $\mathbf{Z}_k$ (see Appendix A.11 for details):

$$\mathcal{LW}\big(\mathbf{Z}_k;\ m_k-1,\ \mathbf{X}_k\big) = |\mathbf{X}_k|^{-\frac{m_k-1}{2}}\ \mathrm{etr}\big(-\tfrac{1}{2}(\mathbf{Z}_k\mathbf{X}_k^{-1})\big). \tag{2.54}$$

References

1. R.R. Yager, L. Liu (eds.). *Classic works of the Dempster-Shafer theory of belief functions* (Springer, Berlin, 2008)
2. Y. Bar-Shalom, P.K. Willett, X. Tian, Tracking and Data Fusion. in *A Handbook of Algorithms* (YBS Publishing, Storrs, 2011)
3. X. Rong, Survey of maneuvering target tracking. Part I. dynamic models. IEEE Trans. Aerosp. Electron. Syst. **39**(4), 1333–1364 (2003)
4. X.R. Li, V.P. Jilkov, A survey of maneuvering target tracking-part II: ballistic target models. in *Proceedings of the 2001 SPIE Conference on Signal and Data Processing of Small Targets*, vol. 4473, 559–581 (2001)
5. X.R. Li, V.P. Jilkov, A survey of maneuvering target tracking-part III: measurement models, in *Proceedings of the 2001 SPIE Conference on Signal and Data Processing of Small Targets*, vol. 4473, pp. 423–446, 2001

6. X.R. Li and V.P. Jilkov, A survey of maneuvering target tracking-Part IV: decision-based methods. in *Proceedings of the 2002 SPIE Conference on Signal and Data Processing of Small Targets*, vol. 4728, p. 60, 2002
7. X. Rong, Li, V.P. Jilkov, Survey of maneuvering target tracking. Part V. Multiple model methods". in *IEEE Transactions on Aerospace and Electronic Systems* vol. 41(4) (2005), pp.1255–1321
8. G.V. Keuk, Zielvervolgung nach Kalman-anwendung auf elektronisches radar. FFM-Bericht, 173 (1971)
9. H.A.P. Blom. An efficient filter for abruptly changing systems. In *Proceedings of the 23rd IEEE Conference on Decision and Control*, Las Vegas, 1984
10. M. Michaelis, F. Govaers, W. Koch, State dependent mode transition probabilities, in IEEE ISIF Workshop Sensor Data Fusion-Trends, Solutions, Applications, Bonn, Germany (IEEE Xplore), October 2013
11. D. Robinson, W. Seels, Kalman-Filter, European Patent No. EP0452797B1, April 1992
12. R. Klemm, M. Mertens, Tracking of convoys by airborne STAP radar. In IEEE international conference on information fusion, Cologne, 2008
13. D.J. Torrieri, Statistical theory of passive location systems. IEEE Trans. Aerosp. Electron. Syst. **AES-20**(2), 183–198 (1984)
14. D. Musicki, R. Kaune, W. Koch, Mobile Emitter Geolocation and Tracking Using TDOA and FDOA Measurements. In *IEEE Transactions on Signal Processing* vol. 58(3) pp.1863–1874 (2010)
15. K. Becker, Three-dimensional target motion analysis using angle and frequency measurements. IEEE Trans. Aerosp. Electron. Syst. **41**(1), 284–301 (2005)
16. C.R. Berger, M. Daun, W. Koch. Low complexity track initialization from a small set of non-invertible measurements. EURASIP J. Adv. Sig. Process. special issue on track-before-detect techniques, 15, (756414) **2008** doi:10.1155/2008/756414
17. R. Baltes, A triangulation system for tracking multiple targets with passive sensors. in *Proceedings of the DGON/ITG International Radar Symposium IRS'98*, Muenchen, 1998
18. M. Daun, U. Nickel, W. Koch, Tracking in multistatic passive radar systems using DAB/DVB-T illumination. In *Proceedings of Signal Processing* pp 1365–1386 (2012)
19. W. Koch, Ground target tracking with STAP radar: selected tracking aspects. Chapter 15 in: R. Klemm (ed.), The Applications of Space-Time Adaptive Processing, (IEE Publishers, London, 2004)
20. W. Koch, M. Blanchette, Tracking in densely cluttered environment—concepts and experimental results. In *Proceedings of ISATA Robotics, Motion, and Machine Vision*, p.215 ff, Aachen, 1994
21. W. Koch, Experimental results on Bayesian MHT for maneuvering closely-spaced objects in a densely cluttered environment. *In IEE International Radar Conference RADAR'97*, Edinburgh, 729 (1999)
22. W. Koch, Generalized smoothing for multiple model/multiple hypothesis filtering: experimental results, in *Proceedings of European Control Conference ECC'99*, Karlsruhe, CD ROM 1999
23. W.-D. Wirth, *Radar Techniques Using Array Antennas*. (IEE Press, London, 2001)
24. P.L. Bogler, *Radar Principles with Applications to Tracking Systems* (Wiley, New York, 1990)

Chapter 3
Bayesian Knowledge Propagation

Within the general framework of Bayesian reasoning and based on object evolution models and sensor likelihood functions, such as those previously discussed, we proceed along the following lines.

1. *Basis.* In the course of time, one or several sensors produce measurements of one or more objects of interest. The accumulated sensor data are an example of a time series. Each object is characterized by its current state.
2. *Objective.* Learn as much as possible about the object states X_l at each time of interest t_l by exploiting the sensor data collected in the times series Z^k, i.e. for past ($l < k$), present ($l = k$), or future ($l > k$) states.
3. *Problem.* The sensor information is usually imperfect, i.e. imprecise, of uncertain origin, false or corrupted, possibly unresolved, ambiguous etc. Moreover, the objects' temporal evolution is usually not well-known.
4. *Approach.* Interpret sensor measurements and object states as random variables and describe what is known about them by using conditional probability densities functions. In particular, information on an object state at time t_l obtained from the sensor data Z^k is represented by $p(X_l|Z^k)$.
5. *Solution.* Based on Bayes' Theorem, derive iteration formulae for calculating the probability density functions $p(X_l|Z^k)$ and develop a mechanism for initiating the iteration process. Derive state estimates from the probability densities along with appropriate quality measures for the estimates.

3.1 Bayesian Tracking Paradigm

A Bayesian tracking algorithm is an iterative updating scheme for calculating conditional probability density functions $p(X_l|Z^k)$ that represent all available knowledge on the object states X_l at discrete instants of time t_l. The densities are explicitly conditioned on the sensor data Z^k accumulated up to some time t_k, typically the present time. Implicitly, however, they are also determined by all available context

W. Koch, *Tracking and Sensor Data Fusion*,
Mathematical Engineering, DOI: 10.1007/978-3-642-39271-9_3,

knowledge on the sensor characteristics, the dynamical object properties, the environment of the objects, topographical maps, or tactical rules governing the objects' overall behavior.

With respect to the instant of time t_l at which estimates of the object states X_l are required, the related density iteration process is referred to as *prediction* ($t_l > t_k$), *filtering* ($t_l = t_k$), or *retrodiction* ($t_l < t_k$). The propagation of the probability densities involved is given by three basic update equations, which will be derived and discussed and illustrated by examples.

Prediction

The prediction density $p(X_k|Z^{k-1})$ is obtained by combining the evolution model $p(X_k|X_{k-1})$ with the previous filtering density $p(X_{k-1}|Z^{k-1})$:

$$p(X_{k-1}|Z^{k-1}) \xrightarrow[\text{constraints}]{\text{evolution model}} p(X_k|Z^{k-1}) \tag{3.1}$$

$$p(X_k|Z^{k-1}) = \int \mathrm{d}X_{k-1}\; \underbrace{p(X_k|X_{k-1})}_{\text{evolution model}}\; \underbrace{p(X_{k-1}|Z^{k-1})}_{\text{previous filtering}}. \tag{3.2}$$

Filtering

The filtering density $p(X_k|Z^k)$ is obtained by combining the sensor model $p(Z_k, m_k|X_k)$ with the prediction density $p(X_k|Z^{k-1})$ according to:

$$p(X_k|Z^{k-1}) \xrightarrow[\text{sensor model}]{\text{current sensor data}} p(X_k|Z^k) \tag{3.3}$$

$$p(X_k|Z^k) = \frac{p(Z_k, m_k|X_k)\; p(X_k|Z^{k-1})}{\int \mathrm{d}X_k\; \underbrace{p(Z_k, m_k|X_k)}_{\text{sensor model}}\; \underbrace{p(X_k|Z^{k-1})}_{\text{prediction}}}. \tag{3.4}$$

Retrodiction

The retrodiction density $p(X_l|Z^k)$ is obtained by combining the previous retrodiction density $p(X_{l+1}|Z^k)$ with the object evolution model $p(X_{l+1}|X_l)$ and the previous prediction and filtering densities $p(X_{l+1}|Z^l)$, $p(X_l|Z^l)$ according to:

$$p(X_{l-1}|Z^k) \xleftarrow[\text{evolution model}]{\text{filtering, prediction}} p(X_l|Z^k) \tag{3.5}$$

$$p(X_l|Z^k) = \int \mathrm{d}X_{l+1}\; \frac{\overbrace{p(X_{l+1}|X_l)}^{\text{evolution}}\; \overbrace{p(X_l|Z^l)}^{\text{prev. filtering}}}{\underbrace{p(X_{l+1}|Z^l)}_{\text{prev. prediction}}}\; \underbrace{p(X_{l+1}|Z^k)}_{\text{prev. retrodiction}}. \tag{3.6}$$

The natural antonym of "prediction", the technical term "retrodiction" was introduced by Oliver Drummond in a series of papers [1–3]. According to his definition, "The process of computing estimates of states, probability densities, or discrete probabilities for a prior time (or over a period of time) based on data up to and including some subsequent time, typically, the current time." [1, p. 255], this term comprises not only standard smoothing, but also the concept of a retrodicted discrete probability that is analogous to a smoothed estimate in usual Kalman filtering. For this reason, the notion of "retrodiction" is general enough as well as adequate for the type of algorithms proposed above. Adopting the classical standard terminology [4], we could speak of *fixed-interval* retrodiction.

The Notion of a Track

According to this paradigm, an *object track* represents all relevant knowledge on a time-varying object state of interest, including its history and measures that describe the quality of this knowledge. As a technical term, 'track' is therefore either a synonym for the collection of densities $p(X_l|Z^k)$, $l = 1, \ldots, k, \ldots$, or of suitably chosen parameters characterizing them, such as estimates related to appropriate risk functions and the corresponding estimation error covariance matrices.

If possible, a one-to-one association between the objects in the sensors' field of view and the produced tracks is to be established and has to be preserved as long as possible (*track continuity*). In many applications, track continuity is even more important than track accuracy. Obviously, the achievable track quality does not only depend on the performance of the underlying sensors, but also on the object properties and the operational conditions within the scenario to be observed.

In this context, the notion of *track consistency* is important. It describes the degree of compliance between the inherent measures of track quality provided by the fusion process itself and the "real" tracking errors involved. Track consistency can be verified in experiments with an established ground truth or in Monte-Carlo-simulations (see the discussion on fusion performance measures in Sect. 1.3.3).

Graphical Illustration

In Fig. 3.1a the conditional probability densities $p(\mathbf{x}_{k-1}|Z^{k-1})$, $p(\mathbf{x}_k|Z^k)$, and $p(\mathbf{x}_{k+1}|Z^{k+1})$ resulting from filtering at time instants t_{k-1}, t_k, and t_{k+1} are displayed along with the predicted density $p(\mathbf{x}_{k+2}|Z^{k+1})$. While at time t_{k-1} one sensor measurement has been processed, no measurement could be associated to it at time t_k. Hence, a missing detection is assumed. Due to the missing sensor information, the density $p(\mathbf{x}_k|Z^k)$ is broadened, since object maneuvers may have occurred. This in particular implies an increased region, where data at the subsequent time t_{k+1} are expected (gates). According to this effect, at time t_{k+1} three correlating sensor measurements are to be processed, leading to a multi-modal probability density function. The multiple modes reflect the ambiguity of the origin of the sensor data and also characterize the predicted density $p(\mathbf{x}_{k+2}|Z^{k+1})$. By this, the data-driven adaptivity of the Bayesian updating scheme is indicated.

In Fig. 3.1b, the density $p(\mathbf{x}_{k+2}|Z^{k+2})$, resulting from processing a single correlating sensor measurement at time t_{k+2}, along with the retrodicted densities

Fig. 3.1 Scheme of Bayesian density iteration: conditional probability density functions resulting from the *prediction*, *filtering*, and *retrodiction steps* at different instants of times. **a** Forward iteration. **b** Backward iteration

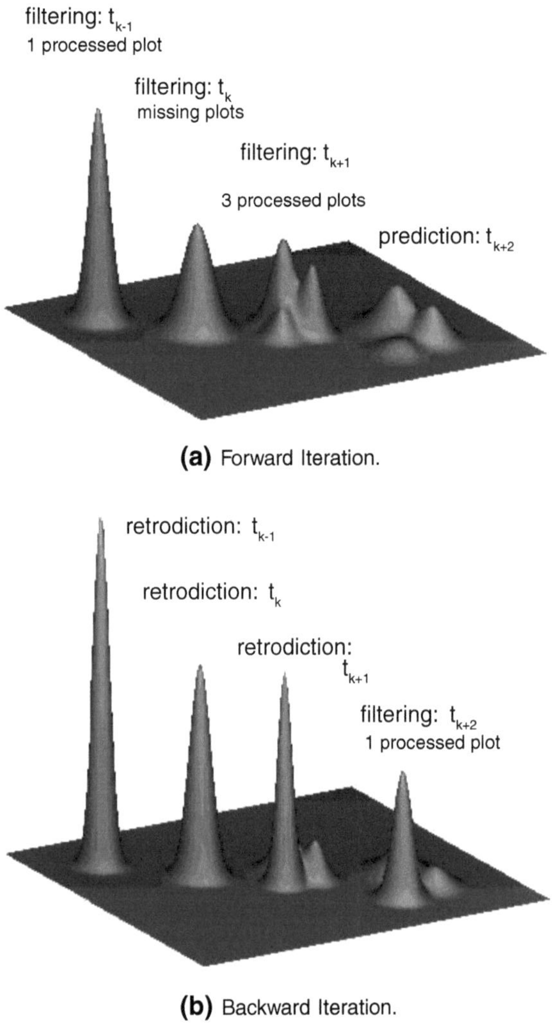

$p(\mathbf{x}_{k+1}|Z^{k+2})$, $p(\mathbf{x}_k|Z^{k+2})$, and $p(\mathbf{x}_{k-1}|Z^{k+2})$ are shown. Obviously, available sensor data at the present time can significantly improve the estimates of the object states in the past.

3.1.1 Characteristic Aspects

The sensor data fusion process, i.e. the iterative calculation of conditional probability densities $p(X_l|Z^k)$ from multiple sensor data and context information on sensors, objects, and the environment, can be characterized by several aspects. The emphasis, which is given to a particular aspect in a concrete application, has a strong impact on the design and architecture of a sensor data fusion system and on the requirements related to the underlying infrastructure, such as the bandwidth and reliability of communication links or navigation systems for sensor registration and alignment.

- In *optimal data fusion*, the conditional probability density functions involved are correctly calculated.
- In *centralized fusion*, only one fusion center exists, where the sensor data or object tracks are fused.
- In *distributed fusion*, the sensor data or object tracks are distributed and fused at several fusion centers.
- In *measurement fusion*, the sensor data and all relevant likelihood parameters are communicated to the fusion center(s).
- In *track-to-track fusion*, the local state estimates and covariances are communicated to the fusion center(s).
- In *full communication rate fusion*, all local sensor data or tracks are communicated to the fusion centers.
- In *reduced-rate communication fusion*, only selected sensor measurements or local tracks are communicated.

The likelihood function discussed in Sect. 2.3.2 is a typical example of centralized or distributed measurement fusion, while the fusion algorithms discussed in Sect. 6 can be characterized as distributed, full or reduced-rate track-to-track fusion. In general, measurement fusion architectures provide better approximations of optimal fusion. The choice of a fusion strategy depends on the particular requirements of a given application. See [5] for more a detailed discussion on benefits and problems of alternative fusion system architectures.

3.1.2 Remarks on Approximations

Under more realistic conditions, the probability densities involved typically have the structure of finite mixtures, i.e. they can be represented by weighted sums of individual probability densities that assume particular data interpretations or model hypotheses to be true. This general structure is a direct consequence of the uncertain origin of the sensor data and/or of the uncertainty related to the underlying object evolution. In concrete implementations, however, it is always necessary to apply certain approximations to handle such mixtures efficiently. Provided the densities $p(X_l|Z^k)$ are calculated at least approximately correctly, "good" estimators can be derived related to various risk functions adapted to the applications. What "good" means depends on the application considered and must often be verified by extensive Monte-Carlo-simulations and experiments.

Gaußian Mixtures

At least approximately correct closed-formula solutions for the Bayesian tracking paradigm can be derived if the prediction, filtering, and retrodiction densities as well as the sensor and evolution models belong to certain families of probability densities, so-called *mutually conjugate densities*. A wide and mathematically comfortable family of conjugate densities for random vectors $\mathbf{x}$ is provided by *Gaußian mixtures* [6], i.e. by weighted sums of Gaußian probability densities, $p(\mathbf{x}) = \sum_i p_i \mathcal{N}(\mathbf{x};\ \mathbf{x}_i,\ \mathbf{P}_i)$ with *mixture coefficients* $p_i \in \mathbb{R}$ that sum up to One, $\sum_i p_i = 1$, but need not necessarily be positive. A Gaußian mixture density is thus completely represented by a relatively small number of parameters $\{p_i, \mathbf{x}_i, \mathbf{P}_i\}_i$. As an early example see [7]. Other examples of families, which lead to at least approximately correct update formulae and are relevant to the work considered here, are Wishart and inverted Wishart mixtures or Gamma and inverted Gamma mixtures (see Appendix A.11).

For many real-world applications, it has been shown that even more sophisticated functional relationships describing the physics of the measurement process within a sensor system can be modeled by likelihood functions of the Gaußian mixture type. Of course, the accuracy of the sensor model, i.e. the number of mixture components that are actually to be taken into account to approximately describe the underlying phenomena, depends on the requirements of the underlying application. The same arguments are valid if the incorporation of context information, such as road-maps, is to be considered. They are also valid in the case of more complex dynamics models, such as those with a state dependent model transition matrix given by Eq. 2.12. Many examples of this type are discussed in Chap. 3.

It is the author's conviction that a large variety of relevant problems still exists in real-world applications of sensor data fusion and sensor management, which can efficiently be solved by using appropriately defined Gaußian mixtures. A particularly interesting indication of this general tendency seems to be the very fact that even in recent approaches, such as in Probability Hypothesis Density filtering (PHD, [8]), Gaußian mixture realizations provide the state-of-the-art solutions (GM-CPHD: Gaußian Mixture Cardinalized PHD). In view of practicality, these realizations are preferable compared to alternative approximation schemes, such as particle filtering. Moreover, explicit calculations in exploiting realistic sensor and evolution models are possible when using Gaußian mixture techniques, which provide a better understanding of the underlying physical and technical phenomena.

Particle Filtering

For implementing the Bayesian tracking paradigm, alternative approximation schemes are applicable that deal with the probability densities involved numerically. The most prominent method among these, *particle filtering*, was first introduced for tracking applications by Neil Gordon [9], who initiated and inspired a stormy development in this field (see [10] and the literature cited therein). Another early example of using particle filtering in a position estimation application for mobile robots is the work of D. Fox, W. Burgard, F. Dellaert, and S. Thrun [11].

Particle filtering techniques numerically represent probability density functions by random samples (called "particles") drawn from them by using random number generators. The method is thus closely related to the random Monte-Carlo techniques, developed for problems in quantum field theory, for instance (see the discussion in [12]). For this reason, particle filtering techniques are computationally intensive. Their main advantage is the fact that they, in principle, provide "numerically exact" solutions at the cost of long computation times. These solutions can serve as benchmarks to test alternatives, such as Gaußian mixture realizations, which are often much less computation time consuming. In the context of the work presented here, performance comparisons using particle filters were done for "road-map assisted tracking" [13].

Particle filtering is a valuable approximation scheme for probability densities especially in applications, where the likelihood function $\ell(X_k; Z_k)$ can only be calculated pointwisely by an algorithm and no functional closed-formula expression is available. In the context of the work presented here, emitter localization and tracking in an urban environment is discussed (see [14] and the discussion in Sect. 3.2.3). Since this scenario is dominated by propagation phenomena, the key to the solution of this tracking problem lies in dealing with multipath phenomena appropriately. This can be done by using ray tracing algorithms for evaluating the most likely propagation channels for randomly chosen candidate emitter positions. Similar examples can be found wherever sophisticated propagation models can be exploited for localization and tracking (ionospheric propagation such as in communications or over-the-horizon radar, shallow-water sonar, indoor navigation) [15].

For advanced approximation techniques beyond classical particle filtering, which combines elements of Gaußian mixture reasoning with intelligent non-random sampling techniques, see the work of Uwe Hanebeck and his group (see [16], for example, and the literature cited herein).

3.1.3 On Track-to-Track Fusion

In certain applications, track-to-track fusion (see e.g. [17–20]) has considerable advantages:

- The communication channels are less overloaded with false tracks, provided these can be suppressed by local data processing.
- We may profit from reduced sensibility to sensor registration errors as local tracking is inherently robust regarding these effects. In this case, the problem is transfered to track-to-track fusion, but on this level its solution profits from efficient track-to-track correlation algorithms in situations that are not too dense.
- Disturbances of individual sensor sites and their corresponding local processors do not lead to the loss of the total system function.

Disadvantages result from suboptimal performance with respect to reaction time, track quality, lacking profit from redundancy, and the lower data rate for sensor individual tracking, which particularly affects track initiation, e.g. Moreover, track-to-track fusion is problematic if data collected by active and passive sensors have to be fused (e.g. position data and bearings), since the production of local, sensor individual tracks may be difficult in non-trivial situations.

We speak of *optimal* track-to-track fusion in a Bayesian sense if the conditional probability density functions $p(X_k|Z^k) = p(X_k|\{Z_s^k\}_{s=1}^S)$, conditioned on all measurements of all sensors, can be correctly reconstructed from the locally produced tracks $p(X_k|Z_s^k)$, obtained by processing the data of the sensors $s = 1, \ldots, S$ individually:

$$\{p(\mathbf{x}_l|Z_s^l)\}_{s,l=1}^{S,k} \xrightarrow[\text{fusion}]{\text{track-to-track}} p(\mathbf{x}_k|\{Z_s^k\}_{s=1}^S). \tag{3.7}$$

In Sect. 6 selected aspects of track-to-track fusion are discussed and exact update formulae for certain special cases are derived.

3.1.4 A First Look at Initiation

At time t_0, the probability density $p(X_0|Z^0)$ describes the initial knowledge of the object state. As an example let us consider state vectors $\mathbf{x}_k = (\mathbf{r}_k^\top, \dot{\mathbf{r}}_k^\top)^\top$, consisting of the object position and velocity, and a first position measurement $\mathbf{z}_0$ with a measurement error covariance matrices $\mathbf{R}_0$. Based on $\mathbf{z}_0$ and the context information on the maximum object speed $v_{\max}$ to be expected, a reasonable initiation is given by $p(\mathbf{x}_0|\mathbf{z}_0) = \mathcal{N}(\mathbf{x}_0;\ \mathbf{x}_{0|0},\ \mathbf{P}_{0|0})$ with:

$$\mathbf{x}_{0|0} = (\mathbf{z}_0^\top, \mathbf{0}^\top)^\top, \quad \mathbf{P}_{0|0} = \text{diag}[\mathbf{R}_0, v_{\max}^2 \mathbf{1}]. \tag{3.8}$$

In the case of an IMM evolution model, we consider the probability density $p(\mathbf{x}_0, i_0|Z^0) = p_{0|0}^{i_0}\ \mathcal{N}(\mathbf{x}_0;\ \mathbf{x}_{0|0}^{i_0},\ \mathbf{P}_{0|0}^{i_0})$ with $p_{0|0}^{i_0} = \frac{1}{r}$. For a numerically robust and quick initiation scheme even from incomplete measurements see [21, 22] and the literature cited therein.

3.2 Object State Prediction

The probability density function $p(X_k|Z^{k-1})$ describes the predicted knowledge of the object state X_k referring to the instant of time t_k based on all the measurements received in the past up to and including the time t_{k-1}. According to the Chapman-Kolmogorov Equation, the prediction density can be calculated by combining the available knowledge on the object state at the past time t_{k-1}, given by $p(X_{k-1}|Z^{k-1})$ with the available knowledge on the object evolution, given by the evolution model $p(X_k|X_{k-1})$. Marginalization and the Markov assumption directly yield:

$$p(X_k|Z^{k-1}) = \int \mathrm{d}X_{k-1}\; p(X_k, X_{k-1}|Z^{k-1}) \tag{3.9}$$

$$= \int \mathrm{d}X_{k-1}\; p(X_k|X_{k-1})\; p(X_{k-1}|Z^{k-1}). \tag{3.10}$$

3.2.1 Kalman Prediction

Let us consider a Gauß-Markov evolution model, such as provided by van Keuk's model (Eq. 2.8), where its deterministic part is characterized by the evolution matrix $\mathbf{F}_{k|k-1}$ and the stochastic part by the evolution covariance matrix $\mathbf{D}_{k|k-1}$, and a Gaußian previous filtering density, given by $p(\mathbf{x}_{k-1}|Z^{k-1}) = \mathcal{N}(\mathbf{x}_{k-1}; \mathbf{x}_{k-1|k-1}, \mathbf{P}_{k-1|k-1})$. Then the prediction density is also provided by a Gaußian:

$$p(\mathbf{x}_k|Z^{k-1}) = \int \mathrm{d}\mathbf{x}_{k-1}\; \mathcal{N}\big(\mathbf{x}_k;\; \mathbf{F}_{k|k-1}\mathbf{x}_{k-1},\; \mathbf{D}_{k|k-1}\big)$$

$$\times\; \mathcal{N}\big(\mathbf{x}_{k-1};\; \mathbf{x}_{k-1|k-1},\; \mathbf{P}_{k-1|k-1}\big) \tag{3.11}$$

$$= \mathcal{N}\big(\mathbf{x}_k;\; \mathbf{x}_{k|k-1},\; \mathbf{P}_{k|k-1}\big) \tag{3.12}$$

with an expectation vector $\mathbf{x}_{k|k-1}$ and a covariance matrix $\mathbf{P}_{k|k-1}$ given by the Kalman prediction update equations:

$$\mathbf{x}_{k|k-1} = \mathbf{F}_{k|k-1}\mathbf{x}_{k-1|k-1} \tag{3.13}$$

$$\mathbf{P}_{k|k-1} = \mathbf{F}_{k|k-1}\mathbf{P}_{k-1|k-1}\mathbf{F}_{k|k-1}^{\top} + \mathbf{D}_{k|k-1}. \tag{3.14}$$

This directly results from a product formula for Gaußians stated and proven in Appendix A.5, Eq. A.28. Note that after applying this formula, the integration variable $\mathbf{x}_{k-1}$ in Eq. 3.11 is no longer contained in the first Gaußian of the product and can be drawn in front of the integral. The integration thus becomes trivial since probability densities are normalized by definition.

3.2.2 Expectation Gates

As a by-product of the prediction process, the statistical properties of object measurements Z_k that are expected at time t_k can be calculated on the basis of previously obtained measurements Z^{k-1}:

$$p(Z_k|Z^{k-1}) = \int \mathrm{d}X_k\; p(Z_k, X_k|Z^{k-1}) \tag{3.15}$$

$$= \int \mathrm{d}X_k\; p(Z_k|X_k)\; p(X_k|Z^{k-1}). \tag{3.16}$$

In the special case of Kalman prediction and with a Gaußian likelihood function, we obtain:

$$p(\mathbf{z}_k|Z^{k-1}) = \int \mathrm{d}\mathbf{x}_k \ \mathcal{N}\big(\mathbf{z}_k;\ \mathbf{H}_k\mathbf{x}_k,\ \mathbf{R}_k\big)\ \mathcal{N}\big(\mathbf{x}_k;\ \mathbf{x}_{k|k-1},\ \mathbf{P}_{k|k-1}\big) \tag{3.17}$$

$$= \mathcal{N}\big(\mathbf{z}_k;\ \mathbf{H}_k\mathbf{x}_{k|k-1},\ \mathbf{S}_{k|k-1}\big). \tag{3.18}$$

Via the product formula (Eq. A.28), the matrix $\mathbf{S}_{k|k-1}$ results from the previous filtering covariance matrix $\mathbf{P}_{k-1|k-1}$ exploiting both the evolution and the sensor model:

$$\mathbf{S}_{k|k-1} = \mathbf{H}_k\mathbf{P}_{k|k-1}\mathbf{H}_k^\top + \mathbf{R}_k. \tag{3.19}$$

This means in particular that the *innovation vector* $\boldsymbol{\nu}_{k|k-1} = \mathbf{z}_k - \mathbf{H}_k\mathbf{x}_{k|k-1}$ is a normally distributed zero mean random variable characterized by the covariance matrix $\mathbf{S}_{k|k-1}$, which is thus called *innovation covariance matrix*. For this reason, the quadratic form

$$|\nu_{k|k-1}|^2_{\mathbf{S}_{k|k-1}} = (\mathbf{z}_k - \mathbf{H}_k\mathbf{x}_{k|k-1})^\top\mathbf{S}^{-1}_{k|k-1}(\mathbf{z}_k - \mathbf{H}_k\mathbf{x}_{k|k-1}), \tag{3.20}$$

called *innovation square* or *Mahalanobis distance* between predicted and actually produced measurements, is a χ^2_n-distributed random variable with n degrees of freedom where n is the dimension of the measurement vector $\mathbf{z}_k$. The ellipsoid defined by:

$$|\nu_{k|k-1}|^2_{\mathbf{S}_{k|k-1}} \leq \chi^2_n(1 - P_c) \tag{3.21}$$

thus contains the expected measurement with a *correlation probability* P_c. The concrete value of $\chi^2_n(1 - P_c)$ can be looked up in a χ^2-table. Such expectation gates are useful to exclude measurements from the fusion process that are very unlikely to belong to a given object. Figure 3.2 schematically illustrates the use of expectation gates in an object tracking example.

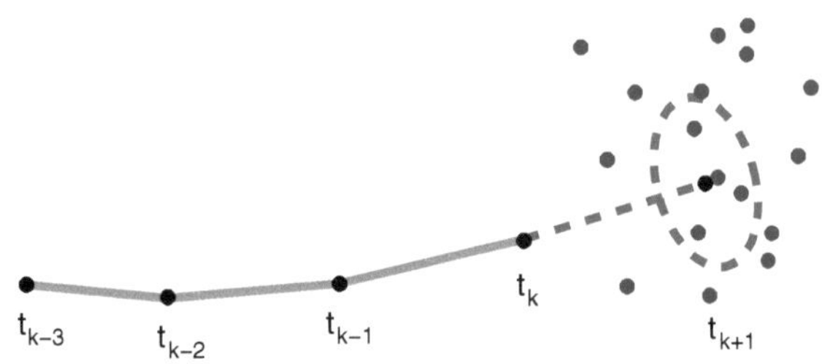

Fig. 3.2 Schematic illustration of expectation gates as a means of excluding measurements belonging to an object with a low probability $1 - P_c$

3.2.3 IMM Prediction

According to the discussion in the Sects. 2.1 and 2.2.2, let the filtering density $p(X_{k-1}|Z^{k-1})$ at time t_{k-1} be given by

$$p(X_{k-1}|Z^{k-1}) = p(\mathbf{x}_{k-1}, i_{k-1}|Z^{k-1}) \tag{3.22}$$
$$= p(\mathbf{x}_{k-1}|i_{k-1}, Z^{k-1})\, p(i_{k-1}|Z^{k-1}) \tag{3.23}$$
$$= p^{i_{k-1}}_{k-1|k-1}\, \mathcal{N}\big(\mathbf{x}_{k-1};\, \mathbf{x}^{i_{k-1}}_{k-1|k-1},\, \mathbf{P}^{i_{k-1}}_{k-1|k-1}\big), \tag{3.24}$$

i.e. by a weighted Gaußian. In this case, the prediction update according to Eq. 3.10 and the product formula for Gaußians (Eq. A.28) yield:

$$p(\mathbf{x}_k, i_k|Z^{k-1}) = \sum_{i_{k-1}} \int \mathrm{d}\mathbf{x}_{k-1}\; p(\mathbf{x}_k, i_k, \mathbf{x}_{k-1}, i_{k-1}|Z^{k-1}) \tag{3.25}$$
$$= \sum_{i_{k-1}} \int \mathrm{d}\mathbf{x}_{k-1}\; p(\mathbf{x}_k, i_k|\mathbf{x}_{k-1}, i_{k-1})\; p(\mathbf{x}_{k-1}, i_{k-1}|Z^{k-1}) \tag{3.26}$$
$$= \sum_{i_{k-1}} p^{i_k i_{k-1}}_{k|k-1}\, \mathcal{N}\big(\mathbf{x}_k;\, \mathbf{x}^{i_k i_{k-1}}_{k|k-1},\, \mathbf{P}^{i_k i_{k-1}}_{k|k-1}\big), \tag{3.27}$$

where the parameters $p^{i_k i_{k-1}}_{k|k-1}$, $\mathbf{x}^{i_k i_{k-1}}_{k|k-1}$, $\mathbf{P}^{i_k i_{k-1}}_{k|k-1}$ of the density are given by:

$$p^{i_k i_{k-1}}_{k|k-1} = p_{i_k i_{k-1}}\, p^{i_{k-1}}_{k-1|k-1} \tag{3.28}$$
$$\mathbf{x}^{i_k i_{k-1}}_{k|k-1} = \mathbf{F}^{i_k}_{k|k-1}\mathbf{x}^{i_{k-1}}_{k-1|k-1} \tag{3.29}$$
$$\mathbf{P}^{i_k i_{k-1}}_{k|k-1} = \mathbf{F}^{i_k}_{k|k-1}\mathbf{P}^{i_{k-1}}_{k-1|k-1}\mathbf{F}^{i_k\top}_{k|k-1} + \mathbf{D}^{i_k}_{k|k-1}. \tag{3.30}$$

In standard IMM applications, $p(\mathbf{x}_k, i_k|Z^{k-1})$ is approximated via moment matching (see [23, p. 56 ff] and the discussion in Appendix A.6) yielding

$$p(\mathbf{x}_k, i_k|Z^{k-1}) \approx p^{i_k}_{k|k-1}\, \mathcal{N}\big(\mathbf{x}_k;\, \mathbf{x}^{i_k}_{k|k-1},\, \mathbf{P}^{i_k}_{k|k-1}\big) \tag{3.31}$$

with parameters $p^{i_k}_{k|k-1}$, $\mathbf{x}^{i_k}_{k|k-1}$ and $\mathbf{P}^{i_k}_{k|k-1}$ given by:

$$p^{i_k}_{k|k-1} = \sum_{i_{k-1}=1}^{r} p^{i_k i_{k-1}}_{k|k-1} \tag{3.32}$$

$$\mathbf{x}^{i_k}_{k|k-1} = \frac{1}{p^{i_k}_{k|k-1}} \sum_{i_{k-1}=1}^{r} p^{i_k i_{k-1}}_{k|k-1}\, \mathbf{x}^{i_k i_{k-1}}_{k|k-1} \tag{3.33}$$

$$\mathbf{P}^{i_k}_{k|k-1} = \frac{1}{p^{i_k}_{k|k-1}} \sum_{i_{k-1}=1}^{r} p^{i_k i_{k-1}}_{k|k-1} \Big(\mathbf{P}^{i_k i_{k-1}}_{k|k-1} \tag{3.34}$$

$$+ (\mathbf{x}^{i_k i_{k-1}}_{k|k-1} - \mathbf{x}^{i_k}_{k|k-1})(\mathbf{x}^{i_k i_{k-1}}_{k|k-1} - \mathbf{x}^{i_k}_{k|k-1})^\top\Big). \tag{3.35}$$

If the predictions of the kinematic state variables $\mathbf{x}_k$ are the only objects of interest, $p(\mathbf{x}_k|Z^{k-1})$ is a direct result from marginalization and is given by a Gaußian sum with r mixture components:

$$p(\mathbf{x}_k|Z^{k-1}) = \sum_{i_k=1}^{r} p^{i_k}_{k|k-1}\, \mathcal{N}\big(\mathbf{x}_k;\, \mathbf{x}^{i_k}_{k|k-1},\, \mathbf{P}^{i_k}_{k|k-1}\big). \tag{3.36}$$

3.3 Data Update: Filtering

The filtering update equation for the conditional probability density function $p(X_k|Z^k)$ that represents our knowledge of the present object state given all available information can be represented according to Bayes' Theorem by:

$$p(X_k|Z^k) = p(X_k|Z_k, m_k, Z^{k-1}) \tag{3.37}$$

$$= \frac{p(Z_k, m_k|X_k)\; p(X_k|Z^{k-1})}{\int \mathrm{d}X_k\; p(Z_k, m_k|X_k)\; p(X_k|Z^{k-1})}. \tag{3.38}$$

This equation states how $p(X_k|Z^k)$ is obtained by combining the prediction density $p(X_k|Z^{k-1})$ with the sensor model $p(Z_k, m_k|X_k)$. As the sensor model appears both in the denominator and the numerator, the conditional densities $p(Z_k, m_k|X_k)$ need to be known up to a factor independent of the object state X_k only. Each function $\ell(X_k; Z_k, m_k) \propto p(Z_k, m_k|X_k)$ provides the same result. This observation is the reason for introducing the term "likelihood function" for denoting functions that are proportional to the conditional probability density $p(Z_k, m_k|X_k)$.

3.3.1 Kalman Filtering

Let us consider kinematic object states only, $X_k = \mathbf{x}_k$, and sensors that produce measurements related to them. Under conditions where the data-to-object associations are unambiguous, e.g. for well-separated objects without false sensor data ($\rho_F = 0$), and in the case of a Gauß-Markov evolution model, such as given by Eq. 2.8, and a Gaußian sensor model (Eq. 2.19), the Bayesian approach leads to the well-known Kalman filter update equations. Kalman filtering can thus be considered as a simple straight-forward realization of the more general Bayesian filtering paradigm.

Equation 3.38 yields according to the product formula for Gaußians (Eq. A.28):

$$p(\mathbf{x}_k|Z^k) = \frac{\mathcal{N}\big(\mathbf{z}_k;\ \mathbf{H}_k\mathbf{x}_k,\ \mathbf{R}_k\big)\ \mathcal{N}\big(\mathbf{x}_k;\ \mathbf{x}_{k|k-1},\ \mathbf{P}_{k|k-1}\big)}{\int \mathrm{d}\mathbf{x}_k\ \mathcal{N}\big(\mathbf{z}_k;\ \mathbf{H}_k\mathbf{x}_k,\ \mathbf{R}_k\big)\ \mathcal{N}\big(\mathbf{x}_k;\ \mathbf{x}_{k|k-1},\ \mathbf{P}_{k|k-1}\big)} \tag{3.39}$$

$$= \frac{\mathcal{N}\big(\mathbf{z}_k;\ \mathbf{H}_k\mathbf{x}_{k|k-1},\ \mathbf{S}_{k|k-1}\big)\ \mathcal{N}\big(\mathbf{x}_k;\ \mathbf{x}_{k|k},\ \mathbf{P}_{k|k}\big)}{\int \mathrm{d}\mathbf{x}_k\ \mathcal{N}\big(\mathbf{z}_k;\ \mathbf{H}_k\mathbf{x}_{k|k-1},\ \mathbf{S}_{k|k-1}\big)\ \mathcal{N}\big(\mathbf{x}_k;\ \mathbf{x}_{k|k},\ \mathbf{P}_{k|k}\big)} \tag{3.40}$$

$$= \mathcal{N}\big(\mathbf{x}_k;\ \mathbf{x}_{k|k},\ \mathbf{P}_{k|k}\big), \tag{3.41}$$

where the parameters $\mathbf{x}_{k|k}$, $\mathbf{P}_{k|k}$ of the resulting Gaußian are alternatively given by:

$$\mathbf{x}_{k|k} = \begin{cases} \mathbf{x}_{k|k-1} + \mathbf{W}_{k|k-1}(\mathbf{z}_k - \mathbf{H}_k\mathbf{x}_{k|k-1}) \\ \mathbf{P}_{k|k}(\mathbf{P}_{k|k-1}^{-1}\mathbf{x}_{k|k-1} + \mathbf{H}_k^\top\mathbf{R}_k^{-1}\mathbf{z}_k) \end{cases} \tag{3.42}$$

$$\mathbf{P}_{k|k} = \begin{cases} \mathbf{P}_{k|k-1} - \mathbf{W}_{k|k-1}\mathbf{S}_{k|k-1}\mathbf{W}_{k|k-1}^\top \\ (\mathbf{P}_{k|k-1}^{-1} + \mathbf{H}_k^\top\mathbf{R}_k^{-1}\mathbf{H}_k)^{-1} \end{cases}. \tag{3.43}$$

Note that there are equivalent formulations of the Kalman update formulae according to various versions of the product formula (Eq. A.28). The innovation covariance matrix $\mathbf{S}_{k|k-1}$ has already appeared earlier in our considerations (see Eq. 3.19), while the *Kalman Gain* matrix is given by:

$$\mathbf{W}_{k|k-1} = \mathbf{P}_{k|k-1}\mathbf{H}_{k|k-1}^\top\mathbf{S}_{k|k-1}^{-1}. \tag{3.44}$$

In Eq. 3.41, the factor $\mathcal{N}\big(\mathbf{z}_k;\ \mathbf{H}_k\mathbf{x}_{k|k-1},\ \mathbf{S}_k\big)$ does not depend on the integration variable $\mathbf{x}_k$ and can be drawn in front of the integral, thus canceling the corresponding quantity in the numerator. Note that the matrix $\mathbf{S}_{k|k-1}$ to be inverted when calculating the Kalman gain matrix has the same dimension as the measurement vector $\mathbf{z}_k$, i.e. $\mathbf{S}_{k|k-1}$ is a low-dimensional matrix in general.

3.3.2 IMM Filtering

This update philosophy can easily be generalized to apply to situations where IMM evolution models are used, i.e. if the object state is given by $X_k = (\mathbf{x}_k, i_k)$. We immediately obtain:

$$p(\mathbf{x}_k, i_k|Z^k)$$

$$= \frac{p_{k|k-1}^{i_k}\ \mathcal{N}\big(\mathbf{z}_k;\ \mathbf{H}_k\mathbf{x}_k,\ \mathbf{R}_k\big)\ \mathcal{N}\big(\mathbf{x}_k;\ \mathbf{x}_{k|k-1}^{i_k},\ \mathbf{P}_{k|k-1}^{i_k}\big)}{\sum_{i_k} p_{k|k-1}^{i_k}\ \int \mathrm{d}\mathbf{x}_k\ \mathcal{N}\big(\mathbf{z}_k;\ \mathbf{H}_k\mathbf{x}_k,\ \mathbf{R}_k\big)\ \mathcal{N}\big(\mathbf{x}_k;\ \mathbf{x}_{k|k-1}^{i_k},\ \mathbf{P}_{k|k-1}^{i_k}\big)} \tag{3.45}$$

$$= \frac{p_{k|k-1}^{i_k}\ \mathcal{N}\big(\mathbf{z}_k;\ \mathbf{H}_k\mathbf{x}_{k|k-1}^{i_k},\ \mathbf{S}_{k|k-1}^{i_k}\big)\ \mathcal{N}\big(\mathbf{x}_k;\ \mathbf{x}_{k|k}^{i_k},\ \mathbf{P}_{k|k}^{i_k}\big)}{\sum_{i_k} p_{k|k-1}^{i_k}\ \mathcal{N}\big(\mathbf{z}_k;\ \mathbf{H}_k\mathbf{x}_{k|k-1}^{i_k},\ \mathbf{S}_{k|k-1}^{i_k}\big)\ \int \mathrm{d}\mathbf{x}_k\ \mathcal{N}\big(\mathbf{x}_k;\ \mathbf{x}_{k|k}^{i_k},\ \mathbf{P}_{k|k}^{i_k}\big)} \tag{3.46}$$

$$= p_{k|k}^{i_k}\ \mathcal{N}\big(\mathbf{x}_k;\ \mathbf{x}_{k|k}^{i_k},\ \mathbf{P}_{k|k}^{i_k}\big), \tag{3.47}$$

where the parameters $\mathbf{x}_{k|k}^{i_k}$, $\mathbf{P}_{k|k}^{i_k}$ are given by a Kalman filtering update based on $\mathbf{x}_{k|k-1}^{i_k}$, $\mathbf{P}_{k|k-1}^{i_k}$ and $\mathbf{S}_{k|k-1}^{i_k} = \mathbf{H}_k\mathbf{P}_{k|k-1}^{i_k}\mathbf{H}_k^{\top} + \mathbf{R}_k$, while the weighting factors $p_{k|k}^{i_k}$ depend on the sensor data $\mathbf{z}_k$ and are given by:

$$p_{k|k}^{i_k} = \frac{p_{k|k-1}^{i_k}\ \mathcal{N}\big(\mathbf{z}_k;\ \mathbf{H}_k\mathbf{x}_{k|k-1}^{i_k},\ \mathbf{S}_{k|k-1}^{i_k}\big)}{\sum_{i_k} p_{k|k-1}^{i_k}\ \mathcal{N}\big(\mathbf{z}_k;\ \mathbf{H}_k\mathbf{x}_{k|k-1}^{i_k},\ \mathbf{S}_{k|k-1}^{i_k}\big)}. \tag{3.48}$$

If only the kinematic state variables $\mathbf{x}_k$ are of interest, $p(\mathbf{x}_k|Z^k)$ is given by a Gaußian sum with r mixture components via marginalization:

$$p(\mathbf{x}_k|Z^k) = \sum_{i_k} p_{k|k}^{i_k}\ \mathcal{N}\big(\mathbf{x}_k;\ \mathbf{x}_{k|k}^{i_k},\ \mathbf{P}_{k|k}^{i_k}\big). \tag{3.49}$$

An Alternative

So-called 'Generalized Pseudo-Bayesian' realizations of the IMM filtering problem (GPB, [24, 25]) fit well into this framework. The difference between GPB and standard IMM filtering is simply characterized by the instant of time when the moment-matching step is performed. While in standard IMM this is done after the prediction step and before the new sensor data are processed, GPB filtering approximates the probability density

$$\begin{aligned}
p(\mathbf{x}_k, i_k|Z^k) &\propto \mathcal{N}\big(\mathbf{z}_k;\ \mathbf{H}_k\mathbf{x}_k,\ \mathbf{R}_k\big) \sum_{i_{k-1}} p_{k|k-1}^{i_k i_{k-1}}\ \mathcal{N}\big(\mathbf{x}_k;\ \mathbf{x}_{k|k-1}^{i_k i_{k-1}},\ \mathbf{P}_{k|k-1}^{i_k i_{k-1}}\big)\\
&= \sum_{i_{k-1}} p_{k|k-1}^{i_k i_{k-1}}\ \mathcal{N}\big(\mathbf{z}_k;\ \mathbf{H}_k\mathbf{x}_{k|k-1}^{i_k i_{k-1}},\ \mathbf{S}_{k|k-1}^{i_k i_{k-1}}\big)\ \mathcal{N}\big(\mathbf{x}_k;\ \mathbf{x}_{k|k}^{i_k i_{k-1}},\ \mathbf{P}_{k|k}^{i_k i_{k-1}}\big)\\
&\approx p_{k|k}^{\prime i_k}\ \mathcal{N}\big(\mathbf{x}_k;\ \mathbf{x}_{k|k}^{\prime i_k},\ \mathbf{P}_{k|k}^{\prime i_k}\big)
\end{aligned}$$

with appropriately defined mixture parameters that are directly given by the product formula A.28. Since the moment matching is done with the updated weighting factors, GPB methods show a better reaction to abrupt maneuvers. For a more rigorous discussion of these topics see [26].

3.3.3 MHT Filtering

In the case of ambiguous sensor data, likelihood functions such as in Eq. 2.40 are essentially characterized by taking different data interpretation hypotheses into account. They are the basis for *Multiple Hypothesis Tracking* techniques (MHT, [27, 28]). In such situations, the origin of a time series $Z^k = \{Z_k, m_k, Z^{k-1}\}$ of sensor data accumulated up to the time t_k can be interpreted by interpretation histories $\mathbf{j}_k = (j_k, \ldots, j_1), 0 \leq j_k \leq m_k$ that assume a certain data interpretation j_l to be true at each data collection time t_l, $1 \leq l \leq k$.

Via marginalization, for kinematic object states $X_{k-1} = \mathbf{x}_{k-1}$, the previous filtering density $p(\mathbf{x}_{k-1}|Z^{k-1})$ can be written as a mixture over the interpretation histories $\mathbf{j}_{k-1}$. Let us furthermore assume that its components are given by Gaußians:

$$p(\mathbf{x}_{k-1}|Z^{k-1}) = \sum_{\mathbf{j}_{k-1}} p(\mathbf{x}_{k-1}|\mathbf{j}_{k-1}, Z^{k-1})\, p(\mathbf{j}_{k-1}|Z^{k-1}) \tag{3.50}$$

$$= \sum_{\mathbf{j}_{k-1}} p_{\mathbf{j}_{k-1}}\, \mathcal{N}\big(\mathbf{x}_{k-1};\, \mathbf{x}^{\mathbf{j}_{k-1}}_{k-1|k-1},\, \mathbf{P}^{\mathbf{j}_{k-1}}_{k-1|k-1}\big). \tag{3.51}$$

With a Gauß-Markov evolution model such as in Eq. 2.8, the prediction densities obey a similar representation:

$$p(\mathbf{x}_k|Z^{k-1}) = \sum_{\mathbf{j}_{k-1}} p_{\mathbf{j}_{k-1}}\, \mathcal{N}\big(\mathbf{x}_k;\, \mathbf{x}^{\mathbf{j}_{k-1}}_{k|k-1},\, \mathbf{P}^{\mathbf{j}_{k-1}}_{k|k-1}\big), \tag{3.52}$$

where $\mathbf{x}^{\mathbf{j}_{k-1}}_{k|k-1}$, $\mathbf{P}^{\mathbf{j}_{k-1}}_{k|k-1}$ result from the Eqs. 3.13 and 3.14. By making use of the likelihood function for uncertain data discussed earlier (Eq. 2.40) and according to Bayes' Theorem, we obtain:

$$p(\mathbf{x}_k|Z^k) = \frac{\sum_{j_k, \mathbf{j}_{k-1}} \ell_{j_k}(\mathbf{x}_k)\, p_{\mathbf{j}_{k-1}}\, \mathcal{N}\big(\mathbf{x}_k;\, \mathbf{x}^{\mathbf{j}_{k-1}}_{k|k-1},\, \mathbf{P}^{\mathbf{j}_{k-1}}_{k|k-1}\big)}{\sum_{j_k, \mathbf{j}_{k-1}} \int d\mathbf{x}_k\, \ell_{j_k}(\mathbf{x}_k)\, p_{\mathbf{j}_{k-1}}\, \mathcal{N}\big(\mathbf{x}_k;\, \mathbf{x}^{\mathbf{j}_{k-1}}_{k|k-1},\, \mathbf{P}^{\mathbf{j}_{k-1}}_{k|k-1}\big)} \tag{3.53}$$

$$= \sum_{\mathbf{j}_k} p_{\mathbf{j}_k}\, \mathcal{N}\big(\mathbf{x}_k;\, \mathbf{x}^{\mathbf{j}_k}_{k|k},\, \mathbf{P}^{\mathbf{j}_k}_{k|k}\big) \tag{3.54}$$

by using the product formula for Gaußians. The weighting factors $p_{\mathbf{j}_k}$ are given by:

$$p_{\mathbf{j}_k} = \frac{p^*_{\mathbf{j}_k}}{\sum_{\mathbf{j}_k} p^*_{\mathbf{j}_k}} \tag{3.55}$$

with the unnormalized weighting update:

$$p^*_{\mathbf{j}_k} = p_{\mathbf{j}_{k-1}} \begin{cases} (1 - P_D)\rho_F & \text{for } j_k = 0 \\ P_D\, \mathcal{N}\big(\mathbf{z}^{j_k}_k;\, \mathbf{H}_k \mathbf{x}^{\mathbf{j}_{k-1}}_{k|k-1},\, \mathbf{S}^{\mathbf{j}_{k-1}}_{k|k-1}\big) & \text{for } j_k \neq 0 \end{cases}, \tag{3.56}$$

while $\mathbf{x}^{\mathbf{j}_k}_{k|k}$ and $\mathbf{P}^{\mathbf{j}_k}_{k|k}$ result from:

$$\mathbf{x}^{\mathbf{j}_k}_{k|k} = \begin{cases} \mathbf{x}^{\mathbf{j}_{k-1}}_{k|k-1} & \text{for } j_k = 0 \\ \mathbf{x}^{\mathbf{j}_{k-1}}_{k|k-1} + \mathbf{W}^{\mathbf{j}_k}_{k|k-1}(\mathbf{z}^{j_k}_k - \mathbf{H}_k \mathbf{x}^{\mathbf{j}_{k-1}}_{k|k-1}) & \text{for } j_k \neq 0 \end{cases} \tag{3.57}$$

$$\mathbf{P}^{\mathbf{j}_k}_{k|k} = \begin{cases} \mathbf{P}^{\mathbf{j}_{k-1}}_{k|k-1} & \text{for } j_k = 0 \\ \mathbf{P}^{\mathbf{j}_{k-1}}_{k|k-1} - \mathbf{W}^{\mathbf{j}_k}_{k|k-1}\mathbf{S}^{\mathbf{j}_k}_{k|k-1}\mathbf{W}^{\mathbf{j}_k\top}_{k|k-1} & \text{for } j_k \neq 0 \end{cases} \tag{3.58}$$

with the corresponding innovation covariance and Kalman gain matrices

$$\mathbf{S}^{\mathbf{j}_k}_{k|k-1} = \mathbf{H}_k \mathbf{P}^{\mathbf{j}_{k-1}}_{k|k-1} \mathbf{H}^\top_k + \mathbf{R}^{j_k}_k \tag{3.59}$$

$$\mathbf{W}^{\mathbf{j}_k}_{k|k-1} = \mathbf{P}^{\mathbf{j}_{k-1}}_{k|k-1} \mathbf{H}^\top_{k|k-1} \big(\mathbf{S}^{\mathbf{j}_k}_{k|k-1}\big)^{-1}, \tag{3.60}$$

which are defined in analogy to the expressions in Eqs. 3.19 and 3.44. This filtering update philosophy can directly be generalized to IMM-MHT-type techniques [29].

According to the previous discussion, each mixture component

$$p^{\mathbf{j}_k}_{k|k}\; p(\mathbf{x}_k|\mathbf{j}_k, Z^k)$$

of the resulting densities $p(\mathbf{x}_k|Z^k)$ represents a *track hypothesis*. The structure of a Gaußian mixture for $p(\mathbf{x}_k|Z^k)$ also occurs if an IMM prediction $p(\mathbf{x}_k|Z^{k-1})$ (see previous subsection) is updated by using a Gaußian likelihood according to Eq. 3.49, where $p(i_k|Z^k)\, p(\mathbf{x}_k|i_k, Z^k)$ can be considered as a *model hypothesis*. IMM filtering may thus be considered as a multiple hypothesis tracking method as well. See [30, 31] for an alternative treatment of the multiple hypothesis tracking problem by exploiting expectation maximization techniques.

Figure 3.3 provides a schematic illustration of MHT filtering. A mixture component p^i of the filtering density at time t_k is predicted to time t_{k+1}. Due to uncertainty in the object evolution, the predicted component is broadened (dashed line). Let us assume that three measurements are in the expectation gate, which can be interpreted by four data interpretation hypotheses. The likelihood function is thus a sum of three Gaußians and a constant. The subsequent filtering thus spawns the predicted component into four filtering components with different weights depending on the innovation square of the sensor measurement belonging to each component.

Approximations

In case of a more severe clutter background or in a multiple object tracking task with expectation gates overlapping for a longer time, Bayesian tracking filters inevitably lead to mixture densities $p(\mathbf{x}_k|Z^k)$ with an exponentially growing number of mixture components involved. In contrast to the rigorous Bayesian reasoning, the choice of a prudent approximation scheme is in some sense an "art" depending on the particular application considered.

Practical experience in many real-world applications (see [32–34], for example) shows, however, that the densities are usually characterized by several distinct modes. By using

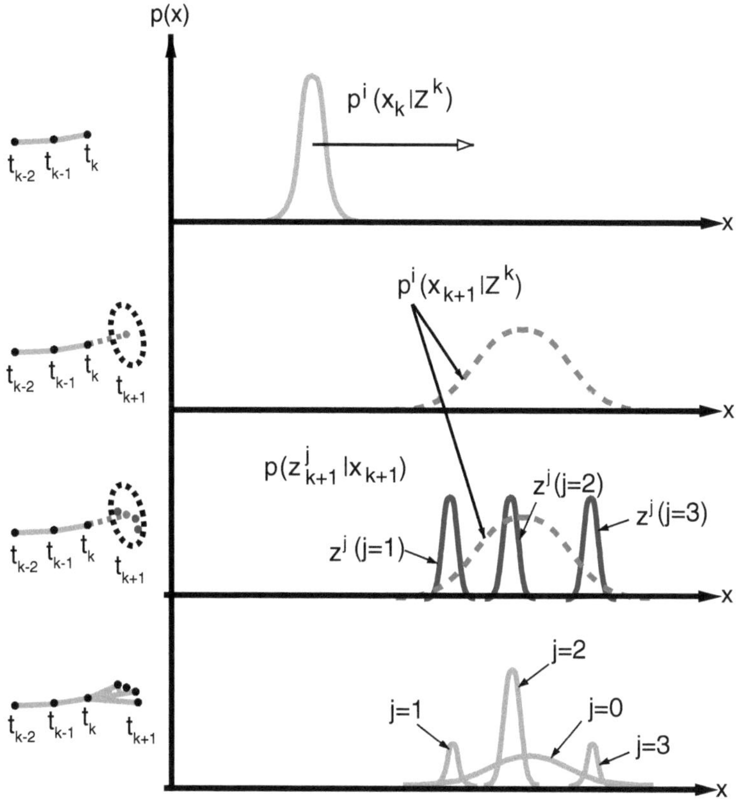

Fig. 3.3 Simplified schematic illustration of the MHT filtering process with 4 measurements to be processed at time t_{k+1}

- *individual gating* for each track hypothesis,
- *local combining* of similar components via moment matching, and
- *pruning* of irrelevant mixture components,

memory explosions can be avoided without destroying the multi-modal structure of the densities $p(\mathbf{x}_k|Z^k)$. Provided this is carefully done with data-driven adaptivity, all statistically relevant information can be preserved, while keeping the number of mixture components under control, i.e. the number of mixture components involved may be fluctuating and even large in critical situations, but does not grow explosively [35–38]. This strategy was first applied by van Keuk et al. [39] and is illustrated in Fig. 3.4.

PDA-type filtering according to Bar Shalom, where *all* mixture components are combined via moment matching, is a limiting case of such techniques [23]. As the phenomenon of distinct modes is inherent in the uncertain origin of the received data, however, relevant statistical information would get lost if global combining

Fig. 3.4 Schematic illustration: local combining of similar components via moment matching and pruning of irrelevant mixture components. **a** Local combining. **b** Pruning

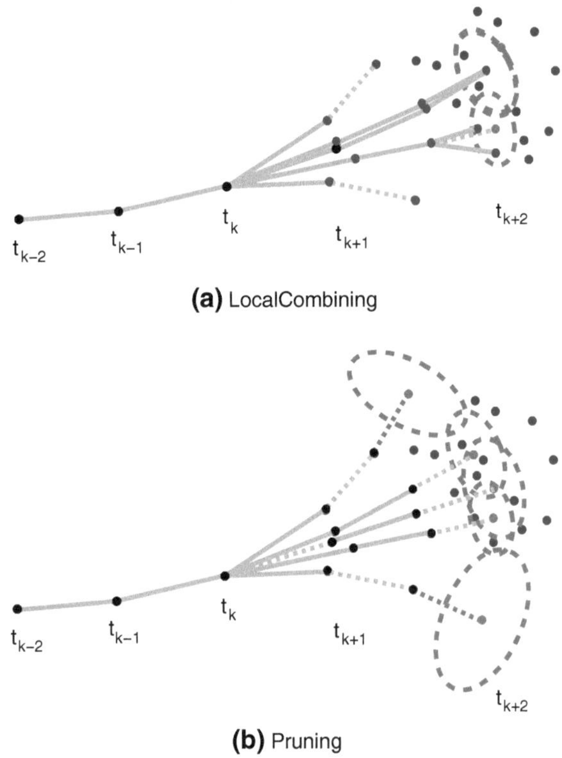

were applied to such cases. The use of PDA-type filtering is thus confined to a relatively restricted area in parameter space (defined by ρ_F, P_D, for instance).

3.4 Object State Retrodiction

For making statements about past object states X_l at time instants t_l with $l < k$, given that sensor information Z^k is available up to the present time t_k, the probability density functions $p(X_l|Z^k)$, i.e. the retrodiction densities, must be calculated. As before in filtering, Bayes Theorem is the key to an iterative calculation scheme, which starts with the knowledge on the object $p(X_k|Z^k)$ available at the present time t_k and is directed towards the past. In deriving a retrodiction update formula, which relates $p(X_l|Z^k)$ to the previously obtained retrodiction density $p(X_{l+1}|Z^k)$ calculated for time t_{l+1}, the object state X_{l+1} at this very time is brought into play via marginalization,

$$p(X_l|Z^k) = \int \mathrm{d}X_{l+1}\; p(X_l, X_{l+1}|Z^k) \tag{3.61}$$

$$= \int \mathrm{d}X_{l+1}\; p(X_l|X_{l+1}, Z^k)\; p(X_{l+1}|Z^k). \tag{3.62}$$

Since in this equation $p(X_{l+1}|Z^k)$ is assumed to be available, it remains to understand the meaning of the density $p(X_l|X_{l+1}, Z^k)$ in the integrand of the previous equation. It seems to be intuitively clear that the knowledge on the object state X_l at time t_l does not depend on the sensor data produced at $t_{l+1}, \ldots, t_k$, i.e. $p(X_l|X_{l+1}, Z^k) = p(X_l|X_{l+1})$. In Appendix A.7, a more formal argumentation is given. A subsequent use of Bayes' Theorem yields:

$$p(X_l|X_{l+1}, Z^k) = p(X_l|X_{l+1}, Z^l) \tag{3.63}$$

$$= \frac{p(X_{l+1}|X_l)\; p(X_l|Z^l)}{\int \mathrm{d}X_l\; p(X_{l+1}|X_l)\; p(X_l|Z^l)}. \tag{3.64}$$

The retrodiction update equation for $p(X_l|Z^k)$ is thus given by:

$$p(X_l|Z^k) = \int \mathrm{d}X_{l+1}\; \frac{p(X_{l+1}|X_l)\; p(X_l|Z^l)}{p(X_{l+1}|Z^l)}\; p(X_{l+1}|Z^k) \tag{3.65}$$

and combines the previously obtained retrodiction, filtering, and prediction densities as well as the object evolution model.

We assemble several characteristic features of retrodiction techniques:

1. In the given formulation, retrodiction applies to single as well as multiple objects, to well-separated objects, object formations, and objects characterized by a more general state.
2. The retrodicted density for time l is completely determined by the filtering density at time l and the following instants of time up to the present ($l \leq k$). Retrodiction is thus decoupled from filtering and prediction and may be switched off without affecting the tracking filter performance (e.g. in overload situations).
3. Accurate filtering and prediction is prerequisite to accurate retrodiction. Provided these processing steps were performed approximately optimally, the retrodiction loop provides an approximately optimal description of the past object states.
4. Besides making use of the underlying evolution model of the objects, retrodiction refers to no other modeling assumption. In particular, the sensor data are not reprocessed by retrodiction.
5. Retrodiction can provide highly precise reconstructions on object trajectory, including their velocity and acceleration histories that may contribute to object classification (see Sect. 1.3.5).
6. The classification of an air target as a helicopter, for example, could be based on precisely retrodicted velocity estimates equal to Zero. Since such retrodiction-based classifications have impact on the evolution model chosen for the future,

the example illustrates in which way retrodiction results may improve available knowledge even on present and future object states.

3.4.1 Fixed Interval Smoothing

Let us consider kinematic object states only, $X_l = \mathbf{x}_l$, and conditions where Kalman filtering is applicable. Under these conditions and using the product formula for Gaußians, Eq. 3.64 can be written as:

$$p(X_l|X_{l+1}, Z^l) = \frac{\mathcal{N}\big(\mathbf{x}_{l+1};\ \mathbf{F}_{l+1|l}\mathbf{x}_l,\ \mathbf{D}_{l+1|l}\big)\ \mathcal{N}\big(\mathbf{x}_l;\ \mathbf{x}_{l|l},\ \mathbf{P}_{l|l}\big)}{\mathcal{N}\big(\mathbf{x}_{l+1};\ \mathbf{x}_{l+1|l},\ \mathbf{P}_{l+1|l}\big)} \tag{3.66}$$

$$= \mathcal{N}\big(\mathbf{x}_l;\ \mathbf{h}_{l|l+1}(\mathbf{x}_{l+1}),\ \mathbf{R}_{l|l+1}\big) \tag{3.67}$$

with the abbreviations:

$$\mathbf{h}_{l|l+1}(\mathbf{x}_{l+1}) = \mathbf{x}_{l|l} + \mathbf{W}_{l|l+1}(\mathbf{x}_{l+1} - \mathbf{x}_{l+1|l}) \tag{3.68}$$

$$\mathbf{R}_{l|l+1} = \mathbf{P}_{l|l} - \mathbf{W}_{l|l+1}\mathbf{P}_{l+1|l}\mathbf{W}_{l|l+1}^{\top} \tag{3.69}$$

and a "retrodiction gain" matrix

$$\mathbf{W}_{l|l+1} = \mathbf{P}_{l|l}\mathbf{F}_{l+1|l}^{\top}\mathbf{P}_{l+1|l}^{-1}. \tag{3.70}$$

Note that $\mathcal{N}\big(\mathbf{x}_l;\ \mathbf{h}_{l|l+1}(\mathbf{x}_{l+1}),\ \mathbf{R}_{l|l+1}\big)$ can be interpreted in analogy to a Gaußian likelihood function with a linear measurement function $\mathbf{h}_{l|l+1}(\mathbf{x}_{l+1})$. For this reason, a second use of the product finally yields:

$$p(\mathbf{x}_l|Z^k) = \int \mathrm{d}\mathbf{x}_{l+1}\ p(\mathbf{x}_l|\mathbf{x}_{l+1}, Z^k)\ \mathcal{N}\big(\mathbf{x}_{l+1};\ \mathbf{x}_{l+1|k},\ \mathbf{P}_{l+1|k}\big) \tag{3.71}$$

$$= \mathcal{N}\big(\mathbf{x}_l;\ \mathbf{x}_{l|k},\ \mathbf{P}_{l|k}\big), \tag{3.72}$$

where the parameters of the retrodicted density $p(\mathbf{x}_l|Z^k)$ are given the *Rauch-Tung-Striebel* [4] retrodiction update equations:

$$\mathbf{x}_{l|k} = \mathbf{x}_{l|l} + \mathbf{W}_{l|l+1}\big(\mathbf{x}_{l+1|k} - \mathbf{x}_{l+1|l}\big) \tag{3.73}$$

$$\mathbf{P}_{l|k} = \mathbf{P}_{l|l} + \mathbf{W}_{l|l+1}\big(\mathbf{P}_{l+1|k} - \mathbf{P}_{l+1|l}\big)\mathbf{W}_{l|l+1}^{\top}. \tag{3.74}$$

The retrodicted state estimates $\mathbf{x}_{l|k}$ do not depend on the corresponding error covariance matrices $\mathbf{P}_{l|k}$. Their computation may thus be omitted in case of limited resources.

3.4.2 Continuous Time Retrodiction

In certain applications, there is a need to produce suitable interpolations between two retrodicted densities $p(X_l|Z^k)$ and $p(X_{l+1}|Z^k)$ at two subsequent data collection times t_l and t_{l+1} by taking full advantage of the available knowledge of the objects' evolution model [40]. For this reason, let us consider probability densities $p(X_{l+\theta}|Z^k)$ with $0 < \theta < 1$, which represent the available knowledge about the object state at an intermediate instant of time $t_l < t_{l+\theta} < t_{l+1}$. In analogy to the previous reasoning, we obtain:

$$p(X_{l+\theta}|Z^k) = \int \mathrm{d}X_{l+1}\; p(X_{l+\theta}|X_{l+1}, Z^k)\; p(X_{l+1}|Z^k) \tag{3.75}$$

with a modified version of the density in Eq. 3.64:

$$p(X_{l+\theta}|X_{l+1}, Z^k) = \frac{p(X_{l+1}|X_{l+\theta})\; p(X_{l+\theta}|Z^l)}{p(X_{l+1}|Z^l)}. \tag{3.76}$$

Note that the prediction densities $p(X_{l+\theta}|Z^l)$, $p(X_{l+1}|Z^l)$ are available according to Eq. 3.10.

Under conditions, where Kalman filtering is applicable, the Markov transition density $p(X_{l+1}|X_{l+\theta})$ is given by the Gaußian:

$$p(\mathbf{x}_{l+1}|\mathbf{x}_{l+\theta}) = \mathcal{N}\big(\mathbf{x}_{l+1};\; \mathbf{F}_{l+1|l+\theta}\mathbf{x}_{l+\theta},\; \mathbf{D}_{l+1|l+\theta}\big), \tag{3.77}$$

yielding as a special case of Eq. 3.67:

$$p(\mathbf{x}_{l+\theta}|\mathbf{x}_{l+1}, Z^k) = \mathcal{N}\big(\mathbf{x}_{l+\theta};\; \mathbf{h}_{l+\theta|l+1}(\mathbf{x}_{l+1}),\; \mathbf{R}_{l+\theta|l+1}\big), \tag{3.78}$$

with the abbreviations:

$$\mathbf{h}_{l+\theta|l+1}(\mathbf{x}_{l+1}) = \mathbf{x}_{l+\theta|l} + \mathbf{W}_{l+\theta|l+1}(\mathbf{x}_{l+1} - \mathbf{x}_{l+1|l}) \tag{3.79}$$

$$\mathbf{R}_{l+\theta|l+1} = \mathbf{P}_{l+\theta|l} - \mathbf{W}_{l+\theta|l+1}\mathbf{P}_{l+1|l}\mathbf{W}^{\top}_{l+\theta|l+1} \tag{3.80}$$

$$\mathbf{W}_{l+\theta|l+1} = \mathbf{P}_{l+\theta|l}\mathbf{F}^{\top}_{l+1|l+\theta}\mathbf{P}^{-1}_{l+1|l+\theta}. \tag{3.81}$$

$p(\mathbf{x}_{l+\theta}|\mathbf{x}_{l+1}, Z^k)$ directly provides an expression for the continuous time retrodiction density $p(\mathbf{x}_{l-\theta}|Z^k)$ according to the product formula:

$$p(\mathbf{x}_{l-\theta}|Z^k) = \frac{p(\mathbf{x}_{l+1}|\mathbf{x}_{l+\theta})\; p(\mathbf{x}_{l+\theta}|Z^l)}{p(\mathbf{x}_{l+1}|Z^l)} \tag{3.82}$$

$$= \mathcal{N}(\mathbf{x}_{l-\theta};\; \mathbf{x}_{l-\theta|k},\; \mathbf{P}_{l-\theta|k}) \tag{3.83}$$

with parameters given by modified Rauch-Tung-Striebel update formulae:

$$\mathbf{x}_{l-\theta|k} = \mathbf{x}_{l-\theta|l-1} + \mathbf{W}_{l|l-\theta}(\mathbf{x}_{l|k} - \mathbf{x}_{l|l-1}) \tag{3.84}$$

$$\mathbf{P}_{l-\theta|k} = \mathbf{P}_{l-\theta|l-1} + \mathbf{W}_{l|l-\theta}(\mathbf{P}_{l|k} - \mathbf{P}_{l|l-1})\mathbf{W}_{l|l-\theta}^{\top} \tag{3.85}$$

$$\mathbf{W}_{l|l-\theta} = \mathbf{P}_{l-\theta|l-1}\mathbf{F}_{l|l-\theta}^{\top}\mathbf{P}_{l|l-1}^{-1}. \tag{3.86}$$

3.4.3 IMM Retrodiction

With an underlying IMM evolution model, we obtain the following expression for the retrodiction density:

$$p(\mathbf{x}_l, i_l|Z^k) = \sum_{i_{l+1}} \int \mathrm{d}\mathbf{x}_{l+1}\; p(\mathbf{x}_l, i_l|\mathbf{x}_{l+1}, i_{l+1}, Z^k)\; p(\mathbf{x}_{l+1}, i_{l+1}|Z^k), \tag{3.87}$$

where we assume that the previous retrodiction density is in analogy to IMM filtering given by a weighted Gaußian:

$$p(\mathbf{x}_{l+1}, i_{l+1}|Z^k) = p_{l+1|k}^{i_{l+1}}\; \mathcal{N}\big(\mathbf{x}_{l+1};\; \mathbf{x}_{l+1|k}^{i_{l+1}},\; \mathbf{P}_{l+1|k}^{i_{l+1}}\big), \tag{3.88}$$

while the remaining factor in the integral results from:

$$p(\mathbf{x}_l, i_l|\mathbf{x}_{l+1}, i_{l+1}, Z^k) = \frac{p(\mathbf{x}_{l+1}, i_{l+1}|\mathbf{x}_l, i_l)\; p(\mathbf{x}_l, i_l|Z^l)}{\sum_{i_l} \int \mathrm{d}\mathbf{x}_l\; p(\mathbf{x}_{l+1}, i_{l+1}|\mathbf{x}_l, i_l)\; p(\mathbf{x}_l, i_l|Z^l)}. \tag{3.89}$$

With $p(\mathbf{x}_l, i_l|Z^l)$ approximately given by Eq. 3.48 and the IMM evolution model in Eq. 2.11, the product formula yields in analogy to Eq. 3.67:

$$p(\mathbf{x}_l, i_l|\mathbf{x}_{l+1}, i_{l+1}, Z^k) = c_{l|l+1}^{i_l}(\mathbf{x}_{l+1})\; \mathcal{N}\big(\mathbf{x}_l;\; \mathbf{h}_{l|l+1}^{i_{l+1},i_l}(\mathbf{x}_{l+1}),\; \mathbf{R}_{l|l+1}^{i_{l+1},i_l}\big) \tag{3.90}$$

with abbreviations $\mathbf{h}_{l|l+1}^{i_{l+1},i_l}(\mathbf{x}_{l+1})$ and $\mathbf{R}_{l|l+1}^{i_{l+1},i_l}$ given by:

$$\mathbf{h}_{l|l+1}^{i_{l+1},i_l}(\mathbf{x}_{l+1}) = \mathbf{x}_{l|l}^{i_l} + \mathbf{W}_{l|l+1}^{i_{l+1},i_l}(\mathbf{x}_{l+1} - \mathbf{F}_{l+1|l}^{i_{l+1}}\mathbf{x}_{l|l}^{i_l}) \tag{3.91}$$

$$\mathbf{R}_{l|l+1}^{i_{l+1},i_l} = \mathbf{P}_{l|l}^{i_l} - \mathbf{W}_{l|l+1}^{i_{l+1},i_l}\; \mathbf{S}_{l|l+1}^{i_{l+1},i_l}\; \mathbf{W}_{l|l+1}^{i_{l+1},i_l\top}, \tag{3.92}$$

where we used:

$$\mathbf{S}_{l|l+1}^{i_{l+1},i_l} = \mathbf{F}_{l+1|l}^{i_{l+1}}\; \mathbf{P}_{l|j_l}^{i_l}\; \mathbf{F}_{l+1|l}^{i_{l+1}\top} + \mathbf{D}_{l+1|l}^{i_{l+1}} \tag{3.93}$$

$$\mathbf{W}_{l|l+1}^{i_{l+1},i_l} = \mathbf{P}_{l|j_l}^{\mathbf{i}_l}\; \mathbf{F}_{l+1|l}^{i_{l+1}\top}\big(\mathbf{S}_{l|l+1}^{i_{l+1},i_l}\big)^{-1} \tag{3.94}$$

and factors $c^{i_l}_{l|j_l}(\mathbf{x}_{l+1})$, which can be interpreted as normalized weighting factors depending on the object state $\mathbf{x}_{l+1}$:

$$c^{i_l}_{l|l+1}(\mathbf{x}_{l+1}) = \frac{p_{i_{l+1}i_l}\, p^{i_l}_{l+1|l}\ \mathcal{N}\big(\mathbf{x}_{l+1};\ \mathbf{F}^{i_{l+1}}_{l+1|l}\,\mathbf{x}^{i_l}_{l+1|l},\ \mathbf{S}^{i_{l+1}i_l}_{l|l+1}\big)}{\sum_{i_l} p_{i_{l+1}i_l}\, p^{i_l}_{l+1|l}\ \mathcal{N}\big(\mathbf{x}_{l+1};\ \mathbf{F}^{i_{l+1}}_{l+1|l}\,\mathbf{x}^{i_l}_{l+1|l},\ \mathbf{S}^{i_{l+1}i_l}_{l|l+1}\big)}. \tag{3.95}$$

According to these considerations, $p(\mathbf{x}_l, i_l|\mathbf{x}_{l+1}, i_{l+1}, Z^k)$ can no longer be interpreted in analogy to a Gaußian likelihood and be evaluated by exploiting the product formula. The problems are caused by the weighting factors $c^{i_l}_{l|l+1}(\mathbf{x}_{l+1|k})$, which explicitly depend on the kinematic object state at time t_{l+1} in a rather complicated way. The product formula would be directly applicable only if they were constant. The best knowledge on $\mathbf{x}_{l+1}$ available at time t_k, however, is given by the expectation $\mathbf{x}_{l+1|k}$ calculated in the previous retrodiction step. We thus consider the approximation:

$$c^{i_l}_{l|l+1}(\mathbf{x}_{l+1}) \approx c^{i_l}_{l|l+1}(\mathbf{x}_{l+1|k}), \tag{3.96}$$

which leads to an approximate expression for the retrodicted density:

$$\begin{aligned} p(\mathbf{x}_l, i_l|Z^k) &\approx \sum_{i_{l+1}} \int d\mathbf{x}_{l+1}\ c^{i_l}_{l|l+1}\, p^{i_{l+1}}_{l+1|k}\ \mathcal{N}\big(\mathbf{x}_l;\ \mathbf{h}^{i_{l+1},i_l}_{l|l+1}(\mathbf{x}_{l+1}),\ \mathbf{R}^{i_{l+1},i_l}_{l|l+1}\big) \\ &\qquad \times \mathcal{N}\big(\mathbf{x}_{l+1};\ \mathbf{x}^{i_{l+1}}_{l+1|k},\ \mathbf{P}^{i_{l+1}}_{l+1|k}\big) \qquad (3.97) \\ &= \sum_{i_{l+1}} p^{i_{l+1}i_l}_{l|k}\ \mathcal{N}\big(\mathbf{x}_l;\ \mathbf{x}^{i_{l+1}i_l}_{l|k},\ \mathbf{P}^{i_{l+1}i_l}_{l|k}\big) \qquad (3.98) \end{aligned}$$

with $p^{i_{l+1}i_l}_{l|k} = c^{i_l}_{l|l+1}\, p^{i_{l+1}}_{l+1|k}$, while the parameters $\mathbf{x}^{i_{l+1}i_l}_{l|k}$ and $\mathbf{P}^{i_{l+1}i_l}_{l|k}$ are obtained by the Rauch-Tung-Striebel formulae 3.73, 3.74. As in standard IMM prediction, $p(\mathbf{x}_k, i_k|Z^{k-1})$ is approximated via moment matching ([23, p. 56 ff], Appendix A.6) yielding

$$p(\mathbf{x}_l, i_l|Z^k) \approx p^{i_l}_{l|k}\ \mathcal{N}\big(\mathbf{x}_l;\ \mathbf{x}^{i_l}_{l|k},\ \mathbf{P}^{i_l}_{l|k}\big) \tag{3.99}$$

with parameters $p^{i_l}_{l|k}$, $\mathbf{x}^{i_l}_{l|k}$, and $\mathbf{P}^{i_l}_{l|k}$ given by:

$$\begin{aligned} p^{i_l}_{l|k} &= \sum_{i_{l+1}=1}^{r} p^{i_{l+1}i_l}_{l|k} \\ \mathbf{x}^{i_l}_{l|k} &= \frac{1}{p^{i_l}_{l|k}} \sum_{i_{l+1}=1}^{r} p^{i_{l+1}i_k}_{l|k}\, \mathbf{x}^{i_{l+1}i_l}_{l|k} \qquad (3.100) \\ \mathbf{P}^{i_l}_{l|k} &= \frac{1}{p^{i_l}_{l|k}} \sum_{i_{l+1}=1}^{r} p^{i_{l+1}i_l}_{l|k}\, \big(\mathbf{P}^{i_k i_{k-1}}_{k|k-1} + (\mathbf{x}^{i_{l+1}i_l}_{l|k} - \mathbf{x}^{i_l}_{l|k})(\mathbf{x}^{i_{l+1}i_l}_{l|k} - \mathbf{x}^{i_l}_{l|k})^\top\big). \end{aligned}$$

If only the retrodictions of the kinematic state variables $\mathbf{x}_l$ are of interest, $p(\mathbf{x}_l|Z^k)$ is given by a Gaußian sum with r mixture components:

$$p(\mathbf{x}_l|Z^k) = \sum_{i_l=1}^{r} p(\mathbf{x}_l, i_l|Z^k) \tag{3.101}$$

$$\approx \sum_{i_l=1}^{r} p_{l|k}^{i_l} \, \mathcal{N}\big(\mathbf{x}_l;\, \mathbf{x}_{l|k}^{i_l},\, \mathbf{P}_{l|k}^{i_l}\big). \tag{3.102}$$

3.4.4 MHT Retrodiction

As discussed before in the case of MHT filtering, data interpretation histories $\mathbf{j}_k$ provide possible explanations of the origin of a time series Z^k consisting of ambiguous sensor data. The notion of retrodiction can also be applied to those situations. Due to the total probability theorem and under the conditions discussed in Sect. 2.3.3, the retrodiction $p(\mathbf{x}_l|Z^k)$ may be represented by a mixture:

$$p(\mathbf{x}_l|Z^k) = \sum_{\mathbf{j}_k} p(\mathbf{x}_l, \mathbf{j}_k|Z^k) \tag{3.103}$$

$$= \sum_{\mathbf{j}_k} p_{\mathbf{j}_k} \, \mathcal{N}\big(\mathbf{x}_l;\, \mathbf{x}_{l|k}^{\mathbf{j}_k},\, \mathbf{P}_{l|k}^{\mathbf{j}_k}\big). \tag{3.104}$$

Since for any given data interpretation history $\mathbf{j}_k$ the conditional probability densities $p(\mathbf{x}_l|\mathbf{j}_k, Z^k)$ are unambiguous, the parameters $\mathbf{x}_{l|k}^{\mathbf{j}_k}$, $\mathbf{P}_{l|k}^{\mathbf{j}_k}$ of the retrodiction density $p(\mathbf{x}_l|Z^k)$ directly result from the Rauch-Tung-Striebel formulae, while the weighting factors $p_{\mathbf{j}_k}$ are those obtained in the filtering step. In other words, the retrodiction process proceeds along the branches of the data interpretation hypotheses tree.

In the following, we assemble several aspects of MHT retrodiction.

1. For well-separated objects and a single evolution model under ideal operational conditions, i.e. without false measurements and assuming perfect detection, the approach comes down to the Rauch-Tung-Striebel fixed-interval smoothing as a limiting case. Hence, the Rauch-Tung-Striebel formulae play a role in MHT retrodiction that is completely analogous to the Kalman update formulae in MHT filtering.
2. MHT retrodiction can be combined with IMM evolution models. Under ideal conditions with well-separated objects, we obtain a hierarchy of approximations to the original retrodiction problem. Adopting the standard terminology [4], the Fraser-Potter-type algorithms in [41, 42] are approximations to optimal retrodiction in the two-filter form insofar as the results of backward and forward filters are combined. In our view, however, the Rauch-Tung-Striebel-type formulation of approximate IMM-smoothing offers advantages over Fraser-Potter-type

algorithms insofar as computational effort is concerned (e.g. matrix inversions involved). In addition, Rauch-Tung-Striebel-type algorithms are initialized by the filtering results (no diffuse prior density).

3. In order to avoid memory explosions such as mentioned in Sect. 3.3.3, those mixture components in the filtering process are neglected ($p_{\mathbf{j}_k} \to 0$) that are either statistically irrelevant or can be combined with other mixture components. This has useful consequences: If all hypotheses with the same prehistory $\mathbf{j}_{k-1}$ are canceled, $\mathbf{j}_{k-1}$ is irrelevant itself ($p_{\mathbf{j}_{k-1}} \to 0$). This scheme may be applied repeatedly to all subsequent prehistories $\mathbf{j}_l$, $l < k-1$, finally leading to a unique track. This process is called *reconstruction of histories* [28, 29, 32]. The work reported in [43] also points in that direction. As observed in [28, 32], we assemble the following facts.

   (a) The history is correctly reconstructed with high probability.
   (b) The number of relevant hypotheses to be stored can be drastically reduced.
   (c) The number of missed detections in a reconstructed history provides on-line estimations of the detection probability that are otherwise not easily obtainable.

4. Oliver Drummond's *Retrodiction of Probabilities* [2, 3] is an approximation of the retrodiction density $p(\mathbf{x}_k|Z^k)$ that omits the Rauch-Tung-Striebel update of the retrodicted expectation vector and the corresponding covariance matrix. In other words, we assume:

$$\mathcal{N}\big(\mathbf{x}_l;\ \mathbf{x}^{\mathbf{j}_k}_{l|k},\ \mathbf{P}^{\mathbf{j}_l}_{l|k}\big) \approx \mathcal{N}\big(\mathbf{x}_l;\ \mathbf{x}^{\mathbf{j}_k}_{l|l},\ \mathbf{P}^{\mathbf{j}_l}_{l|l}\big). \tag{3.105}$$

As a direct consequence, we yield approximations to the density functions

$$p(\mathbf{x}_l|Z^k) \approx \sum_{\mathbf{j}_l} p^*_{\mathbf{j}_l}\, \mathcal{N}\big(\mathbf{x}_l;\ \mathbf{x}^{\mathbf{j}_l}_{l|l},\ \mathbf{P}^{\mathbf{j}_l}_{l|l}\big) \tag{3.106}$$

with mixture coefficients $p^*_{\mathbf{j}_l}$ that are recursively defined by

$$p^*_{\mathbf{j}_l} = \begin{cases} p_{\mathbf{j}_k} & \text{for } l = k \\ \sum_{\mathbf{j}_{l+1} \in J_{\mathbf{j}_{l+1}|\mathbf{j}_l}} p^*_{\mathbf{j}_{l+1}} & \text{for } l < k, \end{cases} \tag{3.107}$$

where the sum is taken over all histories $\mathbf{j}_{l+1}$ with the same prehistory $\mathbf{j}_l$. True hypotheses that incidentally have had a small weight in at the time, when they were originally created, may well increase in weight incrementally during this procedure as time goes by. Retrodiction of discrete probabilities is computationally cheap since only weighting factors are to be re-processed, leaving the state estimates and their error covariance matrices unchanged.

5. In principle, retrodiction methods do not affect the description of the current object states provided the filtering was done correctly. As previously discussed (Sect. 3.3.3), however, in any practical realization approximations must be applied

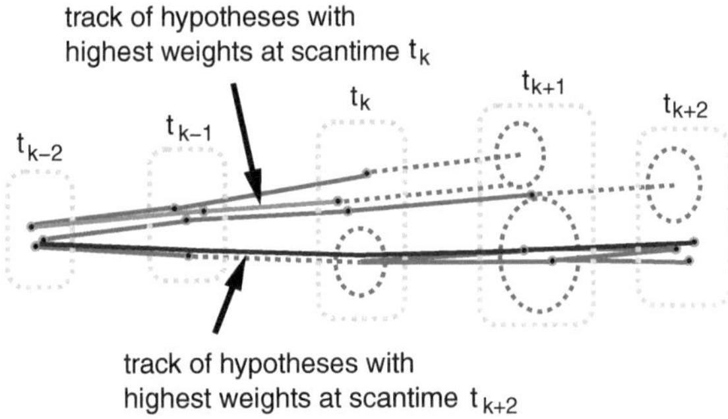

Fig. 3.5 Schematic overview of retrodiction within a multiple hypothesis framework

to avoid memory explosions. In this context, retrodiction-based pruning offers the potential of improved approximations to optimal filtering. The scheme generalizes the strategy in [1, 3] in that, as each set of sensor data becomes available, the modification of earlier track hypotheses has impact on subsequent tracks (Multiple Frame Data Association).

(a) In *retrodiction-based cut-off*, we first permit hypotheses with even very small weights at present. By retrodiction of probabilities, retrospectively some past hypotheses increase in weight, while others decrease. Then, starting at a certain time in the past, hypotheses with insignificant weights are neglected. This has impact up to the present scan since all descending track hypotheses vanish themselves, while the remaining weighting factors are renormalized. This scheme may be applied repeatedly over several data collection times.
(b) In close analogy to retrodiction-based cut-off, we might also delay the decision if two hypotheses are to be combined, thus leading to *retrodiction-based local combining*.

Retrodiction-based pruning seems to be particularly useful in track initiation/ extraction [44], an issue addressed below (Sect. 4).

Figure 3.5 provides a schematic overview of retrodiction within a multiple hypothesis framework.

3.4.5 Discussion of an Example

The following aspects are illustrated by an example with real radar data:

1. Data association conflicts arise even for well-separated objects if a high false return background is to be taken into account, which cannot completely be suppressed by clutter filtering at the signal processing level.
2. Even in the absence of unwanted sensor reports, ambiguous correlations between newly received sensor data and existing tracks are an inherent problem for objects

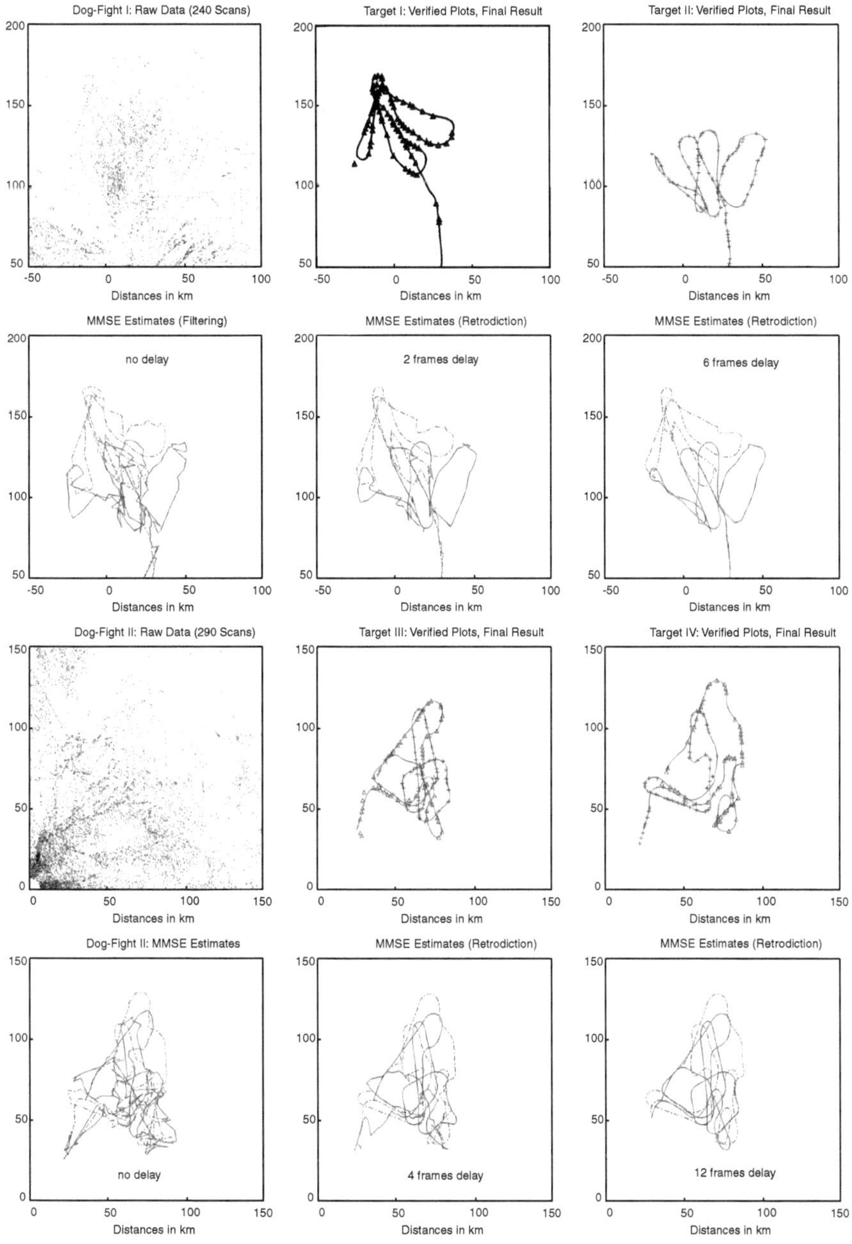

Fig. 3.6 An example taken from wide-area ground-based air surveillance: two pairs of highly maneuvering aircraft in a training situation (high residual clutter background)

moving closely-spaced for some time. Furthermore, resolution phenomena make the data association task even harder.

3. Additional problems arise from poor quality sensor data, due to large measurement errors, low signal-to-noise ratios, or fading phenomena (i.e. successively missing plots). Besides that, the scan rates may be low (especially in long-range surveillance applications).

Figure 3.6 shows a radar data set accumulated over about 240 and 290 scans, respectively. As well as many false alarms, the data of two pairs of highly maneuvering aircraft in a training situation were recorded. The detection probability is between 40 and 60 %. The data were collected from a rotating S-band long-range radar measuring target range and azimuth (scan period: 10 s, range accuracy: 350 ft, bearing accuracy: 0.22 °, range resolution: 1,600 ft, bearing resolution: 2.4 °). Information on the real object position is crucial for evaluating tracking filters. Here a secondary radar was used. The verified primary plots are indicated by $\triangle$ and $+$ in the figures right to the raw data along with the final tracking result (i.e. tracking output according to multiple data association hypotheses and subsequent retrodiction until no further improvement is obtained).

The 2nd and 4th row in 3.6 show for both scenarios the MMSE-estimates of the objects' positions are displayed for a retrodiction delay of zero, 2, 4, 6, and 12 scans. The estimates with no delay are simply obtained by MHT-type filtering. The resulting trajectories seem to be of small value for assessing the air situation. The related variances (very large) are not indicated. The high inaccuracy observed reflects the complex hypothesis tree resulting from ambiguous data interpretations. Multiple dynamics model filtering (IMM) does outperform single model filtering in some particular situations that are characterized by fewer data association conflicts and at least one non-maneuvering target. Aside from those situations, however, the overall impression of the pure filtering result is similar for both cases.

By using MHT-retrodiction, even a delay of two frames significantly improves the filtering output. We displayed the MMSE estimates derived from $p(\mathbf{x}_l|Z^k)$ for $l = 2, 4, 6, 12$. A delay of 6 frames (i.e. 1 min) provides easily interpretable trajectories, while the maximum gain by retrodiction is obtained after 12 frames delay. Evidently the final retrodiction results fit the verified primary plots very well. If IMM-retrodiction is used, we essentially obtain the same final trajectory. However, in certain flight phases (not too many false returns, no maneuvers) it is obtained by a shorter delay (about 1–3 frames less). i.e. Under certain circumstances, accurate speed and heading information is available earlier than in case of a single dynamics model.

References

1. O.E. Drummond, Target tracking with retrodicted discrete probabilities, in *Proceedings of SPIE 3163, Signal and Data Processing of Small Targets*,1997, p. 249

2. O.E. Drummond, Multiple sensor tracking with multiple frame, probabilistic data association, in *SPIE 2561, Signal and Data Processing of Small Targets*, 1995, p. 322
3. O.E. Drummond, Multiple target tracking with multiple frame, probabilistic data association, in *Proceedings of SPIE 1954, Signal and Data Processing of Small Targets*, 1993, p. 394
4. A. Gelb (ed.), *Applied Optimal Estimation* (M.I.T Press, Cambridge, 1974)
5. M.E Liggins, D.L Hall, J. Llinas (eds.), *Handbook of Multisensor Data Fusion - Theory and Practice*, 2nd edn. (CRC Press, Boca Raton, 2008)
6. D.M. Titterington, A.F.M. Smith, U.E. Makov, *Statistical Analysis of Finite Mixture Distributions* (Wiley, New York, 1985)
7. D.L. Alspach, H.W. Sorenson, Nonlinear Bayesian estimation using gaussian sum approximation. IEEE Trans. Autom. Control **20**, 439–447 (1972)
8. M. Ulmke, O. Erdinc, and P. Willett, Ground moving target tracking with cardinalized Gaussian mixture PHD filtering, in *Proceedings of the 10th International Conference on Information Fusion FUSION 2007*, Quebec, Canada, July 2007
9. N. Gordon, A hybrid bootstrap filter for target tracking in clutter, IEEE Trans. Aerosp. Electron. Syst. **33**, 353–358 (1997)
10. B. Ristic, S. Arulampalam, N. Gordon, *Beyond the Kalman Filter: Particle Filters for Tracking Applications* (Artech House Radar Library, London, 2004)
11. D. Fox, W. Burgard, F. Dellaert, S. Thrun, Monte Carlo localization: Efficient position estimation for mobile robots, in *Proceedings of AAAI-99.*
12. A. Doucet, N. de Freitas, N. Gordon (eds.), *Sequential Monte Carlo Methods in Practice* (Springer, Berlin, 2001)
13. M. Ulmke, W. Koch, Road-map assisted ground moving target tracking. IEEE Trans. Aerosp. Electron. Syst. AES-42 **4**, 1264–1274 (2006)
14. V. Algeier, B. Demissie, W. Koch, R. Thomä, State space initiation for blind mobile terminal position tracking. EURASIP J. Adv. Sig. Process **2008**(394219), 14. doi:10.1155/2008/394219
15. W. Koch, Perspectives for a Backbone technology: tracking and fusion in real-world applications, in *11th International Conference on Information Fusion, Cologne, Germany (2008).* Under review: IEEE AESS Magazine
16. V. Klumpp, F. Sawo, U.D. Hanebeck, D. Franken, The sliced Gaussian mixture filter for efficient nonlinear estimation, in *11th International Conference on Information Fusion*, Cologne, Germany, 2008
17. K.-C. Chang, T. Zhi, R.K. Saha, Performance evaluation of track fusion with information matrix filter. IEEE Trans. Aerosp. Electron. Syst. **38**(2), 455 (2002)
18. H. Chen, T. Kirubarajan, Y. Bar-Schalom, Performance limits of track-to-track fusion versus centralized fusion: Theory and applications. IEEE Trans. Aerosp. Electron. Syst. **39**(2), 386 (2003)
19. C.-Y. Chong, S. Mori, H. Barker, K.-C. Chang, Architectures and algorithms for track association and fusion, in *IEEE Aerospace and Electronic Systems Society Magazine*, Jan, 2000
20. S. Mori, W.H. Barker, C.-Y. Chong, K.-C. Chang, Track association and track fusion with nondeterministic target dynamics. IEEE Trans. Aerosp. Electron. Syst. **38**(2), 659 (2002)
21. C.R. Berger, M. Daun, W. Koch, Low complexity track initialization from a small set of non-invertible measurements. EURASIP J. Adv. Signal Process., special issue on track-before-detect techniques **2008**(756414), 15 (2008)
22. Ch. Berger, M. Daun, W. Koch, Track initialization from incomplete measurements, in *Proceedings of 10th ISIF International Conference on Information Fusion*, Quebec, Canada, 2007
23. Y. Bar-Shalom, X.-R. Li, T. Kirubarajan, *Estimation with Applications to Tracking and Navigation* (Wiley, New York, 2001)
24. Rong X. Li, V.P. Jilkov, Survey of maneuvering target tracking. Part V. Multiple model methods. IEEE Trans. Aerosp. Electron. Syst. **41**(4), 1255–1321 (2005)
25. X.R. Li, Hybrid Estimation Techniques, in *Control and Dynamic Systems*, ed. by C.T. Leondes, vol. 76 (Academic Press, San Diego, 1996)
26. H.A.P. Blom, J. Jygeros (Eds.). *Stochastic Hybrid Systems: Theory and Safety Critical Applications*, (Springer, Berlin, 2006)

27. S.S. Blackman, Multiple Hypothesis Methods. in *IEEE AES Magazine***19**(1), 41–52 (2004)
28. W. Koch, On Bayesian MHT for well-separated targets in a densely cluttered environment, in *Proceedings of the IEEE International Radar Conference, RADAR'95*, Washington, May 1995, p. 323
29. W. Koch, Fixed-interval retrodiction approach to Bayesian IMM-MHT for maneuvering multiple targets, IEEE Trans. Aerosp. Electron. Syst. AES-36 **1**, (2000)
30. R. Streit, T. E. Luginbuhl, *Probabilistic Multi-Hypothesis Tracking*. Naval Undersea Warfare Center Dahlgren Division, *Research Report NUWC-NPT/10/428* (1995)
31. M. Wieneke, W. Koch, The PMHT: Solutions to some of its problems *In Proceedings of SPIE Signal and Data Processing of Small Targets*, San Diego, USA, Aug 2007
32. W. Koch, M. Blanchette, Tracking in densely cluttered environment - concepts and experimental results, in *Proceedings of ISATA Robotics, Motion, and Machine Vision*, Aachen, Germany, 1994, p. 215 ff
33. W. Koch, Experimental results on Bayesian MHT for maneuvering closely-spaced objects in a densely cluttered environment, in *IEE International Radar Conference RADAR'97*, Edinburgh, UK, 1999, p. 729
34. W. Koch, Generalized smoothing for multiple model/multiple hypothesis filtering: Experimental results, in *Proceedings of European Control Conference ECC'99*, Karlsruhe, Germany, CD ROM, 1999
35. D. Salmond, Mixture reduction algorithms for target tracking in clutter. SPIE signal Data Process. Small Targets **1305**, 434–445 (1990)
36. L.Y. Pao, Multisensor Multitarget Mixture Reduction Algorithms for Tracking. J. Guidance Control Dyn. **17**, 1205 (1994)
37. R. Danchick, G.E. Newnam, A fast method for finding the exact N-best hypotheses for multitarget tracking, IEEE Trans. Aerosp. Electron. Syst. AES-29 **2**, (1993)
38. I.J. Cox, M.L. Miller, On finding ranked assignments with application to multitarget tracking and motion correspondence. IEEE Trans. Aerosp. Electron. Syst. AES-31, **1**, (1995)
39. W. Fleskes, G. van Keuk, On single target tracking in dense clutter environment - Quantitative results, in *Proceedings of IEE International Radar Conference RADAR 1987*, 1987, pp. 130–134
40. W. Koch, J. Koller, M. Ulmke, Ground target tracking and road-map extraction. ISPRS J. Photogram. Remote Sens. **61**, 197–208 (2006)
41. H.A.P. Blom, Y. Bar-Shalom, Time-reversion of a hybrid state stochastic difference system with a jump-linear smoothing application, IEEE Trans. Inf. Theory **36**(4), 1990
42. R.E. Helmick, W.D. Blair, S.A. Hoffman, Fixed-interval smoothing for Markovian switching systems. IEEE Trans. Info. Technol. **41**(6) (1995)
43. I. Cox, J.J. Leonard, Modeling a dynamic environment using a Bayesian multiple hypothesis approach. Artif. Intell. **66**, 311 (1994)
44. G. van Keuk, Sequential track extraction. IEEE Trans. Aerosp. Electron. Syst. **34**, 1135 (1998)

Chapter 4
Sequential Track Extraction

Iterative tracking algorithms must be initiated appropriately. Under simple conditions, this is not a difficult task, as has been shown above (Eq. 3.8). For low observable objects, i.e. objects embedded in a high clutter background [1–5] or in case of incomplete measurements [6, 7], more than a single set of observations at particular data collection times are usually necessary for detecting all objects of interest moving in the sensors' fields of view. Only then, the probability density iteration can be initiated based on 'extracted' object tracks, i.e. by tentative tracks, whose existence is 'detected' by a detection process working on a higher level of abstraction. This process makes use of a time series of accumulated sensor data $Z^k = \{Z_i\}_{i=1}^k$.

4.1 Well-Separated Objects

Assuming at first that the objects are well-separated, for the sake of simplicity, we thus have to decide between two alternatives before a tracking process can be initiated:

- h_1: Besides false data, Z^k also contains real object measurements.
- h_0: There is no object in the FoV; all sensor data in Z^k are false.

As a special case of the more general theory of statistical decision processes, the performance of a track extraction algorithm is characterized by two probabilities related to the decision errors of first and second kind:

1. $P_1 = P(\text{accept } h_1|h_1)$, i.e. the conditional probability that h_1 is accepted given h_1 is actually true (corresponding to the detection probability P_D of a sensor discussed in Sect. 2.3.4).
2. $P_0 = P(\text{accept } h_1|h_0)$: the conditional probability that h_1 is accepted given it is actually false (corresponding to the false alarm probability P_F of a sensor).

W. Koch, *Tracking and Sensor Data Fusion*,
Mathematical Engineering, DOI: 10.1007/978-3-642-39271-9_4,

4.1.1 Sequential Likelihood Ratio Test

In typical tracking applications, the decisions between the alternatives must be made as quickly as possible on average for given decision probabilities P_0, P_1. The decision algorithm discussed below fulfills this requirement and is of enormous practical importance. It is called *Sequential Likelihood Ratio Test* and was first proposed by Abraham Wald [2–4, 8, 9].

The starting point for sequential decision-making in the context of track extraction is the ratio of the conditional probabilities $p(h_1|Z^k)$ of h_1 being true given all data have been processed appropriately and $p(h_0|Z^k)$ of h_0 being true given the sensor data. If $p(h_1|Z^k)$ is close to One and $p(h_0|Z^k)$ close to Zero, the ratio is large, while it is small if $p(h_1|Z^k)$ is close to Zero and $p(h_0|Z^k)$ close to One. If both hypotheses are more or less equally probable, the ratio is of an intermediate size. According to Bayes' Theorem, we obtain:

$$\frac{p(h_1|Z^k)}{p(h_0|Z^k)} = \frac{p(Z^k|h_1)}{p(Z^k|h_0)}\,\frac{p(h_1)}{p(h_0)}. \tag{4.1}$$

Since the a priori probabilities $p(h_1)$ and $p(h_0)$ are in most applications assumed to be equal, this defines a test function, which is called *likelihood ratio*:

$$\mathrm{LR}(k) = \frac{p(Z^k|h_1)}{p(Z^k|h_0)} \tag{4.2}$$

and can be calculated iteratively by exploiting the underlying object evolution and sensor models $p(X_k|X_{k-1})$ and $p(Z_k|X_k)$.

An intuitively plausible sequential test procedure starts with a time window of length $k = 1$ and iteratively calculates the test function $\mathrm{LR}(k)$ until a decision can be made. At each step of this iteration the likelihood ratio is compared with two thresholds A and B:

$\mathrm{LR}(k) < A$, accept the hypothesis h_0 (i.e. no object existent)
for $\mathrm{LR}(k) > B$, accept the hypothesis h_1 (i.e. an object exists)
$A < \mathrm{LR}(k) < B$, expect new data Z_{k+1}, repeat the test with $\mathrm{LR}(k+1)$.

4.1.2 Properties Relevant to Tracking

Note that the iterative calculation of likelihood ratios has a meaning, which is completely different from the iterative calculation of probability density functions, although similar formulae and calculations are implied, as will become clear below. By iteratively calculated likelihood ratios we wish to decide, whether an iterative tacking process should be initiated or not.

1. The most important theoretical result on sequential likelihood ratio tests is the fact that the test has a *minimum decision length on average* given predefined statistical

decision errors of first and second kind, which have to be specified according the the requirements in a given application.

2. Furthermore, the thresholds A, B can be expressed as functions of the decision probabilities P_0, P_1, i.e. they can be expressed as functions of the statistical decision errors of first and second kind and are thus not independent test parameters to be chosen appropriately. A useful approximation in many applications is given by:

$$A \approx \frac{1 - P_1}{1 - P_0}, \quad B \approx \frac{P_1}{P_0}. \tag{4.3}$$

4.1.3 Relation to MHT Tracking

Likelihood ratios $\mathrm{LR}(k)$ can be calculated iteratively as a by-product of the standard Bayesian tracking methodology previously discussed, provided we look upon it from a different perspective. This can be seen directly:

$$\mathrm{LR}(k) = \frac{p(Z^k|h_1)}{p(Z^k|h_0)} \tag{4.4}$$

$$= \frac{\int d\mathbf{x}_k\, p(Z_k, m_k, \mathbf{x}_k, Z^{k-1}|h_1)}{p(Z_k, m_k, Z^{k-1}, h_0)} \tag{4.5}$$

$$= \frac{\int d\mathbf{x}_k\, \overbrace{p(Z_k, m_k|\mathbf{x}_k, h_1)}^{\text{likelihood}}\, \overbrace{p(\mathbf{x}_k|Z^{k-1}, h_1)}^{\text{prediction}}}{\underbrace{|\mathrm{FoV}|^{-m_k}\, p_F(m_k)}_{\text{clutter model}}} \mathrm{LR}(k-1). \tag{4.6}$$

According to these considerations, the likelihood ratio is in general a sum of a temporally increasing number of individual likelihood ratios,

$$\mathrm{LR}(k) = \sum_i \lambda_k^i. \tag{4.7}$$

In order to avoid memory explosion in calculating the likelihood ratio, the same type of mixture approximation techniques as discussed in Sect. 3.3.3 can be applied (merging of similar, pruning of summands λ_k^i that are too small). Figure 4.1 provides a schematic illustration of the hypothesis tree structure, which is created by sequentially calculating the likelihood ratio test function. As soon as a decision in favor of object existence is made, e.g. at time t_k, the normalized individual likelihood ratios can be used for initializing the tracking process:

$$p(\mathbf{x}_k|Z^k) = \sum_i \frac{\lambda_k^i}{\sum_j \lambda_k^j}\, \mathcal{N}(\mathbf{x}_k;\, \mathbf{x}_{k|k}^i,\, \mathbf{P}_{k|k}^i), \tag{4.8}$$

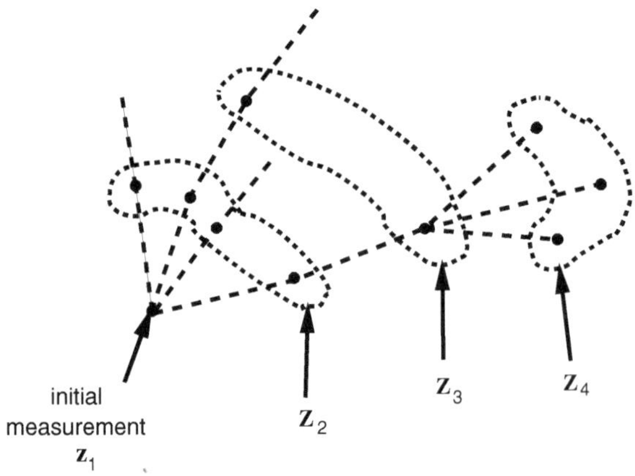

Fig. 4.1 Schematic illustration of the hypothesis tree structure created by sequentially calculating the likelihood ratio test function

where $\mathbf{x}^i_{k|k}$ and $\mathbf{P}^i_{k|k}$ are by-products of the calculation of λ^i_k. As soon as the track has been initiated, the calculation of the likelihood ratio can be restarted as it is a by-product of track maintenance. The output of these subsequent sequential ratio tests can serve to re-confirm track existence or track deletion, depending on the test output. See [1, 2] for details. So far, the problem of multiple well-separated object track extraction, track maintenance, and track deletion, i.e. the full life cycle of a track, is solved in principle. See [5] for an alternative calculation of $\text{LR}(k)$ by using PMHT techniques and [10] for a proof that for well-separated objects, this scheme is identical with Gaußian Mixture Cardinalized PHD filtering (GM-CPHD). Careful quantitative performance evaluations can be found in [11].

4.2 Object Clusters

Sequential likelihood testing can be extended to the problem of extracting object clusters with an unknown number of objects involved. To this end let us assume that the number n of objects involved is limited by N (not too large). The discussed method is confined to N being less than around 10. This means that it can be applied to aircraft formations and convoys of ground moving objects, which are practically relevant examples of object clusters. It is not applicable to larger object clouds or swarms.

4.2.1 Generalized Likelihood Ratio

The ratio of the probability $p(h_1 \vee h_2 \vee \ldots \vee h_N | Z^k)$ that a cluster consisting of at least one and at most N objects exists versus the probability of having false returns only can be written as:

$$\frac{p(h_1 \vee \ldots \vee h_N|Z^k)}{p(h_0|Z^k)} = \frac{\sum_{n=1}^{N} p(h_n|Z^k)}{p(h_0|Z^k)} \tag{4.9}$$

$$= \sum_{n=1}^{N} \frac{p(Z^k|h_n)}{p(Z^k|h_0)} \frac{p(h_n)}{p(h_0)}. \tag{4.10}$$

We thus very naturally obtain a generalized test function

$$\mathrm{LR}(k) = \frac{1}{N} \sum_{n=1}^{N} \mathrm{LR}_n(k) \quad \text{with} \quad \mathrm{LR}_n(k) = \frac{p(Z^k|h_n)}{p(Z^k|h_0)} \tag{4.11}$$

to be calculated in analogy to the case $n = 1$. In practical applications the finite resolution capabilities of the sensors involved have to be taken into account (see section IV.A). For the sake of simplicity this has been omitted here.

4.2.2 On Cluster Cardinality

It seems to be reasonable to interpret the normalized individual likelihood ratios as a 'cardinality', i.e. as a measure of the probability of having n objects in the cluster.

$$c_k(n) = \frac{\mathrm{LR}_n(k)}{\sum_{n=1}^{N} \mathrm{LR}_n(k)}. \tag{4.12}$$

An estimator for the number of objects within the cluster is thus given by

$$\bar{n} = \sum_{i=1}^{N} n\, c_k(n). \tag{4.13}$$

See [2, 3] for a more detailed description of the iterative calculation of the likelihood ratios, practical implementation issues and quantitative results.

References

1. W. Koch, J. Koller, M. Ulmke, Ground target tracking and road-map extraction. ISPRS J. Photogram. Remote Sens. **61**, 197–208 (2006)
2. G. van Keuk, Sequential track extraction. IEEE Trans. Aerosp. Electron. Syst. **34**, 1135 (1998)
3. G. van Keuk, MHT extraction and track maintenance of a target formation. IEEE Trans. Aerosp. Electron. Syst. **38**, 288 (2002)
4. Y. Boers, Hans Driessen, in *IEE Proceedings on Radar, Sonar, Navigation*. A particle filter multi target track before detect application, vol. 151, no. 6 (2004)

5. M. Wieneke, W. Koch, On Sequential Track Extraction within the PMHT Framework. EURASIP J. Adv. Signal Process. **2008**, Article ID 276914, 13 (2008). doi:10.1155/2008/276914
6. C.R. Berger, M. Daun, W. Koch, Low complexity track initialization from a small set of non-invertible measurements. EURASIP J. Adv. Signal Process. special issue on track-before-detect techniques **2008**, Article ID 756414, 15 (2008). doi:10.1155/2008/756414
7. Ch. Berger, M. Daun, W. Koch, Track initialization from incomplete measurements, in *Proceedings of 10th ISIF International Conference on Information Fusion*, Quebec, Canada (2007)
8. A. Wald, *Sequential Analysis*, (John Wiley and Sons, New York, 1947)
9. A. Wald, *Statistical Decision Functions*, (John Wiley and Sons, New York, 1950)
10. M. Ulmke, O. Erdinc, and P. Willett, Ground moving target tracking with cardinalized gaussian mixture PHD filtering, in *Proceedings of the 10th International Conference on Information Fusion FUSION 2007*, Quebec, Canada, July (2007)
11. J. Koller, H. Bos, M. Ulmke, Track extraction and multiple hypothesis tracking for GMTI sensor data. FKIE, Report 90 (2005)

Chapter 5
On Recursive Batch Processing

Most target tracking algorithms aim at calculating the conditional probability densities $p(X_l|Z^k)$ of target states X_l, which describe the available knowledge on the target properties at a certain instant of time t_l, given a time series Z^k of imperfect sensor data accumulated to time t_k. In certain applications, however, the kinematic target states $\mathbf{x}_k, \dots, \mathbf{x}_n$, $n \leq k$, accumulated over a certain time window from a past instant of time t_n up to the present time t_k are of interest. The statistical properties the accumulated state vectors are completely described the joint probability density function of them, given the time series Z^k, $p(\mathbf{x}_k, \dots, \mathbf{x}_n|Z^k)$. These densities may be called *Accumulated State Densities (ASD)*. By marginalization, the standard filtering and retrodiction densities directly result. In addition, ASDs fully describe the correlations between the state estimates produced for different instants of time. For this reason ADSs are useful in recursive batch processing, e.g. in case of out-of sequence measurements or when Expectation Maximization methods are used for solving data association problems.

5.1 Accumulated Object State Densities

ASDs are particularly useful in tracking applications, where out-of-sequence (OoS) measurements are to be processed in a centralized measurement fusion architecture, i.e. when the sensor data do not arrive in the temporal order, in which they have been produced. The OoS problem is unavoidable in any real-world multiple sensor tracking application. To the author's knowledge, Yaakov Bar-Shalom was the first, who picked up the problem and provided an exact solution in the case of Kalman filtering [1]. For the subsequent development and generalizations see [2–6]. To avoid storing and reprocessing of the entire time series of sensor data as well as to avoid the temporal delay related to it, OoS measurement have to be inserted into the ongoing tracking process.

W. Koch, *Tracking and Sensor Data Fusion*,
Mathematical Engineering, DOI: 10.1007/978-3-642-39271-9_5,

All information on the object states accumulated over a time window $t_k, t_{k-1}, \ldots, t_n$ of length $n + 1$,

$$X_{k:n} = (X_k, \ldots, X_n), \tag{5.1}$$

that can be extracted from the time series of accumulated sensor data Z^k up to and including time t_k is contained in the joint density function $p(X_{k:n}|Z^k)$, which may be called *Accumulated State Density* (*ASD*). Via marginalization over $X_k, \ldots, X_{l+1}, X_{l-1}, \ldots, X_n$,

$$p(X_l|Z^k) = \int \mathrm{d}X_k, \ldots, \mathrm{d}X_{l+1}, \mathrm{d}X_{l-1}, \ldots, \mathrm{d}X_n \; p(X_k, \ldots, X_n|Z^k), \tag{5.2}$$

we obtain from the accumulated state density the filtering density $p(X_k|Z^k)$ for $l = k$ and the retrodiction densities $p(X_l|Z^k)$ for $l < k$. In addition, the accumulated state density contains all mutual correlations between the individual object state estimates. Bayes' Theorem and a Markovian object evolution model directly provides a recursion formula for calculating it:

$$p(X_{k:n}|Z^k) = \frac{p(Z_k, m_k|X_k)\; p(X_k|X_{k-1})\; p(X_{k-1:n}|Z^{k-1})}{\int \mathrm{d}X_{k-1}\; p(Z_k, m_k|X_k)\; p(X_k|X_{k-1})\; p(X_{k-1:n}|Z^{k-1})}. \tag{5.3}$$

The sensor data Z_k explicitly appear in this representation. A little formalistically speaking, "sensor data processing" means nothing else than to achieve by certain reformulations that the sensor data are no longer be explicitly present in the resulting expressions.

In practical applications, we expect that the index n describing the length of the accumulated state $X_{k:n}$ will not be large. For a more detailed discussion of the material in this section see [7, 8].

5.1.1 Closed-Form Representations

Under conditions, where Kalman filtering is applicable (perfect data sensor-data-to-track association, linear Gaußian sensor and evolution model), a closed-form representation of $p(X_{k:n}|Z^k)$ can be derived, with $X_{k:n} = \mathbf{x}_{k:n}$, i.e. kinematic state vectors are considered. In this case, let the likelihood function be given by:

$$p(Z_k, m_k|X_k) = \mathcal{N}\big(\mathbf{z}_k;\; \mathbf{H}_k\mathbf{x}_k,\; \mathbf{R}_k\big), \tag{5.4}$$

where $Z_k = \mathbf{z}_k$ denotes the vector of sensor measurements at time t_k, $X_k = \mathbf{x}_k$ the kinematic state vector of the object, $\mathbf{H}_k$ the measurement matrix, and $\mathbf{R}_k$ the measurement error covariance matrix, while the evolution model of the target is represented by:

$$p(X_k|X_{k-1}) = \mathcal{N}\big(\mathbf{x}_k;\ \mathbf{F}_{k|k-1}\mathbf{x}_{k-1},\ \mathbf{D}_{k|k-1}\big), \tag{5.5}$$

where $\mathbf{F}_{k|k-1}$ denotes the evolution matrix and $\mathbf{D}_{k|k-1}$ the evolution covariance matrix according to the previous discussion.

A repeated use of the product formula for Gaußians (see Appendix, Eq. A.28) directly yields a product representation of the augmented state density:

$$p(\mathbf{x}_{k:n}|Z^k) = \mathcal{N}\big(\mathbf{x}_k;\ \mathbf{x}_{k|k},\ \mathbf{P}_{k|k}\big) \prod_{l=n}^{k-1} \mathcal{N}\big(\mathbf{x}_l;\ \mathbf{h}_{l|l+1}(\mathbf{x}_{l+1}),\ \mathbf{R}_{l|l+1}\big), \tag{5.6}$$

where the auxiliary quantities $\mathbf{h}_{l|l+1}$, $\mathbf{R}_{l|l+1}$, $l \leq k$, are defined by:

$$\mathbf{h}_{l|l+1}(\mathbf{x}_{l+1}) = \mathbf{x}_{l|l} + \mathbf{W}_{l|l+1}(\mathbf{x}_{l+1} - \mathbf{x}_{l+1|l}) \tag{5.7}$$

$$\mathbf{R}_{l|l+1} = \mathbf{P}_{l|l} - \mathbf{W}_{l|l+1}\mathbf{P}_{l|l+1}\mathbf{W}_{l|l+1}^{\top} \tag{5.8}$$

and a "retrodiction gain" matrix

$$\mathbf{W}_{l|l+1} = \mathbf{P}_{l|l}\mathbf{F}_{l+1|l}^{\top}\mathbf{P}_{l+1|l}^{-1}. \tag{5.9}$$

Note that $\mathcal{N}\big(\mathbf{x}_l;\ \mathbf{h}_{l|l+1}(\mathbf{x}_{l+1}),\ \mathbf{R}_{l|l+1}\big)$ can be interpreted in analogy to a Gaußian likelihood function with a linear measurement function $\mathbf{h}_{l|l+1}(\mathbf{x}_{l+1})$. These quantities are defined by the parameters of the filtering density $p(\mathbf{x}_l|Z^l) = \mathcal{N}(\mathbf{x}_l;\ \mathbf{x}_{l|l},\ \mathbf{P}_{l|l})$ and given by Eqs. 3.42 and 3.43. Note that there exist equivalent formulations of the Kalman update formulae according to the various versions of the product formula (Eq. A.28). Also the parameters of the prediction density $p(\mathbf{x}_{l+1}|Z^l) = \mathcal{N}(\mathbf{x}_{l+1};\ \mathbf{x}_{l+1|l},\ \mathbf{P}_{l+1|l})$, given by Eqs. 3.13 and 3.14, enter into the product representation in Eq. 5.6.

With $\mathbf{x}_{l|k}$, $\mathbf{P}_{l|k}$, $\mathbf{W}_{l|l+1}$ known from the Rauch-Tung-Striebel recursion (Eqs. 3.73 and 3.74), we can rewrite $p(\mathbf{x}_{k:n}|Z^k)$ by the following product:

$$\begin{aligned} p(\mathbf{x}_{k:n}|Z^k) = {} & \mathcal{N}\big(x_k;\ \mathbf{x}_{k|k},\ \mathbf{P}_{k|k}\big) \\ & \times \prod_{l=n}^{k-1} \mathcal{N}\big(x_l - \mathbf{W}_{l|l+1}x_{l+1};\ \mathbf{x}_{l|k} - \mathbf{W}_{l|l+1}\mathbf{x}_{l+1|k},\ \mathbf{Q}_{l|k}\big), \end{aligned} \tag{5.10}$$

where the matrix $\mathbf{Q}_{l|k}$ is defined by:

$$\mathbf{Q}_{l|k} = \mathbf{P}_{l|k} - \mathbf{W}_{l|l+1}\mathbf{P}_{l+1|k}\mathbf{W}_{l|l+1}^{\top} \tag{5.11}$$

$$= \big(\mathbf{P}_{l|k}^{-1} + \mathbf{P}_{l|k}^{-1}\mathbf{W}_{l|l+1}(\mathbf{P}_{l+1|k}^{-1} - \mathbf{W}_{l|l+1}^{\top}\mathbf{P}_{l|k}^{-1}\mathbf{W}_{l|l+1})^{-1}\mathbf{W}_{l|l+1}^{\top}\mathbf{P}_{l|k}^{-1}\big)^{-1}. \tag{5.12}$$

With projectors $\boldsymbol{\Pi}_l$ defined by:

$$
\begin{aligned}
\boldsymbol{\Pi}_l \mathbf{x}_{k:n} &= \begin{cases} (\mathbf{1}, \mathbf{O}, \dots, \mathbf{O})\mathbf{x}_{k:n}, & l = k \\ (\mathbf{O}, \dots, -\mathbf{W}_{l|l+1}, \mathbf{1}, \dots, \mathbf{O})\, \mathbf{x}_{k:n}, & n \le l < k \end{cases} \\
&= \begin{cases} \mathbf{x}_k, & l = k \\ \mathbf{x}_l - \mathbf{W}_{l|l+1}\mathbf{x}_{l+1}, & n \le l < k. \end{cases}
\end{aligned} \tag{5.13}
$$

the accumulated state density $p(\mathbf{x}_{k:n}|Z^k)$ can be rewritten:

$$
p(\mathbf{x}_{k:n}|Z^k) = \prod_{l=n}^{k} \mathcal{N}\big(\mathbf{x}_{k:n};\ \mathbf{x}_{k:n}^k,\ (\boldsymbol{\Pi}_l^\top \mathbf{Q}_{l|k}^{-1} \boldsymbol{\Pi}_l)^{-1}\big) \tag{5.14}
$$

$$
= \mathcal{N}\big(\mathbf{x}_{k:n};\ \mathbf{x}_{k:n}^k,\ \mathbf{P}_{k:n}^k\big) \tag{5.15}
$$

with an accumulated expectation vector $\mathbf{x}_{k:n}^k$ defined by:

$$
\mathbf{x}_{k:n}^k = (\mathbf{x}_{k|k}^\top, \mathbf{x}_{k-1|k}^\top, \dots, \mathbf{x}_{n+1|k}^\top, \mathbf{x}_{n|k}^\top)^\top, \tag{5.16}
$$

and a covariance matrix $\mathbf{P}_{k:n}^k$, which is given by an harmonic mean according to the product formula for Gaußians:

$$
\mathbf{P}_{k:n}^k = \Big(\sum_{l=n}^{k} \boldsymbol{\Pi}_l^\top \mathbf{Q}_{l|k}^{-1} \boldsymbol{\Pi}_l\Big)^{-1}. \tag{5.17}
$$

With the auxiliary matrices $\mathbf{T}_{l|k}$, $n \le l \le k$ defined by:

$$
\mathbf{T}_{l|k} = \begin{cases} \mathbf{Q}_{n|k}^{-1} & \text{for } l = n \\ \mathbf{P}_{k|k}^{-1} + \mathbf{W}_{l-1|l}^\top \mathbf{Q}_{l-1|k}^{-1} \mathbf{W}_{l-1|l} & \text{for } l = k \\ \mathbf{Q}_{l|k}^{-1} + \mathbf{W}_{l-1|l}^\top \mathbf{Q}_{l-1|k}^{-1} \mathbf{W}_{l-1|l} & \text{else,} \end{cases} \tag{5.18}
$$

we obtain:

$$
\mathbf{P}_{k:n}^k = \begin{pmatrix} \mathbf{T}_{k|k} & -\mathbf{W}_{k-1|k}^\top \mathbf{Q}_{k-1|k}^{-1} & \mathbf{O} & \cdots & \mathbf{O} \\ -\mathbf{Q}_{k-1|k}^{-1} \mathbf{W}_{k-1|k} & \mathbf{T}_{k-1|k} & * & * & \vdots \\ \mathbf{O} & -\mathbf{Q}_{k-2|k}^{-1} \mathbf{W}_{k-2|k-1} & * & * & \mathbf{O} \\ \vdots & * & * & \mathbf{T}_{n+1|k} & -\mathbf{W}_{n|n+1}^\top \mathbf{Q}_{n|k} \\ \mathbf{O} & \cdots & \mathbf{O} & -\mathbf{Q}_{n|k} \mathbf{W}_{n|n+1} & \mathbf{T}_{n|k} \end{pmatrix}^{-1}. \tag{5.19}
$$

i.e. $\mathbf{P}_{k:n}^k$ is given by the inverse of a tridiagonal block matrix. This structure is a consequence of the Markov property of the underlying evolution model. This representation of $\mathbf{P}_{k:n}^{k\ -1}$ is useful in practical calculations.

In appendix A.8 it is shown that the inverse of this tridiagonal block matrix can be calculated and is given by:

$$\mathbf{P}^k_{k:n} = \begin{pmatrix} \mathbf{P}_{k|k} & \mathbf{P}_{k|k}\mathbf{W}^\top_{k-1|k} & \cdots & \mathbf{P}_{k|k}\mathbf{W}^\top_{n+1|k} & \mathbf{P}_{k|k}\mathbf{W}^\top_{n|k} \\ \mathbf{W}_{k-1|k}\mathbf{P}_{k|k} & \mathbf{P}_{k-1|k} & \ddots & \vdots & \mathbf{P}_{k-1|k}\mathbf{W}^\top_{n|k-1} \\ \mathbf{W}_{k-2|k}\mathbf{P}_{k|k} & \mathbf{W}_{k-2|k-1}\mathbf{P}_{k-1|k} & \ddots & \mathbf{P}_{n+1|k}\mathbf{W}^\top_{n+1|n+2} & \vdots \\ \vdots & \vdots & \ddots & \mathbf{P}_{n+1|k} & \mathbf{P}_{n+1|k}\mathbf{W}^\top_{n|n+1} \\ \mathbf{W}_{n|k}\mathbf{P}_{k|k} & \mathbf{W}_{n|k-1}\mathbf{P}_{k-1|k} & \cdots & \mathbf{W}_{n|n+1}\mathbf{P}_{n+1|k} & \mathbf{P}_{n|k} \end{pmatrix} \tag{5.20}$$

with the auxiliary matrices $\mathbf{W}_{l|l'+1}$, $n \leq l, l' \leq k-1, l < l'$, defined by:

$$\mathbf{W}_{l|l'+1} = \prod_{\lambda=l}^{l'-1} \mathbf{W}_{\lambda|\lambda+1} = \prod_{\lambda=l}^{l'-1} \mathbf{P}_{\lambda|\lambda}\mathbf{F}^\top_{\lambda+1|\lambda}\mathbf{P}^{-1}_{\lambda+1|\lambda}, \tag{5.21}$$

The densities $\{\mathcal{N}(\mathbf{x}_l;\ \mathbf{x}_{l|k},\ \mathbf{P}_{l|k})\}_{l=m}^k$ are directly obtained via marginalization, since the covariance matrices $\mathbf{P}_{l|k}$, $m \leq l \leq k$, appear on the diagonal of this block matrix. Note that the ASD is completely defined by the results of prediction, filtering, and retrodiction obtained for the time window $t_k, \ldots, t_n$, i.e. it is a by-product of Kalman filtering and Rauch-Tung-Striebel smoothing.

5.1.2 Discussion of Generalizations

These considerations can directly be generalized to the case of ambiguity with respect to the origin of the sensor data or with respect to the evolution model currently being in effect, i.e. to Multiple Hypothesis Tracking (MHT) and Interacting Multiple Model filters (IMM).

ASDs for MHT Filtering

As previously discussed, a sensor output at time t_k, consisting of m_k measurements collected in the set Z_k, can be ambiguous, i.e. the origin of the sensor data has to be explained by a set of data interpretations, which are assumed to be exhaustive and mutually exclusive. As an example, let us consider likelihood functions such as those defined by Eq. 2.40:

$$p(Z_k, m_k|\mathbf{x}_k) = \sum_{j_k=0}^{m_k} p(Z_k, m_k|\, j_k, \mathbf{x}_k)\; p(j_k). \tag{5.22}$$

According to the previous discussion on MHT tracking, the marginalization using the data interpretation histories $\mathbf{j}_k$ yields for the accumulated state density:

$$p(\mathbf{x}_k, \ldots, \mathbf{x}_n|Z^k) = \sum_{\mathbf{j}_k} p(\mathbf{x}_k, \ldots, \mathbf{x}_n, \mathbf{j}_k|Z^k) \tag{5.23}$$

$$= \sum_{\mathbf{j}_k} p(\mathbf{j}_k|Z^k)\; p(\mathbf{x}_{k:n}|\mathbf{j}_k, Z^k) \tag{5.24}$$

$$= \sum_{\mathbf{j}_k} p_{k|k}^{\mathbf{j}_k}\; \mathcal{N}\big(\mathbf{x}_{k:n};\; \mathbf{x}_{k:n}^{\mathbf{j}_k},\; \mathbf{P}_{k:n}^{\mathbf{j}_k}\big), \tag{5.25}$$

i.e. the ASD for MHT applications is simply a weighted sum of individual ASDs, which are completely defined by the results of prediction, filtering, and retrodiction along a certain branch of the hypothesis tree defined by a particular interpretation history $\mathbf{j}_k$. The corresponding weighting factor is given by the probability of $\mathbf{j}_k$ being true at time t_k given the data: $p(\mathbf{j}_k|Z^k) = p_{k|k}^{\mathbf{j}_k}$. The ASD for MHT application is thus a by-product of MHT tracking and retrodiction.

ASDs for IMM Filtering

In applications where it is uncertain, which evolution model out of a set of r possible alternatives is currently being in effect, Markovian IMM evolution models should be used (see Sect. 2.2.2):

$$p(x_k, i_k|x_{k-1}, i_{k-1}) = p_{i_k i_{k-1}}\; \mathcal{N}\big(\mathbf{x}_k;\; \mathbf{F}_{k|k-1}^{i_k}\mathbf{x}_{k-1},\; \mathbf{D}_{k|k-1}^{i_k}\big), \tag{5.26}$$

which are characterized r by kinematic linear Gaußian transition densities $p(\mathbf{x}_k|\mathbf{x}_{k-1}, i_k)$ and class transition probabilities $p_{i_k i_{k-1}} = p(i_k|i_{k-1})$.

By making use of marginalization, also the IMM approach can easily be combined with Kalman or MHT filtering. Let us denote the dynamics histories "m scans back" by $\mathbf{i}_k$, an m-tuple of indices. According to these considerations and for the same reasons as before, the accumulated IMM-MHT state density is as mixture of individual ASDs for each data interpretation and model history:

$$p(\mathbf{x}_{k:n}|Z^k) = \sum_{\mathbf{i}_k\mathbf{j}_k} p_{k|k}^{\mathbf{i}_k\mathbf{j}_k}\; \mathcal{N}\big(\mathbf{x}_{k:n};\; \mathbf{x}_{k:n}^{\mathbf{i}_k\mathbf{j}_k},\; \mathbf{P}_{k:n}^{\mathbf{i}_k\mathbf{j}_k}\big). \tag{5.27}$$

Each ASD component is defined by the results of prediction, filtering, and retrodiction given these histories. The corresponding weighting factor results from the filtering step. In the case of standard IMM filtering with r evolution models, the ASD consists of r components.

5.1.3 Out-of-Sequence Measurements

In any real-world application of sensor data fusion, we have to be aware of *out-of-sequence measurements*. Due to latencies in the underlying communication infrastructure, for example, such measurements arrive at a processing node in a distributed data fusion system "too late", i.e. after sensor data with a time stamp newer than the time stamp of an out-of-sequence measurement have already been processed. Accumulated object state densities are useful for dealing with this type of sensor data, which may provide valuable information on an object state of interest in spite of their latency, especially if the sensor involved is of high quality.

Under conditions, where Kalman filtering is applicable, let us consider a measurement $\mathbf{z}_m$ produced at time t_m with $n \leq m < k$, i.e. before the 'present' time t_k, where the time series Z^k is available and has been exploited. We wish to understand the impact this new but delayed sensor information has on the present and the past object states $\mathbf{x}_l$, $l \leq k$, i.e. on the accumulated object state $\mathbf{x}_{k:n}$. Let $\mathbf{z}_m$ be a measurement of the object state $\mathbf{x}_m$ at time t_m characterized by a Gaußian likelihood function, which is defined by a measurement matrix $\mathbf{H}_m$ and a corresponding measurement error covariance matrix $\mathbf{R}_m$ according to Eq. 2.19. We furthermore introduce a projection matrix $\boldsymbol{\Pi}_m$, defined by $\boldsymbol{\Pi}_m \mathbf{x}_{k:n} = \mathbf{x}_m$, which extracts the object state $\mathbf{x}_m$ from the accumulated state vector $\mathbf{x}_{k:n}$. The likelihood function of the out-of-sequence measurement with respect to the accumulated object state is thus given by:

$$p(\mathbf{z}_m|\mathbf{x}_{k:n}) = \mathcal{N}\big(\mathbf{z}_m;\ \mathbf{H}_m \boldsymbol{\Pi}_m \mathbf{x}_{k:n},\ \mathbf{R}_m\big). \tag{5.28}$$

Standard Bayesian reasoning and the product formula directly yields for the accumulated state density:

$$p(\mathbf{x}_{k:n}|\mathbf{z}_m, Z^k) = \frac{p(\mathbf{z}_m|\mathbf{x}^k)\ p(\mathbf{x}_{k:n}|Z^k)}{\int \mathrm{d}\mathbf{x}^k\ p(\mathbf{z}_m|\mathbf{x}^k)\ p(\mathbf{x}_{k:n}|Z^k)} \tag{5.29}$$

$$= \mathcal{N}\big(\mathbf{x}_{k:n};\ \mathbf{x}^k_{k:m:n},\ \mathbf{P}^k_{k:m:n}\big) \tag{5.30}$$

with parameters obtained by a version of the Kalman update equations:

$$\mathbf{x}^k_{k:m:n} = \mathbf{x}^k_{k:n} + \mathbf{W}_{k:m:n}(\mathbf{z}_m - \mathbf{H}_m \mathbf{x}_{m|k}) \tag{5.31}$$

$$\mathbf{P}^k_{k:m:n} = (\mathbf{1} - \mathbf{W}_{k:m:n}\mathbf{H})\mathbf{P}^k_{k:n}, \tag{5.32}$$

where the corresponding Kalman gain and innovation matrices are given by:

$$\mathbf{S}_{m|k} = \mathbf{H}_m \boldsymbol{\Pi}_m \mathbf{P}^k_{k:n} \boldsymbol{\Pi}^\top_m \mathbf{H}^\top_m + \mathbf{R}_m \tag{5.33}$$

$$= \mathbf{H}_m \mathbf{P}_{m|k} \mathbf{H}^\top_m + \mathbf{R}_m \tag{5.34}$$

$$\mathbf{W}_{k:m:n} = \mathbf{P}^k_{k:n} \boldsymbol{\Pi}^\top_m \mathbf{H}^\top_m \mathbf{S}^{-1}_{m|k} \tag{5.35}$$

$$= \begin{pmatrix} \mathbf{P}_{k|k}\mathbf{W}_{m|k}^\top \\ \mathbf{P}_{k-1|k}\mathbf{W}_{m|k-1}^\top \\ \vdots \\ \mathbf{P}_{m|k} \\ \mathbf{W}_{m-1|m}\mathbf{P}_{m|k} \\ \vdots \\ \mathbf{W}_{n|m}\mathbf{P}_{m|k} \end{pmatrix} \mathbf{H}_m^\top \mathbf{S}_{m|k}^{-1} \tag{5.36}$$

Note that the matrix $\mathbf{S}_{m|k}$ to be inverted when calculating the Kalman gain matrix has the same dimension as the measurement vector $\mathbf{z}_m$, i.e. it is a low-dimensional matrix, just as in standard Kalman filtering. Nevertheless, the processing of an out-of-sequence measurement $\mathbf{z}_m$ has impact on all state estimates and the related error covariance matrices in the time window considered. The strongest impact is observed for the time t_m, where the measurement has actually been produced, while it declines the further we proceed to the present time $t_k > t_m$ or deeper into the past. Accumulated state densities are therefore well suited to quantitatively discuss phenomena related to what is sometimes called "information aging" related to the decreasing relevancy of a piece of information. If we are interested in the updated state estimates for the time t_l, $1 \leq l \leq k$, we simply have to consider the density:

$$p(\mathbf{x}_l|\mathbf{z}_m, Z^k) = \mathcal{N}\big(\mathbf{x}_l;\ \boldsymbol{\Pi}_l\mathbf{x}_{k:m:n}^k,\ \boldsymbol{\Pi}_l\mathbf{P}_{k:m:n}^k\boldsymbol{\Pi}_l^\top\big) \tag{5.37}$$

applying the projection matrix $\boldsymbol{\Pi}_l$ previously introduced. In a practical application, we will usually be interested in the effect of out-of-sequence measurements, which were produced not too long ago, on the present time and the most recent past. It is therefore sufficient to consider accumulated state densities $p(\mathbf{x}_{k:n}|Z^k)$ characterized by lower dimensional parameters $\mathbf{x}_{k:n}^k$, $\mathbf{P}_{k:n}^k$.

By using the accumulated state density given by Eq. 5.25, these considerations can directly be generalized to the case of ambiguous sensor data.

5.1.4 Discussion of an Example

Figure 5.1 shows a simulated trajectory of a maneuvering air target defined by $x(t) = \frac{v^2}{q}\sin(\frac{q}{2v}t)$, $y(t) = 2x(t)$, $v = 300\,\frac{\mathrm{m}}{\mathrm{s}}$, $q = 4\,\mathrm{g}$. It is observed by two typical mid-range radars located at $(-50, 0)$ km and $(0, 50)$ km (scan period: $T = 5$ s, measurement error standard deviations: $\sigma_r = 20$ m (range), $\sigma_\varphi = 0.2°$ (azimuth). Let us consider measurement fusion as discussed in Sect. 2.3.2. The fusion center is receiving measurements from sensor one without communication delay, while the measurements are arriving "out-of-sequence", i.e. with a temporal delay of zero, one, two, and five scan periods.

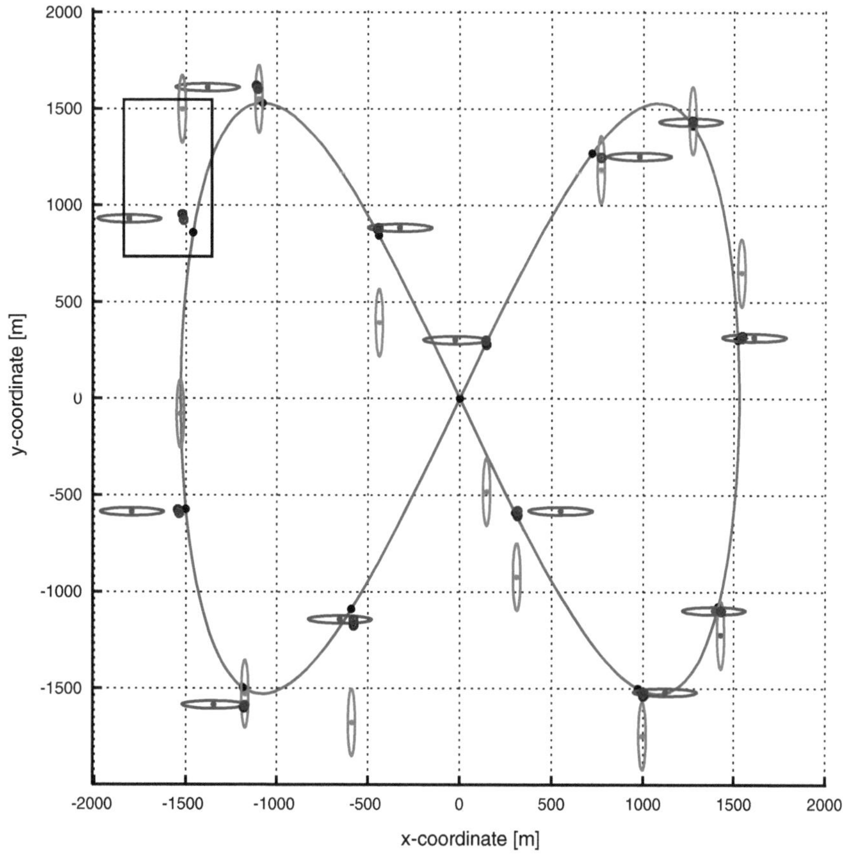

Fig. 5.1 Simulated trajectory of a highly maneuvering object

Figure 5.2 shows numerical results based on 1000 Monte-Carlo-runs (mean error of the expectation vector in position of the filtering steps and the mean trace of corresponding filtering error covariances matrix). For no communication delay and in each sub-figure, the black lines and the region shaded in black represent the mean filtering error in position and the corresponding variance, respectively. The gray lines and the regions shaded in gray show these quantities for different delays as a function of the tracking time. In the right sub-figure in the second row no measurements of sensor 2 are processed at all. Obviously, out-of-sequence measurements produced by sensor 2, which arrive at the fusion center with a delay of five scan periods or more, are nearly useless and do not significantly improve the filtering result at the present time. There may be a significant improvement for retrodicted estimates at the time at which this OoS measurement has actually been produced.

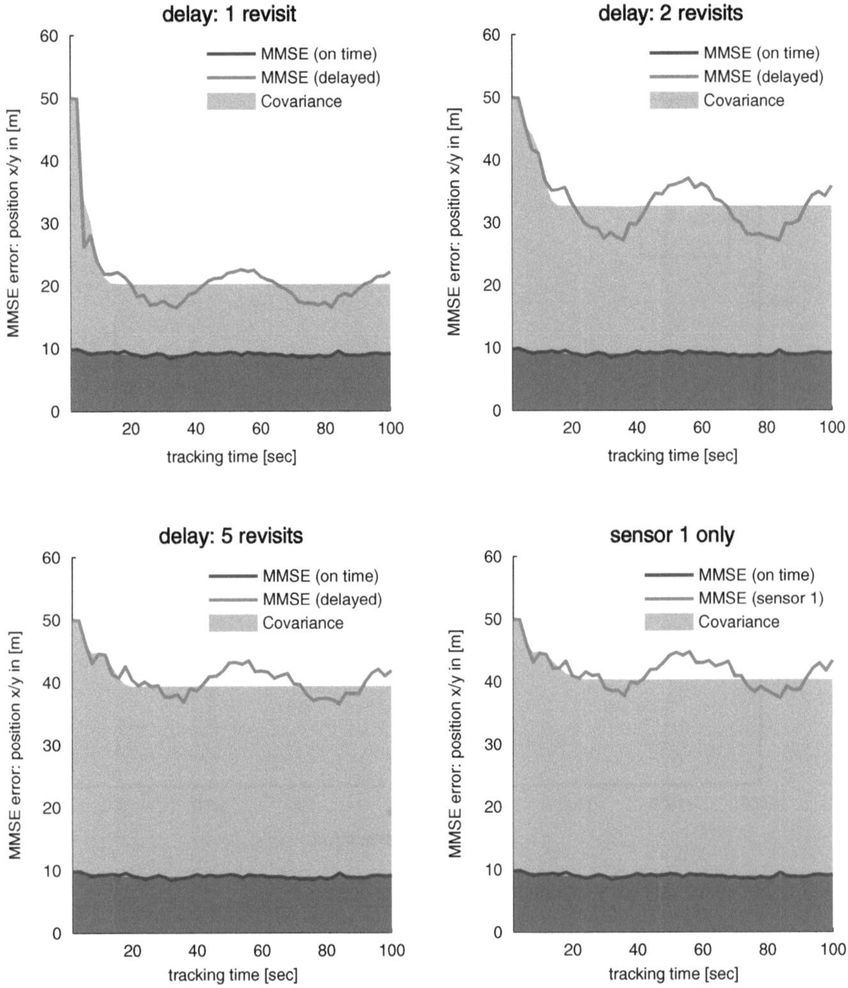

Fig. 5.2 Mean error of the expectation vector in position of the filtering steps and the mean trace of corresponding filtering error covariances matrix for various communication delays

5.2 Solving Tracking Problems by EM

In discussing the data association conflicts in track maintenance, we already dealt with ambiguities in the origin of the sensor data that are caused by unwanted targets and false returns. In this case, an exhaustive and mutually exclusive enumeration of association hypotheses results in MHT- and PDA-type approaches to target tracking as previously discussed. For data association conflicts in a multiple target scenario, these techniques can in principle be applied as well. See the discussion of a multiple target likelihood function in Sect. 2.3.3 and its use in Sect. 7.1.

Tracking in the presence of data association conflicts, however, can also be considered as an "incomplete data problem" [9], which can be solved by the Expectation Maximization methodology (EM). Roy Streit and Tod Ludinbuhl have been the first who applied this approach to target tracking applications [10–12].

5.2.1 Data Augmentation Methods

More generally speaking and in full accordance to the Bayesian approach previously discussed, data augmentation methods, such as Expectation Maximization (EM), intend to make statements about a quantity X given that measurements Z of X are available. Due to missing information, for example, lacking knowledge of the correct measurement-to-target associations, the calculation of the related conditional probability density $p(X|Z)$ may be difficult. If additional information A were known, however, e.g. the data associations, the augmented density $p(X|A, Z)$, could more easily be calculated.

According to this strategy, data augmentation algorithms use the augmented conditional probability density $p(X|A, Z)$ in order to calculate at least certain characteristics of the original probability density $p(X|Z)$. This concept is guided by the general observation that $p(X|Z)$ can be expressed by probability densities involving the additional information A as a direct consequence of $p(A|X, Z) = p(X|A, Z)\, p(A|Z)/p(X|Z)$, i.e. of Bayes rule:

$$p(X|Z) = \frac{p(X|A, Z)\, p(A|Z)}{p(A|X, Z)}. \tag{5.38}$$

The EM algorithm is a particular realization of this more general concept. Let us assume that X^i is a certain preliminary estimate of X. By exploiting the augmented density $p(X|A, Z)$, we wish to calculate a "better" estimate X^{i+1} in the sense that

$$p(X^{i+1}|Z) > p(X^i|Z). \tag{5.39}$$

In other words, we are looking for an iterative algorithm to localize the "posterior mode" of the conditional density $p(X|Z)$. Assuming that $p(A|X_i, Z)$ is available as well, we are looking for an X for which the following is true:

$$\log p(X|Z) > \log p(X^i|Z) \quad \Leftrightarrow \tag{5.40}$$

$$\int \mathrm{d}A \ \log p(X|Z)\ p(A|X^i, Z) > \int \mathrm{d}A \ \log p(X^i|Z)\ p(A|X^i, Z). \tag{5.41}$$

Using as an abbreviation a function Q defined by:

$$Q(X; X^i) = \int \mathrm{d}A \ \log p(X|A, Z)\ p(A|X^i, Z), \tag{5.42}$$

we obtain from Eq. 5.40 by exploiting Eq. 5.38 and Jensen's inequality:

$$Q(X; X_i) - Q(X^i; X^i) > \int \mathrm{d}A \, \log \frac{p(A|X, Z)}{p(A|X^i, Z)} \, p(A|X^i, Z) \tag{5.43}$$

$$\geq \log \int \mathrm{d}A \, \frac{p(A|X, Z)}{p(A|X^i, Z)} \, p(A|X^i, Z) \tag{5.44}$$

$$= 0. \tag{5.45}$$

According to these considerations, an EM algorithm essentially consists of two consecutive steps:

1. *Expectation*: With X^i denoting the current estimate according to $p(X|Z)$ (initialization required!), compute the function $Q(X; X^i)$ defined above.
2. *Maximization*: Find the next update by maximizing $Q(X; X^i)$,

$$X^{i+1} = \mathrm{argmax}_X \, Q(X; X^i), \tag{5.46}$$

and repeat the EM-step until the condition $|Q(X^{i+1}; X^i) - Q(X^i; X^i)| < \epsilon$ holds. A theorem guarantees convergence [9].

In multiple target tracking, auxiliary information on which measurement has been originated by which target is particularly useful. Let

$$a_l^{j \to s_j}, \quad l = k, \ldots, n, \quad j = 1, \ldots, m_l, \quad s_j = 1, \ldots, S, \tag{5.47}$$

denote the hypothesis that at time t_l the measurement $\mathbf{z}_l^j$ is to be associated with a target indexed by s_j. Let a set of all feasible associations that map the measurements in the time window $t_{k:n}$ to each target be denoted by $a_{k:n}$. According to the more general discussion, the function Q is in this case obtained by calculating an expectation over data association probabilities. By using the notion of accumulated state vectors as in the previous section, Q is given by:

$$Q(\mathbf{x}_{k:n}; \mathbf{x}_{k:n}^i) = \sum_{a_{k:n}} \log \left(p(\mathbf{x}_{k:n}|a_{k:n}, Z^k) \right) \, p(a_{k:n}|\mathbf{x}_{k:n}^i, Z^k). \tag{5.48}$$

As will be shown below, under linear Gaußian assumptions regarding the sensor likelihood functions and the transition densities describing the targets' evolution, these functions prove to be given by Gaußian Accumulated State densities as discussed in the previous section (up to a constant irrelevant to maximization). The maximization step in the EM loop is thus trivial and given by the corresponding accumulated vector of expectation vectors obtained by Kalman filtering and retrodiction while processing suitably chosen "synthetic" data.

5.2.2 Expectation and Maximization Steps

By applying Bayes' rule to the argument of the logarithm in the previous function Q,

$$p(\mathbf{x}_{k:n}|a_{k:n}, Z^k) \propto p(Z_{k:n}|a_{k:n}, \mathbf{x}_{k:n})\, p(\mathbf{x}_{k:n}|Z^{n-1}), \tag{5.49}$$

we obtain for Q an expression, which is given up to a constant independent of the state variables by:

$$\begin{aligned} Q(\mathbf{x}_{k:n}; \mathbf{x}^i_{k:n}) =& \log p(\mathbf{x}_{k:n}|Z^{n-1}) \\ &+ \sum_{a_{k:n}} \log\left(p(Z_{k:n}|a_{k:n}, \mathbf{x}_{k:n})\right) p(a_{k:n}|\mathbf{x}^i_{k:n}, Z^k) + \text{const.} \end{aligned} \tag{5.50}$$

Q can be expressed more explicitly by using the known transition densities, sensor likelihood functions, prior information on the object states at time t_{n-1}, and the posterior association probabilities yielding:

$$\begin{aligned} Q(\mathbf{x}_{k:n}; \mathbf{x}^i_{k:n}) =& \sum_{s=1}^{S} \log p(\mathbf{x}^s_{n-1}|Z^{n-1}) + \sum_{l=n}^{k} \sum_{s=1}^{S} \log p(\mathbf{x}^s_l|\mathbf{x}^s_{l-1}) \\ &+ \sum_{l=n}^{k} \sum_{a_l} \log\left(p(Z_l|a_l, \mathbf{x}_l)\right) p(a_l|\mathbf{x}^i_l, Z_l) + \text{const.}, \end{aligned} \tag{5.51}$$

where we assumed $p(a_l|\mathbf{x}^i_{k:n}, Z_k) = p(a_l|\mathbf{x}^i_l, Z_l)$. To proceed, let us first calculate the posterior association probabilities with the additional assumption

$$p(a_l|\mathbf{x}^i_l, Z_l) = p(a_l^{1\to s_1}, \ldots, a_l^{m_l\to s_{m_l}}|\mathbf{x}^i_l, Z_l) \tag{5.52}$$

$$= \prod_{j=1}^{m_l} p(a_l^{j\to s_j}|\mathbf{x}^i_l, Z_l). \tag{5.53}$$

According to Bayes' rule we obtain:

$$p(a_l|\mathbf{x}^i_l, Z_l, m_l) = \frac{p(Z_l|a_l, m_l, \mathbf{x}^i_l)\, \pi_l^{j\to s_j}}{\sum_{a_l} p(Z_l|a_l, m_l, \mathbf{x}^i_l)\, \pi_l^{j\to s_j}} \tag{5.54}$$

with $\pi_l^{j\to s_j}$ denoting the prior probability for associating at time t_l the measurement $\mathbf{z}^j_l$ to the object indexed by s_j:

$$\pi_l^{j\to s_j} = p(a_l^{j\to s_j}|\mathbf{x}^i_l, m_l). \tag{5.55}$$

These quantities will be discussed below in greater detail. The nominator can be rewritten by using an induction argument:

$$\sum_{a_l} p(Z_l|a_l, m_l, \mathbf{x}_l^i)\, p(a_l|\mathbf{x}_l^i, m_l) \tag{5.56}$$

$$= \sum_{s_1=1}^{S} \cdots \sum_{s_{m_l}=1}^{S} \prod_{j=1}^{m_l} p(\mathbf{z}_l^j|\mathbf{x}_l^{i;s_j})\, \pi_l^{j\to s_j} \tag{5.57}$$

$$= \sum_{s_l=1}^{S} \cdots \sum_{s_{m_l-1}=1}^{S} \prod_{j=1}^{m_l-1} p(\mathbf{z}_l^j|\mathbf{x}_l^{i;s_j})\, \pi_l^{j\to s_j}$$

$$\times \left\{ \sum_{s=1}^{S} p(\mathbf{z}_l^{m_l}|\mathbf{x}_l^{i;s})\, \pi_l^{j\to s_j} \right\} \tag{5.58}$$

$$= \prod_{j=1}^{m_l} \sum_{s_j=1}^{S} p(\mathbf{z}_l^j|\mathbf{x}_l^{i;s_j})\, \pi_l^{j\to s_j}. \tag{5.59}$$

The posterior association probabilities are thus given by:

$$p(a_l|\mathbf{x}_l^i, Z_l, m_l) = \prod_{j=1}^{m_l} \frac{\mathcal{N}\big(\mathbf{z}_l^j;\, \mathbf{H}_l\mathbf{x}_l^{i;s_j},\, \mathbf{R}_l^j\big)\, \pi_l^{j\to s_j}}{\sum_{s=1}^{S} \mathcal{N}\big(\mathbf{z}_l^j;\, \mathbf{H}_l\mathbf{x}_l^{i;s},\, \mathbf{R}_l^j\big)\, \pi_l^{j\to s_j}} \tag{5.60}$$

$$= \prod_{j=1}^{m_l} w_l^{i;j\to s_j}, \tag{5.61}$$

where we used as an abbreviation individual weighting factors $w_l^{i;j\to s}$ denoting the posterior association probability for associating in the iteration step i at time t_l the measurement $\mathbf{z}_l^j$ to the object indexed by s. With this result and using

$$\log\big(p(Z_l|a_l, \mathbf{x}_l)\big)\, p(a_l|\mathbf{x}_l^i, Z^k) = \log\left\{ \prod_{j=1}^{m_l} \mathcal{N}\big(\mathbf{z}_l^j;\, \mathbf{H}_l\mathbf{x}_l^{s_j},\, \mathbf{R}_l^j/w_l^{i;j\to s_j}\big) \right\}, \tag{5.62}$$

we can finally calculate the expectation according to the previous considerations:

$$\sum_{a_l} \log\big(p(Z_l|a_l, \mathbf{x}_l)\big)\, p(a_l|\mathbf{x}_l^i, Z_l) \tag{5.63}$$

$$= \sum_{s_1=1}^{S} \cdots \sum_{s_{m_l}=1}^{S} \sum_{j=1}^{m_l} \log \mathcal{N}\big(\mathbf{z}_l^j;\, \mathbf{H}_l\mathbf{x}_l^{s_j},\, \mathbf{R}_l^j/w_l^{i;j\to s_j}\big) \tag{5.64}$$

$$= \sum_{s_1=1}^{S} \cdots \sum_{s_{m_l-1}=1}^{S} \sum_{j=1}^{m_l-1} \sum_{s=1}^{S} \log \mathcal{N}\big(\mathbf{z}_l^{m_l};\ \mathbf{H}_l \mathbf{x}_l^s,\ \mathbf{R}_l^{m_l}/w_l^{i;j\to s}\big) \tag{5.65}$$

$$= \sum_{j=1}^{m_l} \sum_{s=1}^{S} \log \mathcal{N}\big(\mathbf{z}_l^j;\ \mathbf{H}_l \mathbf{x}_l^{s_j},\ \mathbf{R}_l^j/w_l^{i;j\to s}\big). \tag{5.66}$$

Since a sum of logarithms is the logarithm of a product, the resulting product of linear Gaußian likelihood functions is equivalent to a likelihood function characterized by a "fused" measurement as previously discussed,

$$p(\mathbf{z}_l^{*s}|\mathbf{x}_l^s) \propto \prod_{j=1}^{m_l} \mathcal{N}\big(\mathbf{z}_l^j;\ \mathbf{H}_l \mathbf{x}_l^s,\ \mathbf{R}_l^j/w_l^{i;j\to s}\big) \tag{5.67}$$

$$\propto \mathcal{N}\big(\mathbf{z}_l^{*s};\ \mathbf{H}_l \mathbf{x}_l^s,\ \mathbf{R}_l^{*s}\big). \tag{5.68}$$

For each object s, the "synthetic" measurement $\mathbf{z}_l^{*s}$ obtained by combining all measurements at the corresponding time frame t_l according to weighting factors specific for each object s is given by:

$$\mathbf{z}_l^{*s} = \sum_{j=1}^{m_l} w_l^{i;j\to s} \mathbf{R}_l^{j^{-1}} \mathbf{z}_l^j \tag{5.69}$$

$$\mathbf{R}_l^{*s^{-1}} = \sum_{j=1}^{m_l} w_l^{i;j\to s} \mathbf{R}_l^{j^{-1}}. \tag{5.70}$$

Inserting these results into the function Q and taking the logarithm of Q, it is proportional to a double product, which is can be represented by a product of S Accumulated State Densities, according to the discussion in the previous section:

$$\log Q(\mathbf{x}_{k:n};\ \mathbf{x}_{k:n}^i) \propto \prod_{s=1}^{S} \prod_{l=n}^{k} p(\mathbf{z}_l^{*s}|\mathbf{x}_l^s)\ p(\mathbf{x}_l^s|\mathbf{x}_{l-1}^s)\ p(\mathbf{x}_{n-1}^s|Z^{n-1}) \tag{5.71}$$

$$\propto \prod_{s=1}^{S} \mathcal{N}\big(\mathbf{x}_{k:n}^s;\ \mathbf{x}_{k:n}^{*s},\ \mathbf{P}_{k:n}^{*s}\big). \tag{5.72}$$

The maximization of the function Q is thus obtained by running S independent Kalman filters on the synthetic measurements followed by Kalman retrodiction. The resulting expectation vectors $\mathbf{x}_{k:n}^* = \mathbf{x}_{k:n}^{i+1}$ are input for calculating the $Q(\mathbf{x}_{k:n};\ \mathbf{x}_{k:n}^{i+1})$ to be maximized in the next step. The EM loop is repeated until convergence. Instead of enumeration association hypotheses followed by pruning an merging, the EM philosophy thus tries to solve the association problem by an iteration. Linearity in the number of targets and measurements is the main motivation for a further development and extension of this methodology, which is often called Probabilistic Multiple Hypothesis Tracking (PMHT).

5.2.3 Discussion of Problems

Unfortunately, compared with alternatives such as the Probabilistic Data Association Filter (PDAF), PMHT has not yet shown its superiority in terms of track-lost statistics. Furthermore, the problem of track extraction and deletion is apparently not yet satisfactorily solved within this framework. Four properties of PMHT are responsible for its problems in track maintenance: *Non-Adaptivity*, *Hospitality*, *Narcissism*, and *Local Maxima* [13]. Approaches towards a solution for each of these phenomena and derivations of improved PMHT trackers have been proposed such as [14]. Moreover, a sequential Likelihood-Ratio (LR) test for track extraction has been developed, which in the context of the PMHT methodology has the potential for track extraction [15]. As PMHT provides all required ingredients for a sequential LR calculation, the LR is thus a by-product of the PMHT iteration process.

Here, he have considered point-source objects. A generalization to extended objects is straightforward and will in a be discussed in Chap. 8. For introducing missing and false measurements into the framework, let P_D^s be denoting a detection probability related to object s, which produces at most *one* measurement at time t_l per object and sensor. Moreover, $\mathbf{z}_l^0$ indicates a non-informative measurement representing a missing detection, i.e. a measurement with a very large measurement error. The probability of having m_l false measurements is $p_F(m_t)$ given by a Poisson distribution characterized by a spatial clutter density as before. Let us introduce a fictitious target $s = 0$ producing $m_l \geq 0$ false measurements at time t_l, which are equally distributed in space. The detection probability related to this fictitious target $s = 0$ is given by $P_D^0 = 1 - p_F(0)$. As before, we assume that the data associations are independent of each other, which is a critical one since it can be justified as an approximation only.

For performance improvements, much depends on the proper formulation of *prior* probabilities $\pi_l^{j \to s_j}$ under the hypothesis S targets are existent. See [14] for a detailed discussion.

References

1. Y. Bar Shalom, Update with out-of-sequence measurements in tracking: exact solution. IEEE Trans. Aerosp. Electron. Syst. **38**(3), 769 (2002)
2. M. Mallick, Out-of-sequence track filtering using the decorrelated pseudo measurement approach. In *SPIE Signal and Data Processing of Small Targets*, vol. 5428, p. 154 (2004)
3. Y. Bar-Shalom, H. Chen, IMM etimator with out-of-sequence measurements. IEEE Trans. Aerosp. Electron. Syst. **41**(1), 90–98 (2005)

4. M. Orton, A. Marrs, Particle filters for tracking with out-of-sequence measurements. IEEE Trans. Aerosp. Electron. Syst. **41**(2), 693–702 (2005)
5. R.G. Everitt, S.R. Maskell, R. Wright, M. Briers, Multi-target out-of-sequence data association: tracking using graphical models. Elsevier J. Inf. Fusion **7**(4), 434–447 (2007)
6. J. Zhen, A. Balasuriya, S. Challa, Sensor fusion-based visual target tracking for autonomous vehicles with the out-of-sequence measurements solution. Elsevier J. Robot. Auton. Syst. **56**(2), 157–176 (2008)
7. W. Koch, On accumulated state densities and their use for fusing out-of-sequence measurements. In *Proceeding of 12th International Conference on Information Fusion*, Seattle, WA, USA (2009)
8. W. Koch, F. Govaers, On accumulated state densities with applications to out-of-sequence measurement processing. IEEE Trans. Aerosp. Electron. Syst. **47**(4), 2766 (2011)
9. M.A. Tanner, *Tools for Statistical Inference* (Springer, New York, 1996)
10. R.L. Streit (ed.), *Studies in Probabilistic Multi-Hypothesis Tracking and Related Topics*, vol. SES-98-01. Naval Undersea Warfare Center Division, Newport, Rhode Island, USA
11. R.L. Streit, T.E. Luginbuhl, Maximum likelihood method for multi-hypothesis tracking. In *Signal and Data Processing of Small Targets*, SPIE, vol. 2335, April (1994)
12. H. Gauvrit, J.-P. LeCadre, C. Jauffet, A formulation of multitarget tracking as an incomplete data problem. IEEE Trans. Aerosp. Electron. Syst. **AES-33**(4), 1242 (1997)
13. P. Willett, R. Yanhua, R. Streit, PMHT: problems and some solutions. IEEE Trans. Aerosp. Electron. Syst. **38**(3), 738 (2002)
14. M. Wieneke, W. Koch, The PMHT: solutions for some of its problems. In *Signal and Data Processing of Small Targets 2007*, Proc SPIE, vol. 6699, p. 669917 (2007). doi:10.1117/12.734388
15. M. Wieneke, W. Koch, On sequential track extraction within the PMHT framework. EURASIP J. Adv. Signal Process. **2008**, Article ID 276914, 13 (2008). doi:10.1155/2008/276914

Chapter 6
Aspects of Track-to-Track Fusion

In many distributed tracking applications, the quality of a state estimation suffers from communication bandwidth limitations. In particular, scenarios using wireless channels such as HF radio, WLAN, or 3G networks experience link breakdowns and limited capacity constraints. Furthermore, if sensors are involved which have a high update rate, e.g. sonar or lidar, or return much clutter, e.g. radar, network technologies are hardly sufficient to cover all needs with respect to a constantly full transfer of measurements. A common approach for such applications is to transmit only local tracks instead of measurements. This method has important advantages. First and foremost, the load on communication channels is much lower because false tracks and clutter are suppressed already. However, such a paradigm raises the question of an optimal track-to-track fusion (T2TF) in a central fusion center. The pioneering paper in this area has been written by Chee Chong [1]

The goal of T2TF is combining locally preprocessed information of individual sensors optimally, i.e. in a way that is equivalent to fusing all measurements of all sensors directly (measurement fusion, see the discussion in sect. 3.1.1). It is well known that this can be achieved for deterministically moving targets or in situations where the sensor tracks produced at *all* individual data collection times are available in the fusion center. Full-rate communication in this sense, however, is impractical in many applications. We thus propose a distributed Kalman-type processing scheme for maneuvering targets, which provides optimal track-to-track fusion results at arbitrarily chosen instants of time by communicating and combining the local sensor 'tracks' referring to this time. Applications can be found in situations with a highly fluctuating connectivity.

For the sake of simplicity, let us assume conditions where measurement fusion with Kalman filtering is applicable and S synchronized sensors produce measurements at the same instants of time t_l, $l = 1, \ldots, k$ denoted by $Z_l = \{\mathbf{z}_l^s\}_{s=1}^S$. The proposed methodology can be directly extended to asynchronous sensors.

W. Koch, *Tracking and Sensor Data Fusion*,
Mathematical Engineering, DOI: 10.1007/978-3-642-39271-9_6,

6.1 Full-Rate Communication Fusion

It is well-known that optimal track-to-track fusion, i.e. the full reconstruction of $p(\mathbf{x}_k|\{Z_s^k\}_{s=1}^S)$ from the locally produced densities $p(\mathrm{x}_k|Z_s^k)$, $1 \leq s \leq S$, is possible if the local tracks produced at all instants of time t_l, $l = 1, \ldots, k$ of all sensors are available in the fusion center. This can easily be seen by considering the joint probability density:

$$p(Z_k^s, \mathrm{x}_k|Z_s^{k-1}) = p(\mathrm{x}_k|Z_s^k)\, p(Z_k^s|Z_s^{k-1}) \tag{6.1}$$

$$= p(Z_k^s|\mathrm{x}_k)\, p(\mathrm{x}_k|Z_s^{k-1}), \tag{6.2}$$

which allows the representation of the local likelihood function $p(Z_k^s|\mathrm{x}_k) \propto \ell_k^s(\mathrm{x}_k; Z_k^s)$ of the sensor s at each time t_k using the results of prediction and filtering up to a factor independent of the object state:

$$\ell_k^s(\mathrm{x}_k; Z_k^s) \propto \frac{p(\mathrm{x}_k|Z_s^k)}{p(\mathrm{x}_k|Z_s^{k-1})}. \tag{6.3}$$

6.1.1 *Frenkel Tracklets*

Under the conditions where Kalman filtering is applicable, Eq. 6.3 is a quotient of Gaußians, which can be calculated up to a factor independent of x_k according to a product formula for Gaußians (see Appendix A.5, Eq. A.28, 2nd version):

$$\ell_k^s(\mathrm{x}_k; \mathrm{z}_k^s) \propto \frac{\mathcal{N}\big(\mathrm{x}_k;\, \mathrm{x}_{k|k}^s,\, \mathbf{P}_{k|k}^s\big)}{\mathcal{N}\big(\mathrm{x}_k;\, \mathrm{x}_{k|k-1}^s,\, \mathbf{P}_{k|k-1}^s\big)} \tag{6.4}$$

$$\propto \mathcal{N}\big(\mathrm{x}_k;\, \mathrm{x}_{k|k}^{*s},\, \mathbf{P}_{k|k}^{*s}\big) \tag{6.5}$$

with $\mathrm{x}_{k|k}^{*s}$ and $\mathbf{P}_{k|k}^{*s}$ given by:

$$\mathbf{P}_{k|k-1}^{*s\ -1} = \mathbf{P}_{k|k}^{s\ -1} - \mathbf{P}_{k|k-1}^{s\ -1} \tag{6.6}$$

$$\mathrm{x}_{k|k-1}^{*s} = \mathbf{P}_{k|k}^{*s}\big(\mathbf{P}_{k|k}^{s\ -1}\mathrm{x}_{k|k}^{s} - \mathbf{P}_{k|k-1}^{s\ -1}\mathrm{x}_{k|k-1}^{s}\big). \tag{6.7}$$

This representation of the local likelihood function $\ell_k^s(\mathrm{x}_k; \mathrm{z}_k^s)$ has also been called a 'tracklet' according to G. Frenkel [2].

6.1.2 Tracklet Fusion

Along this line of argumentation and with a repeated use of the product formula, the joint likelihood function of the measurements of S independent sensors producing data at the same time t_k can be written as:

$$\ell_k(\mathrm{x}_k;\, Z_k) \propto \prod_{s=1}^{S} \mathcal{N}\big(\mathrm{x}^{*s}_{k|k-1};\, \mathrm{x}_k,\, \mathbf{P}^{*s}_{k|k-1}\big) \tag{6.8}$$

$$\propto \mathcal{N}\big(\mathrm{x}_k;\, \mathrm{x}^{*}_{k|k-1},\, \mathbf{P}^{*}_{k|k-1}\big) \tag{6.9}$$

with $\mathrm{x}^*_{k|k}$ and $\mathbf{P}^*_{k|k}$ given by

$$\mathbf{P}^{*\ -1}_{k|k-1} = \sum_{s=1}^{S} \big(\mathbf{P}^{s\ -1}_{k|k} - \mathbf{P}^{s\ -1}_{k|k-1}\big) \tag{6.10}$$

$$\mathrm{x}^{*}_{k|k-1} = \mathbf{P}^{*}_{k|k-1} \sum_{s=1}^{S} \big(\mathbf{P}^{s\ -1}_{k|k}\mathrm{x}^{s}_{k|k} - \mathbf{P}^{s\ -1}_{k|k-1}\mathrm{x}^{s}_{k|k-1}\big). \tag{6.11}$$

With this representation of the joint likelihood function $\ell_k(\mathrm{x}_k;\, Z_k)$ built up by using the local sensor tracks, formulae for optimal track-to-track fusion are a direct result of Bayes' Theorem and the product formula:

$$p(\mathrm{x}_k|Z^k) = \frac{\ell_k(\mathrm{x}_k;\, Z_k)\; p(\mathrm{x}_k|Z^{k-1})}{\int \mathrm{d}\mathrm{x}_k\; \ell_k(\mathrm{x}_k;\, Z_k)\; p(\mathrm{x}_k|Z^{k-1})} \tag{6.12}$$

$$= \mathcal{N}\big(\mathrm{x}_k;\, \mathrm{x}_{k|k},\, \mathbf{P}_{k|k}\big) \tag{6.13}$$

with $\mathrm{x}_{k|k}$ and $\mathbf{P}_{k|k}$ given by

$$\mathbf{P}^{-1}_{k|k} = \mathbf{P}^{-1}_{k|k-1} + \mathbf{P}^{*\ -1}_{k|k-1} \tag{6.14}$$

$$\mathrm{x}_{k|k} = \mathbf{P}_{k|k}\big(\mathbf{P}^{-1}_{k|k-1}\mathrm{x}_{k|k-1} + \mathbf{P}^{*\ -1}_{k|k-1}\mathrm{x}^{*}_{k|k-1}\big). \tag{6.15}$$

6.2 Arbitrary Communication Rate Fusion

In many applications, the full communication of all local sensor tracks produced at all scan times of the individual sensors along with fusing them as previously sketched is impractical. We therefore propose a distributed Kalman-type processing scheme that provides optimal track-to-track fusion results at an *arbitrarily chosen* instant of time by communicating the local sensor 'tracks' referring to this time only and fusing them in the fusion center.

More concretely speaking, we derive fusion formulae to correctly reconstruct the probability densities $p(\mathrm{x}_l|Z^k)$ for an arbitrary instant of time t_l from local state estimates $\mathrm{x}^s_{l|k}$, $s = 1, \ldots, S$, which are calculated by processing data from sensor s only:

$$\{\mathrm{x}^s_{l|k}\}_{s=1}^S \xrightarrow[\text{fusion}]{\text{track-to-track}} p(\mathrm{x}_l|\{Z^k_s\}_{s=1}^S). \tag{6.16}$$

At each local sensor processing site, the positions $\mathbf{r}_s$, $s = 1, \ldots, S$ of all contributing sensors with reference to the underlying global Cartesian coordinate system are known. For the sake of notational simplicity, let all sensors be equally aligned and synchronized with the same data update rate. These assumptions, however, are not essential and can be relaxed. We furthermore assume that the measurement error covariance matrices of all individual sensors are known for the local processors (e.g. the sensor-specific standard deviations in x and y coordinates, σ^s_x, σ^s_y). The detection probability is assumed to be one.

In calculating the local 'tracks' $\mathrm{x}^s_{l|k}$, the filtering update with the local sensor data is identical with the standard Kalman filtering update step, while the prediction and the retrodiction steps must be modified in order to be able to reconstruct $p(\mathrm{x}_l|\{Z^k_s\}_{s=1}^S)$ after fusing them according to Eq. 6.23. For a distributed application of the standard Kalman prediction and retrodiction steps, i.e. if the object evolution model has to be taken into account, we essentially have to perform a "globalization" of the covariances. By this term we refer to a local adaptation of the covariance matrices to the global knowledge of the distributed system. As a result of this globalization, all estimate covariances of the local sensors will be mutually the same. In consequence, the local 'tracks' $\mathrm{x}^s_{l|k}$ are not identical with the results of local Kalman filtering. As becomes clear below, however, the locally optimal tracks can easily be obtained as by-products when calculating the state estimates $\mathrm{x}^s_{l|k}$.

The proposed distributed Kalman-type processing scheme essentially makes use of the fact that the sensor measurements do *not* enter into the update equation for the estimation error covariance matrices. This means in particular that the covariance matrices of *all* sensors can be calculated at each individual sensor site without any further need of communication (given the relevant parameters of all sensors are known at each sensor site).

A consequence of this fact is, however, that the proposed optimal processing scheme cannot be generalized directly to IMM-or PDA-type tracking, where the covariance matrices are explicitly data-dependent. Moreover, in the proposed distributed processing scheme for multiple sensor Kalman filtering and retrodiction, the local 'tracks' $\mathrm{x}^s_{l|k}$ to be communicated are identical with the state estimates of locally optimal Kalman filtering and retrodiction based on the measurements of sensor s in certain limiting cases only.

6.2.1 *Gaußian Product Densities*

Let us consider probability density functions $p(\mathrm{x}_l|Z^k)$ that are proportional to a *product* of S Gaußians:

$$p(\mathrm{x}_l|Z^k) = c_{l|k} \prod_{s=1}^{S} \mathcal{N}\big(\mathrm{x}_l;\ \mathrm{x}_{l|k}^s,\ \mathbf{P}_{l|k}^s\big). \tag{6.17}$$

The normalizing constant $c_{l|k}$ is a direct result of a repeated use of the product formula for Gaußians (Eq. A.28):

$$c_{l|k}^{-1} = \prod_{s=1}^{S} \int \mathrm{d}\mathrm{x}_l\ \mathcal{N}\big(\mathrm{x}_l;\ \mathrm{x}_{l|k}^s,\ \mathbf{P}_{l|k}^s\big) \tag{6.18}$$

$$= \prod_{s=1}^{S-1} \mathcal{N}\big(\mathrm{x}_{l|k}^{*s};\ \mathrm{x}_{l|k}^{s+1},\ \mathbf{P}_{l|k}^{*s} + \mathbf{P}_{l|k}^{s+1}\big), \tag{6.19}$$

where the parameters $\mathrm{x}_{l|k}^{*s}$ and $\mathbf{P}_{l|k}^{*s}$ are given by

$$\mathbf{P}_{l|k}^{*s\ -1} = \sum_{i=1}^{s-1} \mathbf{P}_{l|k}^{i\ -1}$$

$$\mathrm{x}_{l|k}^{*s} = \mathbf{P}_{l|k}^{*s\ -1} \sum_{i=1}^{s-1} \mathbf{P}_{l|k}^{i\ -1} \mathrm{x}_{l|k}^{i}.$$

In the sequel, as well as in most applications, it is unnecessary to calculate the normalization constant $c_{l|k}$ explicitly.

By virtue of the same product formula, probability densities defined by Gaußian products as in Eq. 6.17 can be transformed into a single Gaußian:

$$p(\mathrm{x}_l|Z^k) = c_{l|k} \prod_{s=1}^{S} \mathcal{N}\big(\mathrm{x}_l;\ \mathrm{x}_{l|k}^s,\ \mathbf{P}_{l|k}^s\big) \tag{6.20}$$

$$= \mathcal{N}\big(\mathrm{x}_l;\ \mathrm{x}_{l|k},\ \mathbf{P}_{l|k}\big), \tag{6.21}$$

with an expectation vector $\mathrm{x}_{l|k}$ and a covariance matrix $\mathbf{P}_{l|k}$ obtained by 'fusing' $\mathrm{x}_{l|k}^s$ and $\mathbf{P}_{l|k}^s$, $s = 1, \ldots, S$ according to the formulae:

$$\mathbf{P}_{l|k}^{-1} = \sum_{s=1}^{S} \mathbf{P}_{l|k}^{s\ -1} \tag{6.22}$$

$$\mathrm{x}_{l|k} = \mathbf{P}_{l|k}\Big(\sum_{s=1}^{S} \mathbf{P}^{s\ -1}_{l|k}\mathrm{x}^{s}_{l|k}\Big). \tag{6.23}$$

'Convex combinations' of this type are fundamental in almost all data fusion applications (see e.g. [3, Chap. 12]).

This simple observation leads to the question whether it is possible to calculate at least the expectation vectors $\mathrm{x}^s_{l|k}$ by using the measurements of sensor s only, i.e. in a way that can be done locally at each node of a distributed sensor network. As previously stated, under conditions where Kalman filtering is appropriate for tracking, the covariance matrices $\mathbf{P}^s_{l|k}$ can be calculated locally for all sensors without exchanging sensor data, provided the measurement error covariance matrices of each individual sensor are known, or if they can be reconstructed at each node of the sensor network.

If the locally produced 'tracks' $\mathrm{x}^s_{k|k}$ are sent at some arbitrary instant of time t_k to a fusion node, they can be fused according to Eq. 6.2, leading to the density $p(\mathrm{x}_k|Z^k)$ from which optimal state estimates based on all sensor measurements of the sensors $s = 1, \ldots, S$ can be derived. According to this processing scheme, the fusion step does not have to be performed at each sensor scan time in order to obtain an optimal result. The fusion result $\mathrm{x}_{k|k}, \mathbf{P}_{k|k}$ may or may not be communicated to the individual nodes of the sensor network. The proposed distributed processing scheme is thus well-suited for applications where reduced or arbitrary rate communication is to be taken into account.

6.2.2 Distributed Filtering Update

Let the distributed multiple sensor fusion procedure be initiated by

$$p(\mathrm{x}_1|Z^0) = \mathcal{N}\big(\mathrm{x}_1;\ \mathrm{x}_{1|0},\ \mathbf{P}_{1|0}\big) \tag{6.24}$$

$$\propto \prod_{s=1}^{S} \mathcal{N}\big(\mathrm{x}_1;\ \mathrm{x}^s_{1|0},\ \mathbf{P}^s_{1|0}\big) \tag{6.25}$$

with $\mathrm{x}^s_{1|0} = \mathrm{x}_{1|0}, \mathbf{P}^s_{1|0} = S\mathbf{P}_{1|0}$. The iterative calculation of the probability densities involved is thus started by distributing an initial track $\mathrm{x}_{0|0}$ obtained from initial measurements in the sensor network along with the covariance matrix $\mathbf{P}_{1|0}$.

The expectation vectors $\mathrm{x}^s_{k|k}$, $s = 1, \ldots, S$ are obtained by a simple Kalman update provided the prediction density $p(\mathrm{x}_k|Z^{k-1})$ also obeys the structure previously discussed,

$$p(\mathrm{x}_k|Z^{k-1}) \propto \prod_{s=1}^{S} \mathcal{N}\big(\mathrm{x}_k;\ \mathrm{x}^s_{k|k-1},\ \mathbf{P}^s_{k|k-1}\big). \tag{6.26}$$

That this representation can actually be achieved is shown below. The joint likelihood function $p(Z_k|\mathrm{x}_k) = \ell_k(Z_k|\mathrm{x}_k)$ given by:

$$\ell_k(\mathrm{x}_k; \mathrm{Z}_k) = \prod_{s-1}^{S} \ell_k^s(\mathrm{x}_k; \mathrm{z}_k^s) \tag{6.27}$$

is essential for the filtering update, with $\ell_k^s(\mathrm{x}_k; \mathrm{z}_k^s) = \mathcal{N}\big(\mathrm{z}_k^s; \mathbf{H}_k^s \mathrm{x}_k, \mathbf{R}_k^s\big)$. Bayes Theorem and a repeated use of the product formula (Eq. A.28) yield:

$$p(\mathrm{x}_k|Z^k) = \frac{\ell_k(Z_k|\mathrm{x}_k)\, p(\mathrm{x}_k|Z^{k-1})}{\int \mathrm{dx}_k\, \ell_k(Z_k|\mathrm{x}_k)\, p(\mathrm{x}_k|Z^{k-1})} \tag{6.28}$$

$$\propto \frac{\prod_{s=1}^{S} \ell_k^s(\mathrm{x}_k; \mathrm{z}_k^s)\, \mathcal{N}\big(\mathrm{x}_k; \mathrm{x}_{k|k-1}^s, \mathbf{P}_{k|k-1}^s\big)}{\int \mathrm{dx}_k \prod_{s=1}^{S} \ell_k^s(\mathrm{x}_k; \mathrm{z}_k^s)\, \mathcal{N}\big(\mathrm{x}_k; \mathrm{x}_{k|k-1}^s, \mathbf{P}_{k|k-1}^s\big)} \tag{6.29}$$

$$\propto \prod_{s=1}^{S} \mathcal{N}\big(\mathrm{x}_k; \mathrm{x}_{k|k}^s, \mathbf{P}_{k|k}^s\big) \tag{6.30}$$

with the parameters $\mathrm{x}_{k|k}^s$ and $\mathbf{P}_{k|k}^s$, $s = 1, \ldots, S$ given by the standard Kalman update equations:

$$\mathrm{x}_{k|k}^s = \mathrm{x}_{k|k-1}^s + \mathbf{W}_{k|k-1}^s(\mathbf{z}_k^s - \mathbf{H}_k^s \mathrm{x}_{k|k-1}^s) \tag{6.31}$$

$$\mathbf{P}_{k|k}^s = \mathbf{P}_{k|k-1}^s - \mathbf{W}_{k|k-1}^{s\top} \mathbf{S}_{k|k-1}^{s\;-1} \mathbf{W}_{k|k-1}^s \tag{6.32}$$

with the corresponding Kalman gain and innovation covariance matrices

$$\mathbf{S}_{k|k-1}^s = \mathbf{H}_k^s \mathbf{P}_{k|k-1}^s \mathbf{H}_k^{s\top} + \mathbf{R}_k^s \tag{6.33}$$

$$\mathbf{W}_{k|k-1}^s = \mathbf{P}_{k|k-1}^s \mathbf{H}_k^{s\top} \mathbf{S}_{k|k-1}^{s\;-1}. \tag{6.34}$$

6.2.3 Distributed Prediction Update

As all sensors observe the same target a standard Kalman filter prediction applied to the local tracks would implicate cross-correlations and therefore the product representation from above would not hold anymore. As a consequence the prediction step has to be modified in order to obtain the product representation. To this end, we first note that the filtering probability density function from time t_{k-1} can be rewritten in the following way.

$$p(\mathrm{x}_{k-1}|Z^{k-1}) \propto \prod_{s=1}^{S} \mathcal{N}\big(\mathrm{x}_{k-1}; \mathrm{x}_{k-1|k-1}^s, \mathbf{P}_{k-1|k-1}^s\big) \tag{6.35}$$

$$= \mathcal{N}\big(\mathrm{x}_{k-1};\ \frac{1}{S}\sum_{s=1}^{S}\overline{\mathrm{x}}^s_{k-1|k-1},\ \overline{\mathbf{P}}_{k-1|k-1}\big), \tag{6.36}$$

$$= \prod_{s=1}^{S}\mathcal{N}\big(\mathrm{x}_{k-1};\ \overline{\mathrm{x}}^s_{k-1|k-1},\ S\overline{\mathbf{P}}_{k-1|k-1}\big), \tag{6.37}$$

where the parameters $\overline{\mathrm{x}}^s_{k-1|k-1}$ and the global filtering covariance $\overline{\mathbf{P}}_{k-1|k-1}$ are given by:

$$\overline{\mathrm{x}}^s_{k-1|k-1} = S\overline{\mathbf{P}}_{k-1|k-1}\mathbf{P}^{s\ -1}_{k-1|k-1}\mathrm{x}^s_{k-1|k-1} \tag{6.38}$$

$$\overline{\mathbf{P}}_{k-1|k-1} = \left(\sum_{s=1}^{S}\mathbf{P}^{s\ -1}_{k-1|k-1}\right)^{-1}. \tag{6.39}$$

As the Markovian propagation density $p(\mathrm{x}_k|\mathrm{x}_{k-1})$ can be written as:

$$p(\mathrm{x}_k|\mathrm{x}_{k-1}) = \mathcal{N}\big(\mathrm{x}_k;\ \mathbf{F}_{k|k-1}\mathrm{x}_{k-1},\ \mathbf{D}_{k|k-1}\big) \tag{6.40}$$

$$\propto \mathcal{N}\big(\mathrm{x}_k;\ \mathbf{F}_{k|k-1}\mathrm{x}_{k-1},\ S\mathbf{D}_{k|k-1}\big)^S, \tag{6.41}$$

we obtain for the prediction density $p(\mathrm{x}_k|Z^{k-1})$:

$$p(\mathrm{x}_k|Z^{k-1}) = \int d\mathrm{x}_{k-1}\ p(\mathrm{x}_k|\mathrm{x}_{k-1})\ p(\mathrm{x}_{k-1}|Z^{k-1}) \tag{6.42}$$

$$\propto \int d\mathrm{x}_{k-1}\prod_{s=1}^{S}\mathcal{N}\big(\mathrm{x}_k;\ \mathbf{F}_{k|k-1}\mathrm{x}_{k-1},\ S\mathbf{D}_{k|k-1}\big)\ \mathcal{N}\big(\mathrm{x}_{k-1};\ \overline{\mathrm{x}}^s_{k-1|k-1},\ S\overline{\mathbf{P}}_{k-1|k-1}\big). \tag{6.43}$$

A repeated use of the product formula for Gaussians now yields

$$p(\mathrm{x}_k|Z^{k-1}) \propto \int d\mathrm{x}_{k-1}\prod_{s=1}^{S}\mathcal{N}\big(\mathrm{x}_k;\ \tilde{\mathrm{x}}^s_{k|k-1},\ \tilde{\mathbf{P}}_{k|k-1}\big)\ \mathcal{N}\big(\mathrm{x}_{k-1};\ \tilde{\mathbf{W}}_{k-1}\mathrm{x}_k + \tilde{\mathrm{z}}^s_{k-1},\ \tilde{\mathbf{R}}_{k-1}\big), \tag{6.44}$$

where the following abbreviations were used:

$$\tilde{\mathrm{x}}^s_{k|k-1} = \mathbf{F}_{k|k-1}\overline{\mathrm{x}}^s_{k-1|k-1} \tag{6.45}$$

$$\tilde{\mathbf{P}}_{k|k-1} = S\left(\mathbf{F}_{k|k-1}\overline{\mathbf{P}}_{k-1|k-1}\mathbf{F}^T_{k|k-1} + \mathbf{D}_{k|k-1}\right) \tag{6.46}$$

$$\tilde{\mathrm{z}}^s_{k-1} = \overline{\mathrm{x}}^s_{k-1|k-1} - \tilde{\mathbf{W}}_{k-1}\tilde{\mathrm{x}}^s_{k|k-1} \tag{6.47}$$

$$\tilde{\mathbf{W}}_{k-1} = S\overline{\mathbf{P}}_{k-1|k-1}\mathbf{F}^T_{k|k-1}\tilde{\mathbf{P}}^{-1}_{k|k-1}. \tag{6.48}$$

$$\tilde{\mathbf{P}}_{k|k-1} = S\overline{\mathbf{P}}_{k-1|k-1} - \tilde{\mathbf{W}}_{k-1}\tilde{\mathbf{P}}_{k|k-1}\tilde{\mathbf{W}}^T_{k-1} \tag{6.49}$$

One might say that $\overline{\mathbf{P}}_{k-1|k-1}$ is the global filtering covariance, while $\tilde{\mathbf{P}}_{k|k-1}$ may be seen as the *globalized local prediction covariance*. Obviously, (6.44) can be written as

$$p(\mathrm{x}_k|\mathcal{Z}^{k-1}) = \ell_{k|k-1} \prod_{s=1}^{S} \mathcal{N}\big(\mathrm{x}_k;\ \tilde{\mathrm{x}}^s_{k|k-1},\ \tilde{\mathbf{P}}_{k|k-1}\big) \tag{6.50}$$

by introducing the function $\ell_{k|k-1}$ defined by

$$\ell_{k|k-1} = \int d\mathrm{x}_{k-1} \prod_{s=1}^{S} \mathcal{N}\big(\mathrm{x}_{k-1} - \tilde{\mathbf{W}}_{k-1}\mathrm{x}_k;\ \tilde{\mathrm{z}}^s_{k-1},\ \tilde{\mathbf{R}}_{k-1}\big). \tag{6.51}$$

Again, applying the product formula repeatedly along the product within $\ell_{k|k-1}$ yields

$$\begin{aligned}\ell_{k|k-1} = &\prod_{s=1}^{S-1} \mathcal{N}\Big(\tilde{\mathrm{z}}^{s+1}_{k-1};\ \frac{1}{s}\sum_{i=1}^{s}\tilde{\mathrm{z}}^i_{k-1},\ \Big(1+\frac{1}{s}\Big)\tilde{\mathbf{R}}_{k-1}\Big)\\ &\cdot \underbrace{\int d\mathrm{x}_{k-1}\,\mathcal{N}\Big(\mathrm{x}_{k-1} - \tilde{\mathbf{W}}_{k-1}\mathrm{x}_k;\ \frac{1}{S}\sum_{i=1}^{S}\tilde{\mathrm{z}}^i_{k-1},\ \frac{1}{S}\tilde{\mathbf{R}}_{k-1}\Big)}_{=1}.\end{aligned} \tag{6.52}$$

As one can see, the integration term becomes trivial, as it is over a single Gaussian. As a result, the function $\ell_{k|k-1}$ is independent of the current state variable $\mathbf{x}_k$ and we obtain the desired product representation:

$$p(\mathrm{x}_k|\mathcal{Z}^{k-1}) \propto \prod_{s=1}^{S} \mathcal{N}\big(\mathrm{x}_k;\ \tilde{\mathrm{x}}^s_{k|k-1},\ \tilde{\mathbf{P}}_{k|k-1}\big). \tag{6.53}$$

Figure 6.1 illustrates under which conditions several known track-to-track fusion approaches (Centralized Kalman Filtering and Retrodiction, Local Kalman Track Fusion, Local Fraenkel Tracklet Fusion, and Local Kalman Track Fusion) appear as limiting cases of the proposed exact distributed Kalman filtering scheme.

Illustration of the Processing Flow

The first processing steps for each sensor (initiation, filtering, local prediction) are nearly identical to standard local Kalman filtering. We recall that in the local prediction step a 'relaxed' dynamics covariance matrix is to be used ($S\mathbf{D}_{k|k-1}$). It is essential to the suggested approach that the individual sensor models and the common target dynamics model be known at each network node (global properties). After the local filtering, the local covariances are globalized. In this step, access to the global properties is required. The final prediction densities result from an application of the relaxed evolution model on the globalized densities. The processing loop is thus closed and restarts with a local filtering step.

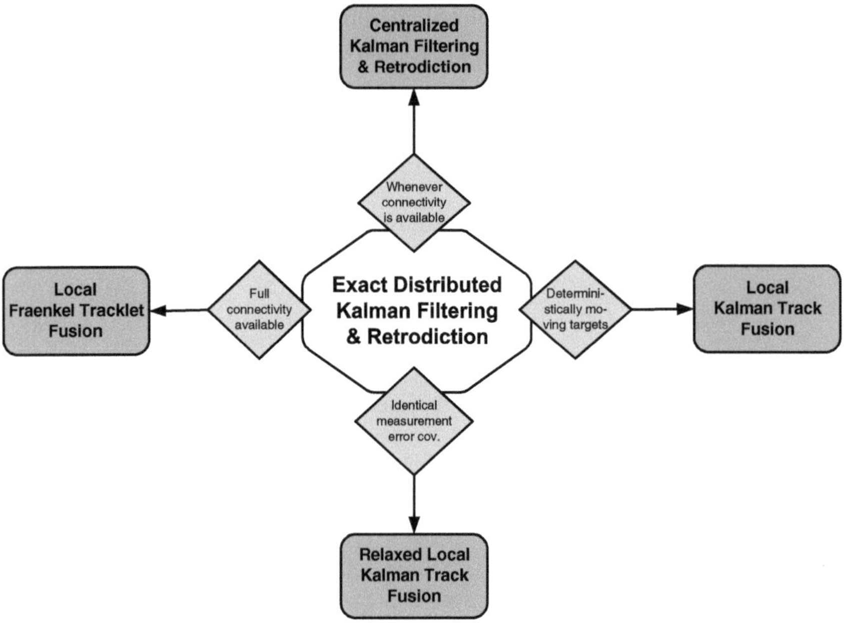

Fig. 6.1 Interrelation of exact distributed Kalman filtering to existing processing schemes

Whenever it is required, a fusion step can be performed for obtaining fused prediction and filtering results, which reproduces the results of full communication-rate measurement fusion at this time, i.e. as if all measurements or local tracks were communicated to the fusion center.

6.2.4 Distributed Retrodiction Update

Essentially the same considerations are valid for distributed retrodiction, if for the retrodicted probability density $p(\mathrm{x}_l|\mathcal{Z}^k)$ a product representation exists,

$$p(\mathrm{x}_l|\mathcal{Z}^k) \propto \prod_{s=1}^{S} \mathcal{N}\big(\mathrm{x}_l;\ \mathrm{x}^s_{l|k},\ \mathbf{P}^s_{l|k}\big), \quad l < k, \tag{6.54}$$

where the retrodicted local state estimates $\mathrm{x}^s_{l|k}$ can be calculated by processing the measurements of sensor s only. To show that this is possible, we assume such a representation for step $l+1$, where $l+1 \leq k$, and derive it for step l. At first, standard probability reasoning yields:

$$p(\mathrm{x}_l|\mathcal{Z}^k) = \int d\mathrm{x}_{l+1}\ p(\mathrm{x}_l, \mathrm{x}_{l+1}|\mathcal{Z}^k) = \int d\mathrm{x}_{l+1}\ p(\mathrm{x}_l|\mathrm{x}_{l+1}, \mathcal{Z}^k)\ p(\mathrm{x}_{l+1}|\mathcal{Z}^k). \tag{6.55}$$

For the integration kernel $p(\mathrm{x}_l|\mathrm{x}_{l+1}, \mathcal{Z}^k)$, we obtain after applying Bayes' rule:

$$p(\mathrm{x}_l|\mathrm{x}_{l+1}, \mathcal{Z}^k) = p(\mathrm{x}_l|\mathrm{x}_{l+1}, \mathcal{Z}^l) = \frac{p(\mathrm{x}_{l+1}|\mathrm{x}_l)\ p(\mathrm{x}_l|\mathcal{Z}^l)}{\int d\mathrm{x}_l\ p(\mathrm{x}_{l+1}|x_l)\ p(\mathrm{x}_l|\mathcal{Z}^l)}. \tag{6.56}$$

As in the prediction step, we rewrite the density $p(\mathbf{x}_l|\mathcal{Z}^l)$. Because of the derivations above, a product representation may be assumed for it. Therefore, by the product formula it holds

$$p(\mathrm{x}_l|\mathcal{Z}^l) \propto \prod_{s=1}^{S} \mathcal{N}\big(\mathrm{x}_l;\ \mathrm{x}_{l|l}^s,\ \mathbf{P}_{l|l}^s\big) \tag{6.57}$$

$$\propto \mathcal{N}\Big(\mathrm{x}_l;\ \frac{1}{S}\sum_{s=1}^{S} \overline{\mathrm{x}}_{l|l}^s,\ \overline{\mathbf{P}}_{l|l}\Big) \tag{6.58}$$

$$\propto \prod_{s=1}^{S} \mathcal{N}\big(\mathrm{x}_l;\ \overline{\mathrm{x}}_{l|l}^s,\ S\overline{\mathbf{P}}_{l|l}\big), \tag{6.59}$$

where we used the following abbreviations:

$$\overline{\mathrm{x}}_{l|l}^s = S\overline{\mathbf{P}}_{l|l}\mathbf{P}_{l|l}^{s\ -1}\mathrm{x}_{l|l}^s \tag{6.60}$$

$$\overline{\mathbf{P}}_{l|l} = \left(\sum_{s=1}^{S} \mathbf{P}_{l|l}^{s\ -1}\right)^{-1}. \tag{6.61}$$

In the same way, we modify the product representation for time t_{l+1} and obtain

$$p(\mathrm{x}_{l+1}|\mathcal{Z}^k) \propto \prod_{s=1}^{S} \mathcal{N}\big(\mathrm{x}_{l+1};\ \overline{\mathrm{x}}_{l+1|k}^s,\ S\overline{\mathbf{P}}_{l+1|k}\big). \tag{6.62}$$

Now, we insert (6.59) into (6.56). This enables us to apply the product formula again:

$$p(\mathrm{x}_l|\mathrm{x}_{l+1}, \mathcal{Z}^l) \propto \prod_{s=1}^{S} \mathcal{N}\big(\mathrm{x}_l;\ \overline{\mathrm{x}}_{l|l}^s,\ S\overline{\mathbf{P}}_{l|l}\big)\ \mathcal{N}\big(\mathrm{x}_{l+1};\ \mathbf{F}_{l+1|l}\mathrm{x}_l,\ S\mathbf{D}_{l+1|l}\big) \tag{6.63}$$

$$\propto \mathcal{N}\big(\mathrm{x}_l;\ \mathbf{h}_{l|l}^s(\mathrm{x}_{l+1}),\ \mathbf{R}_{l|l}\big), \tag{6.64}$$

by using the parameters

$$\mathbf{h}^s_{l|l}(\mathrm{x}_{l+1}) = \overline{\mathrm{x}}^s_{l|l} + \overline{\mathbf{W}}_{l|l+1}\left(\mathrm{x}_{l+1} - \tilde{\mathrm{x}}^s_{l+1|l}\right) \tag{6.65}$$

$$\mathbf{R}_{l|l} = S\left(\overline{\mathbf{P}}_{l|l} - \overline{\mathbf{W}}_{l|l+1}\overline{\mathbf{P}}_{l+1|l}\overline{\mathbf{W}}^\top_{l|l+1}\right) \tag{6.66}$$

$$\overline{\mathbf{W}}_{l|l+1} = S\overline{\mathbf{P}}_{l|l}\mathbf{F}^\top_{l+1|l}\overline{\mathbf{P}}_{l+1|l} \tag{6.67}$$

Inserting $p(\mathrm{x}_l|\mathrm{x}_{l+1}, \mathcal{Z}^k)$ into (6.55) and applying the product formula again repeatedly yields

$$p(\mathrm{x}_l|\mathcal{Z}^k) \propto \ell_{l|k} \prod_{s=1}^{S} \mathcal{N}\big(\mathrm{x}_l;\ \tilde{\mathrm{x}}^s_{l|k},\ \tilde{\mathbf{P}}_{l|k}\big), \tag{6.68}$$

where the parameters $\tilde{\mathrm{x}}^s_{l|k}, \tilde{\mathbf{P}}_{l|k}$ are given by the modified Rauch-Tung-Striebel update equations for Kalman retrodiction

$$\begin{aligned} \tilde{\mathrm{x}}^s_{l|k} &= \overline{\mathrm{x}}^s_{l|l} + \overline{\mathbf{W}}_{l|l+1}(\mathrm{x}^s_{l+1|k} - \tilde{\mathrm{x}}^s_{l+1|l}) \\ \tilde{\mathbf{P}}_{l|k} &= S\left(\overline{\mathbf{P}}_{l|l} + \overline{\mathbf{W}}_{l|l+1}(\overline{\mathbf{P}}_{l+1|k} - \overline{\mathbf{P}}_{l+1|l})\overline{\mathbf{W}}^{s\top}_{l|l+1}\right) \end{aligned} \tag{6.69}$$

and a normalizing function $\ell_{l|k}$, which is defined by

$$\ell_{l|k}(\mathrm{x}_l) = \int \mathrm{d}\mathrm{x}_{l+1} \prod_{s=1}^{S} \mathcal{N}\big(\mathrm{x}_{l+1} - \mathbf{V}_{l|l+1}x_l;\ \mathbf{y}^s_{l|l+1},\ \mathbf{Y}_{l|l+1}\big) \tag{6.70}$$

with the abbreviations

$$\mathbf{y}^s_{l|l+1} = \mathbf{x}^s_{l+1|k} - \mathbf{V}_{l|l+1}\tilde{\mathbf{x}}^s_{l|k}, \tag{6.71}$$

$$\mathbf{Y}_{l|l+1} = S\left(\overline{\mathbf{P}}_{l+1|k} - \mathbf{V}_{l|l+1}\tilde{\mathbf{P}}_{l|k}\mathbf{V}^\top_{l|l+1}\right), \tag{6.72}$$

$$\mathbf{V}_{l|l+1} = S\overline{\mathbf{P}}_{l+1|k}\overline{\mathbf{W}}^\top_{l|l+1}\tilde{\mathbf{P}}^{-1}_{l|k}. \tag{6.73}$$

If we use the product formula along the terms in $\ell_{l|k}$, we obtain

$$\begin{aligned} \ell_{l|k} = &\prod_{s=1}^{S-1} \mathcal{N}\big(y^{s+1}_{l|l+1};\ \frac{1}{s}\sum_{i=1}^{s}\mathbf{y}^i_{l|l+1},\ (1+\frac{1}{s})\mathbf{Y}_{l|l+1}\big) \\ &\cdot \int \mathrm{d}\mathrm{x}_{l+1}\,\mathcal{N}\big(\mathrm{x}_{l+1} - \mathbf{V}_{l|l+1}\mathrm{x}_l;\ \frac{1}{S}\sum_{i=1}^{S}\mathbf{y}^i_{l|l+1},\ \frac{1}{S}\mathbf{Y}_{l|l+1}\big). \end{aligned} \tag{6.74}$$

As in the prediction step the integration term becomes trivial, as it is over a single Gaussian distribution. Therefore, the normalizing function $\ell_{l|k}$ does not depend on $\mathbf{x}_l$, and we have

$$p(\mathrm{x}_l|\mathcal{Z}^k) \propto \prod_{s=1}^{S} \mathcal{N}\big(\mathrm{x}_l;\ \tilde{\mathrm{x}}^s_{l|k},\ \tilde{\mathbf{P}}_{l|k}\big). \tag{6.75}$$

6.3 Discussion of a Simulated Example

Figure 6.2a shows a simulated trajectory of a maneuvering air target defined by $x(t) = \frac{v^2}{q}\sin(\frac{q}{2v}t)$, $y(t) = 2x(t)$, $v = 300\frac{\mathrm{m}}{\mathrm{s}}$, $q = 4g$. It is observed by two typical mid-range radars located at $(-50, 0)$km and $(0, 50)$km (scan period: T=5 s, measurement error standard deviations: $\sigma_r = 20$ m (range), $\sigma_\phi = 0.2°$ (azimuth). This is the same simulated scenario as considered in Sect. 5.1.

On the coarse-grained time scale, defined by $t_k = 2kT, k = 1, 2, \ldots$, the figure also shows the position estimates and the corresponding estimation error ellipses of the local sensor tracks as well as the results of measurement fusion, standard track-to-track fusion, and the results of the proposed new method. To better understand the latter ones, Fig. 6.2b shows a magnification of a characteristic detail of Fig. 6.2a. As expected, the exact distributed Kalman filtering is identical with measurement fusion and somewhat better than local Kalman track fusion. Nevertheless, we can be sure to be able to use a correct distributed data fusion scheme at arbitrary communication times, which is equivalent to measurement fusion.

Figures 6.3 and 6.4 show numerical results based on 1,000 Monte-Carlo-runs. While in Fig. 6.3 full-rate communication is assumed, i.e. the local tracks are fused at each scan, the local tracks of each 4th scan are fused in the situation displayed in Fig. 6.4 (reduced-rate communication). The lines denote the Minimum Mean Squared Error (MMSE), the shadowed areas indicate the corresponding mean tracking error (square-root of the error covariance matrix). The figures compare centralized Kalman filtering (i.e. measurement fusion) with exact distributed Kalman filtering, relaxed local Kalman track fusion, local Fraenkel tracklet fusion, and local Kalman track fusion. As expected, measurement fusion and exact distributed Kalman filtering are numerically identical in both figures.

In the case of full-rate communications, also local Fraenkel tracklet fusion is identical with measurement fusion, while for reduced-rate communications differences occur in the MMSE as well as in the covariance. As reported in the literature, the local Kalman track fusion is over-optimistic in both cases. This effect, however, is not very pronounced. An astonishing result is that relaxed local track fusion is close the exact distributed Kalman track fusion.

These observations are typical in the sense that we obtain very similar tendencies when the target dynamics parameters, the sensor parameters, and the sensor-to-target geometries are varied. See [4] for a more detailed discussion.

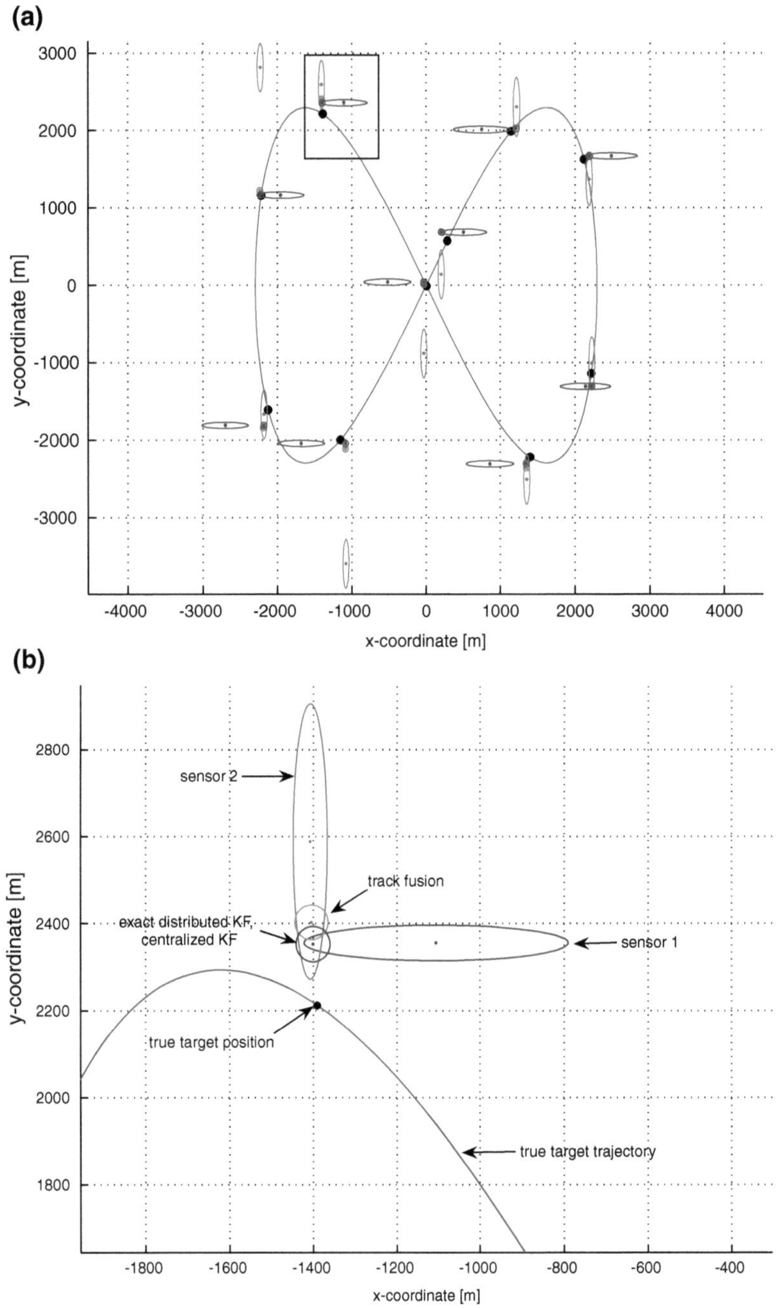

Fig. 6.2 Track-to-track fusion: simulated trajectory of a highly maneuvering object. **a** complete object trajectory. **b** magnification of a detail

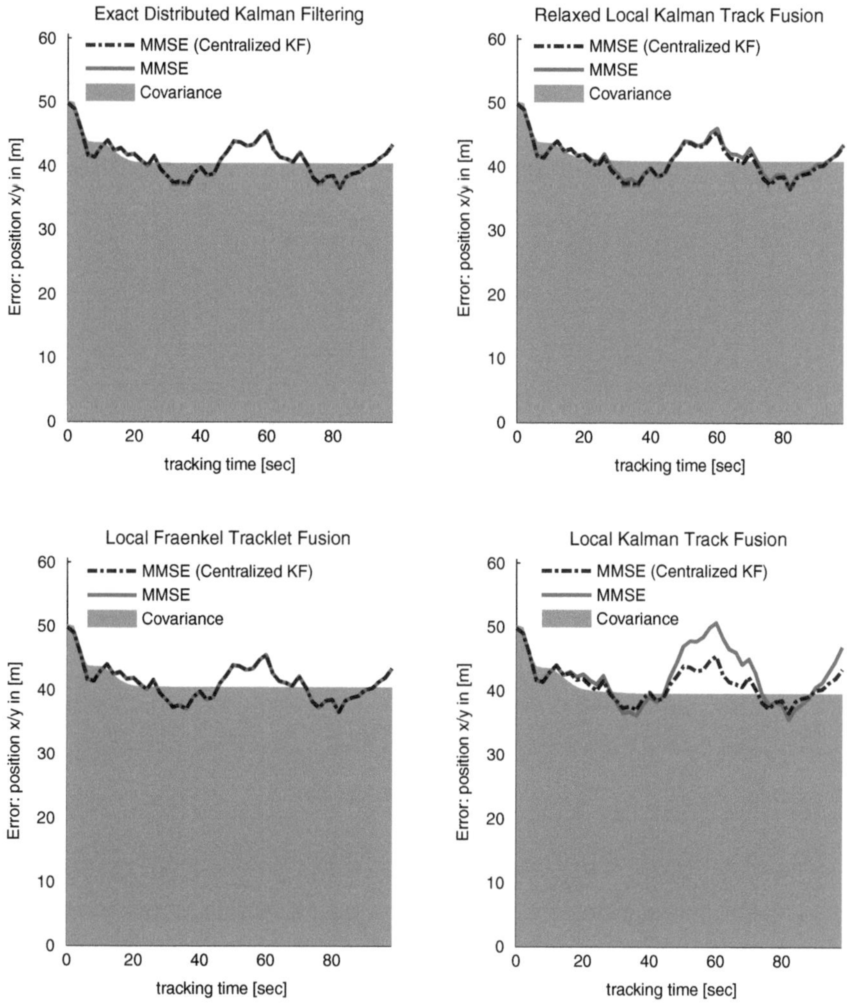

Fig. 6.3 Numerical results (1000 runs): full-rate communication

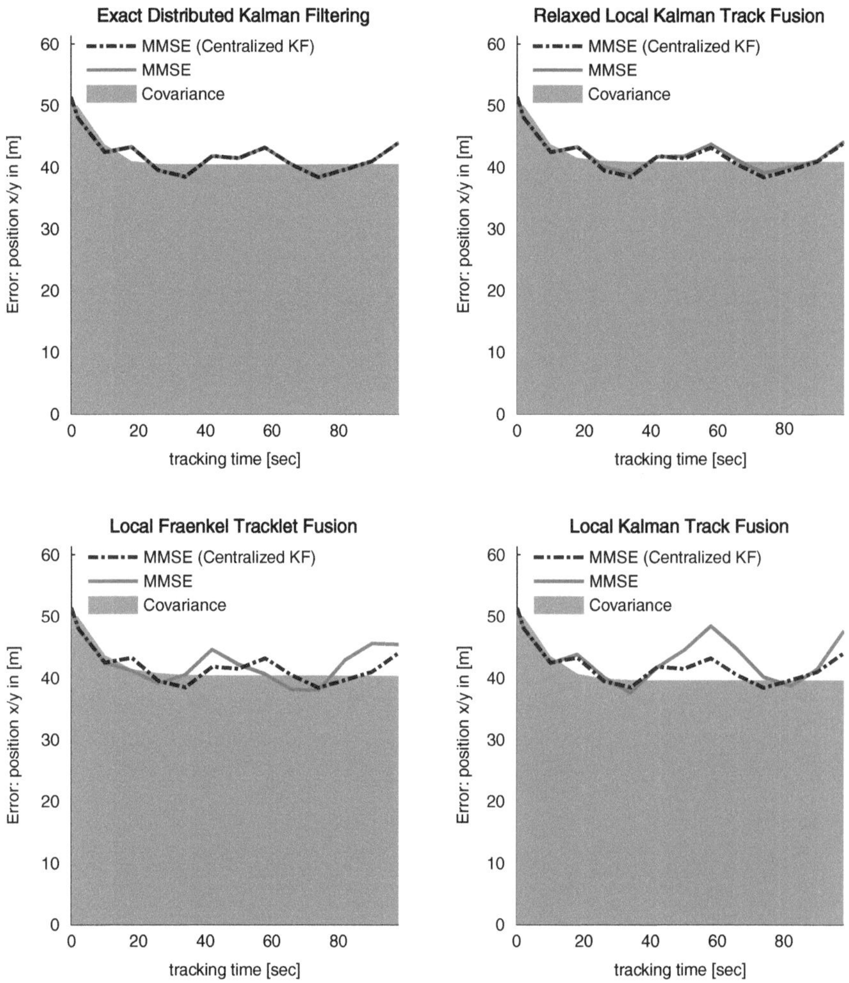

Fig. 6.4 Distributed filtering (reduced-rate communication): numerical results (1000 runs)

References

1. C.-Y. Chong, "Hierarchical Estimation", in 2nd MIT/ONR *Workshop on Distributed Information and Decision Systems* (1979).
2. O. E. Drummond, "Tracklets and a hybrid fusion with process noise". In *SPIE Signal and Data Processing of Small, Targets*, Vol. 3163 (1997)
3. M. E. Liggins, D. L. Hall, J. Llinas (Eds.). *Handbook of multisensor data fusion–theory and practice*. CRC Press, 2nd Edition (2008)
4. F. Govaers, W. Koch, "An exact solution to track-to-track-fusion at arbitrary communication rates". In *IEEE Transactions on Aerospace and Electronic Systems*, Vol. 48, Nr. 3, pp. 2718 (2012)

Part II
Sensor Data Fusion: Selected Applications

Chapter 7
Integration of Advanced Sensor Properties

Advanced signal processing techniques exploit even sophisticated physical phenomena of objects of interest and are fundamental to modern sensor system design. In particular, they have a direct impact on the quantitative and qualitative properties of the sensor data produced and to be fused. This makes a more subtle modeling of the statistical characteristics of the sensor output inevitable. Via constructing appropriate sensor models based on a deeper insight into the physical and technical sensor design principles, the performance of tracking and sensor data fusion systems can be significantly improved.

Chapter 7 is focused on selected physical and technical properties of sensor systems that are used in real-world ISR applications (Intelligence, Surveillance, and Reconnaissance), such as those discussed in [1, Chap. 20]. The analysis of characteristic examples shows that context information on particular performance features of the sensor systems involved is useful, in some cases even inevitable, to fulfill an overall ISR task. The Bayesian methodology discussed in Part I is wide and flexible enough to integrate more sophisticated, appropriately designed, but still mathematically tractable likelihood functions into the process of Bayesian Knowledge Propagation. The discussed examples cover finite sensor resolution, Doppler blindness, and main-lobe jamming.

The possibility to exploit even *negative sensor evidence* is a consequence that is directly connected with the use of more advanced sensor models. This notion covers the conclusions to be drawn from expected, but actually missing sensor measurements for improving the state estimates of objects under track. Even a failed attempt to detect an object of interest is a useful sensor output that is interpretable only if a consistent sensor modeling is available.

7.1 Finite Sensor Resolution

Air surveillance in a dense object/dense clutter environment is a difficult task that requires refined data association and tracking techniques. In this context, tracking

W. Koch, *Tracking and Sensor Data Fusion*,
Mathematical Engineering, DOI: 10.1007/978-3-642-39271-9_7,

for maneuvering groups of objects that join, operate closely spaced for a while, and split off again is confronted with mainly three problems:

1. *Sensor Resolution:* Due to the limited resolution of every radar sensor, closely-spaced targets will continuously transition from being resolved to irrresolved and back again. The importance of resolution phenomena has been addressed in Ref. [2].
2. *Data Association:* Ambiguous data-to-object associations due to overlapping expectation gates are an inherent problem for formations, which is made even more difficult by high false return densities and missed detections.
3. *Maneuvers:* Often distinct maneuvering phases can be identified, as even agile objects will not always make use of their maneuvering capability. Nevertheless, abrupt turns may occur, e.g. if a formation dissolves into well-separated objects.

These problems require the use of multiple hypothesis, multiple model tracking methods as discussed in Part I. The multiple hypothesis character mirrors the uncertain origin of the data, while the multiple models refer to the different maneuvering phases (see discussion in Sect. 8.1). In Sect. 2.3.3, we briefly sketched a model providing a qualitative description of resolution phenomena in terms of the *resolution probability*, by which potentially irresolved measurements can be handled within the Bayesian framework [3]. The data association problem is covered by a likelihood function $p(Z_k, m_k|\mathbf{x}_k)$ that statistically describes what a set of m_k observations $Z_k = \{\mathbf{z}_k^j\}_{j=1}^{m_k}$ can say about the joint state $\mathbf{x}_k$ of the objects to be tracked. Due to the Total Probability Theorem, it can be written as a sum over all possible, mutually exclusive, and exhaustive data interpretations j_k:

$$p(Z_k, m_k|\mathbf{x}_k) = \sum_{j_k} p(Z_k, m_k, j_k|\mathbf{x}_k) \tag{7.1}$$

$$= \sum_{j_k} p(Z_k|m_k, j_k, \mathbf{x}_k)\, p(m_k|j_k, \mathbf{x}_k)\, p(j_k|\mathbf{x}_k). \tag{7.2}$$

Generally, the formulation of such likelihood functions is by no means a trivial task. In many practical cases, however, a given multiple-object tracking problem can be decomposed into independent sub-problems of reduced complexity. The example below is practically important but can still be handled more or less rigorously.

7.1.1 A Radar Resolution Model

For the sake of conciseness, we confine the discussion to non-imaging radar sensors. With some modifications, the results can also be transferred to infrared or electro-optical sensors, for example. Let us consider a medium range radar producing range and azimuth measurements of an object formation consisting of two targets. For physical reasons the resolution in range, azimuth, and range-rate will be independent from each other. In particular, range and cross-range resolution differ significantly in many radar applications. Therefore, the resolution performance of the sensor is expected to depend strongly on the current sensor-to-group geometry and the relative orientation of the targets within the group. The sensor's resolution capability is also determined by the particular signal processing techniques used and the random target fluctuations. As a complete description is rather complicated, we have to look for a simplified, but qualitatively correct and mathematically tractable model. In any case, the radar resolution capability in range and azimuth is limited by the corresponding band- and beam-width. These radar-specific parameters must explicitly enter into any processing of potentially irresolved plots. The typical size of resolution cells in a medium distance is about 50 (range) and 500 m (cross range). As in target formations the mutual distance may well be 50–500 m or even less, the limited sensor resolution is a real problem in object tracking.

Centroid Measurements

Under the hypothesis $j_k = E_k^{ii}$ assuming that the radar plot $\mathbf{z}_k^i$ is an irresolved measurement belonging to two targets with a joint vector $\mathbf{x}_k = (\mathbf{x}_k^{1\top}, \mathbf{x}_k^{2\top})^\top$, the conditional likelihood is given by:

$$p(\mathbf{z}_k^i|\mathbf{x}_k) = \mathcal{N}\big(\mathbf{z}_k^i;\ \mathbf{H}_k^g\mathbf{x}_k,\ \mathbf{R}_k^g\big), \tag{7.3}$$

where the measurement matrix $\mathbf{H}_k^g$ describes a centroid measurement of the group center, characterized by a corresponding measurement error covariance matrix $\mathbf{R}_k^g$:

$$\mathbf{H}_k^g\mathbf{x}_k = \tfrac{1}{2}\mathbf{H}_k(\mathbf{x}_k^1 + \mathbf{x}_k^2), \tag{7.4}$$

where $(r_k, \varphi_k)^\top = \mathbf{H}_k\mathbf{x}_k^i$, $i = 1, 2$, is the measurement of the underlying tracking problem, where resolution phenomena are irrelevant.

Resolution Probability

Resolution phenomena will be observed if the range and angular distances between the objects are small compared with α_r, α_φ: $\Delta r_k/\alpha_r < 1$ and $\Delta\varphi_k/\alpha_\varphi < 1$. The objects within the group are resolvable if $\Delta r_k/\alpha_r \gg 1$ or $\Delta\varphi_k/\alpha_\varphi \gg 1$. Furthermore, we expect a narrow transient region. A more quantitative description is provided by introducing a resolution probability $P_r = P_r(\Delta r, \Delta\varphi)$ depending on the sensor-to-group geometry. It can be expressed by a corresponding probability of being irresolvable $P_r = 1 - P_u(\Delta r_k, \Delta\varphi_k)$. Let us describe P_u by a Gaussian-type function of the relative range and angular distances [3]:

$$P_u(\Delta r_k, \Delta\varphi_k) = \exp\Big[-\log 2\big(\tfrac{\Delta r_k}{\alpha_r}\big)^2\Big]\ \exp\Big[-\log 2\big(\tfrac{\Delta\varphi_k}{\alpha_\varphi}\big)^2\Big]. \tag{7.5}$$

Obviously, this simple model for describing resolution phenomena reflects the previous, more qualitative discussion. We in particular observe that P_u is reduced by a factor of 2 if Δr_k is increased from zero to α_r. Due to the Gaussian character of its dependency on the state vector $\mathbf{x}_k$ the probability P_u can formally be written in terms of a normal density:

$$P_u = \exp\Big[-\log 2\,\big(\mathbf{H}_k(\mathbf{x}_k^1 - \mathbf{x}_k^2)\big)^\top \mathbf{A}^{-1}(\mathbf{H}_k\mathbf{x}_k^1 - \mathbf{H}_k{\mathbf{x}_k}^2)\Big] \tag{7.6}$$

$$= \exp\Big[-\log 2\,(\mathbf{H}_k^u\mathbf{x}_k)^\top \mathbf{A}^{-1}\mathbf{H}_k^u\mathbf{x}_k\Big]. \tag{7.7}$$

Here the *resolution matrix* $\mathbf{A}$ is defined by $\mathbf{A} = \mathbf{diag}(\alpha_r^2, \alpha_\varphi^2)$, while the quantity $\mathbf{H}_k^u\mathbf{x}_k = \mathbf{H}_k(\mathbf{x}_k^1 - \mathbf{x}_k^2)$ can be interpreted as a measurement matrix for distance measurements. Up to a constant factor the resolution probability $P_u(\mathbf{x}_k)$ might formally be interpreted as the fictitious likelihood function of a Zero measurement for the distance $\mathbf{H}_k(\mathbf{x}_k^1 - \mathbf{x}_k^2)$ between the objects with a corresponding *fictitious* measurement error covariance matrix $\mathbf{R}_u$ defined by the resolution parameters α_r, α_φ.

$$P_u(\mathbf{x}_k) = |2\pi\mathbf{R}_u|^{-1/2}\,\mathcal{N}\big(\mathbf{O};\ \mathbf{H}_u\mathbf{x}_k,\ \mathbf{R}_k^u\big). \tag{7.8}$$

with $\mathbf{R}_k^u = \frac{\mathbf{A}}{2\log 2} = \frac{1}{2\log 2}\mathrm{diag}[\alpha_r^2, \alpha_\varphi^2]$. According to a first order Taylor expansion in analogy to the discussion in Sect. 2.3.1, the resolution matrix describing the resolution cells in Cartesian coordinates proves to be time dependent and results from the matrix $\mathbf{A}$ by applying dilatation and a rotation. In the same way as the Cartesian measurement error ellipses, the Cartesian "resolution ellipses" depend on the target range. Suppose we have $\alpha_r = 100$ m and $\alpha_\varphi = 1°$. We then expect that the resolution in a distance of 50 km is about 100 (range) and 900 m (cross range). Since for military targets in a formation their mutual distance may well be 200–500 m or even less, resolution is a real target tracking problem.

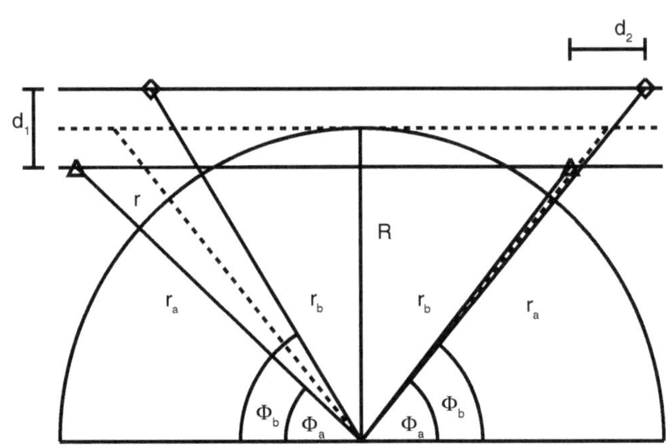

Fig. 7.1 Radar resolution phenomena: simulated object group passing a radar sensor (left: both targets within same range resolution cell; right: both targets within same azimuth resolution cell)

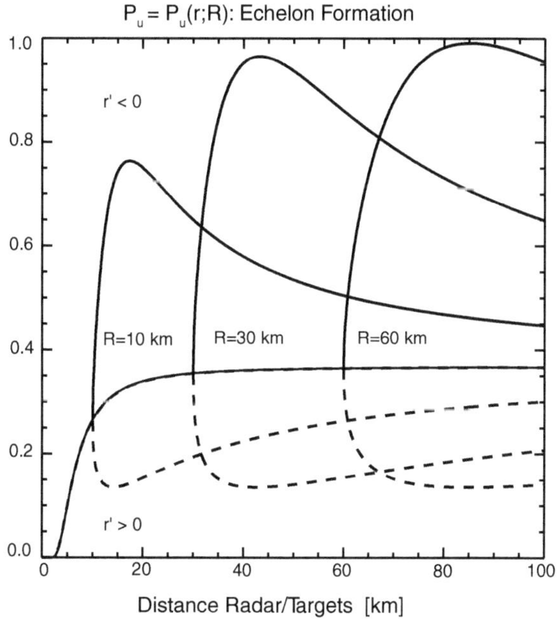

Fig. 7.2 Effect of the underlying sensor-to-group geometry: resolution probability depending on the distance between group center and radar for $R = 0$, 10, 30, 60 km. r' denotes the radial speed

Impact of Sensor-to-Object Geometry

We expect that the resolution performance of the sensor is highly dependent on the current sensor-to-group geometry and the relative orientation of the targets within the group. As an example, let us consider the simplified situation in Fig. 7.1. A formation with two targets is passing a radar. We here consider an echelon formation. R is the minimum distance of the group center from the radar.

Figure 7.2 shows the resulting probability $P_u(r; R)$ parameterized by $R = 0$, 10, 30, 60 km as a function of the distance r between the formation center and the radar. The solid lines refer to a formation approaching the radar ($\dot{r} < 0$), the dashed lines refer to $\dot{r} > 0$. For $R \neq 0$, both flight phases differ substantially. Near R, the probability P_u varies strongly ($0.85 \to 0.15$). For a radial flight ($R = 0$), we observe no asymmetry and P_u is constant over a wide range ($r \gg r_c$).

7.1.2 Resolution-Specific Likelihood

For a cluster of two closely-spaced objects moving in a cluttered environment five different classes of data interpretations exist [3]:

1. E_k^{ii}, $i = 1, \ldots, m_k$: Both objects have not been resolved but detected as a group with probability P_D^u, $\mathbf{z}_k^i \in Z_k$ representing the centroid measurement; all remaining returns are false (m_k data interpretations):

$$p(Z_k|m_k, E_k^{ii}, \mathbf{x}_k) = \frac{\mathcal{N}(\mathbf{z}_k^j; \mathbf{H}_k^g \mathbf{x}_k, \mathbf{R}_k^g)}{|\mathrm{FoV}|^{m_k-1}} \tag{7.9}$$

$$p(m_k|E_k^{ii}, \mathbf{x}_k) = p_F(n_k - 1) \tag{7.10}$$

$$P(E_k^{ii}|\mathbf{x}_k) = \tfrac{1}{m_k} P_u(\mathbf{x}_k)\, P_D^u. \tag{7.11}$$

With P_u as represented in Eq. 7.8, $p(Z_k, m_k, E_k^{ii}|\mathbf{x}_k)$ is up to a constant factor given by:

$$p(Z_k, m_k, E_k^{ii}|\mathbf{x}_k) \propto \mathcal{N}\Big(\begin{pmatrix}\mathbf{z}_k^i\\ 0\end{pmatrix}; \begin{pmatrix}\mathbf{H}_k^g\\ \mathbf{H}_k^u\end{pmatrix}\mathbf{x}_k, \begin{pmatrix}\mathbf{R}_k^g & \mathbf{O}\\ \mathbf{O} & \mathbf{R}_k^u\end{pmatrix}\Big). \tag{7.12}$$

Hence, under the hypothesis E_k^{ii} two measurements are to be processed: the (real) plot $\mathbf{z}_k^i$ of the group center $\mathbf{H}_k^g\mathbf{x}_k = \frac{1}{2}\mathbf{H}_k(\mathbf{x}_k^1 + \mathbf{x}_k^2)$ and a (fictitious) measurement 'zero' of the distance $\mathbf{H}_k^u\mathbf{x}_k = \mathbf{H}_k(\mathbf{x}_k^1 - \mathbf{x}_k^2)$ between the objects. We can thus speak of *'negative' sensor information* [4], as the lack of a second target measurement conveys information on the target position. In the case of a resolution conflict, the relative target distance must be smaller than the resolution.

2. E_k^0: Both objects were neither resolved nor detected as a group, so all returns in Z_k are thus assumed to be false (one interpretation hypothesis):

$$p(Z_k, m_k|E_k^0, x_k) = P_u(\mathbf{x}_k)\,(1 - P_D^u)\, p_F(m_k) \tag{7.13}$$

$$P(E_k^0|\mathbf{x}_k) = P_u(\mathbf{x}_k)\,(1 - P_D^u). \tag{7.14}$$

In analogy to the previous considerations, we can write up to a constant factor:

$$p(Z_k, m_k, E_k^0|\mathbf{x}_k) \propto \mathcal{N}\big(0;\ \mathbf{H}_k^u\mathbf{x},\ \mathbf{R}_k^u\big). \tag{7.15}$$

This means that even under the hypothesis of a missing irresolved plot, at least a fictitious distance measurement 0 is processed with a measurement error given by the sensor resolution.

3. $E_k^{ij}, i, j = 1, \ldots, m_k, i \neq j$: Both objects were resolved and detected, $\mathbf{z}_k^i, z_k^j \in Z_k$ are the measurements, $m_k - 2$ returns are false ($m_k(m_k - 1)$ interpretations):

$$p(Z_k|m_k, E_k^{ij}, \mathbf{x}_k) = \frac{\mathcal{N}(\mathbf{z}_k^i; \mathbf{H}_k\mathbf{x}_k^1, \mathbf{R}_k)\,\mathcal{N}(\mathbf{z}_k^j; \mathbf{H}_k\mathbf{x}_k^2, \mathbf{R}_k)}{|\mathrm{FoV}|^{m_k-2}} \tag{7.16}$$

$$p(m_k|E_k^{ij}, \mathbf{x}_k) = p_F(m_k - 2) \tag{7.17}$$

$$P(E_k^{ij}|\mathbf{x}_k) = \frac{\big(1 - P_u(\mathbf{x}_k)\big)}{m_k(m_k-1)} P_D^2. \tag{7.18}$$

According to the factor $1 - P_u(\mathbf{x}_k) = 1 - |2\pi\mathbf{R}_u|^{\frac{1}{2}}\, \mathcal{N}\big(0;\ \mathbf{H}_k^u\mathbf{x},\ \mathbf{R}_k^u\big)$ the likelihood function becomes a mixture, in which *negative* weighting factors can occur. Nevertheless, the coefficients sum up to one; the density $p(\mathbf{x}_k|Z^k)$ is thus well-defined. This reflects the fact that in case of a resolved group the targets must have a certain minimum distance between each other which is given by the sensor resolution. Otherwise they would not have been resolvable.

4. $E_k^{i0}, E_k^{0i}, i = 1, \ldots, m_k$: Both objects were resolved but only one object was detected, $\mathbf{z}_k^i \in Z_k$ is the measurement, $m_k - 1$ returns in Z_k are false ($2m_k$ interpretations):

$$p(Z_k, m_k|E_k^{i0}, \mathbf{x}_k) = |\text{FoV}|^{1-m_k}\ \mathcal{N}(\mathbf{z}_k^i;\ \mathbf{H}_k\mathbf{x}_k^1,\ \mathbf{R}_k)\ p_F(m_k - 1) \tag{7.19}$$

$$P(E_k^{i0}|\mathbf{x}_k) = \tfrac{1}{m_k}\big(1 - P_u(\mathbf{x}_k)\big)\ P_D\ (1 - P_D). \tag{7.20}$$

5. E_k^{00}: The objects were resolved, but not detected; all m_k plots in Z_k are false (one interpretation):

$$p(Z_k, m_k|E_k^{00}, \mathbf{x}_k) = |\text{FoV}|^{-m_k}\ p_F(m_k) \tag{7.21}$$

$$P(E_k^{00}|\mathbf{x}_k) = \big(1 - P_u(\mathbf{x}_k)\big)\ (1 - P_D)^2. \tag{7.22}$$

Since there exist $(m_k + 1)^2 + 1$ interpretation hypotheses, the ambiguity for even small clusters of closely-spaced objects is much higher than in the case of well-separated objects ($m_k + 1$ each). This means that only small groups can be handled more or less rigorously. For larger clusters (raids of military aircraft, for instance) a collective treatment seems to be reasonable until the group splits off into smaller sub-clusters or individual objects (see the discussion in Sect. 3.3.3).

Up to a factor $\frac{1}{m_k!}\rho_F^{m_k-2}\,|\text{FoV}|^{-m_k}\,\mathrm{e}^{-|\text{FoV}|\rho_F}$ independent of $\mathbf{x}_k$, by using Eq. 2.35, the likelihood function of potentially irresolved sensor data in a clutter background,

$$p(Z_k, m_k|x_k) = p(Z_k, m_k, E_k^0) + \sum_{i,j=0}^{m_k} p(Z_k, E_k^{ij}, m_k|x_k), \tag{7.23}$$

is proportional to a sum of Gaußians and a constant:

$$\begin{aligned} p(Z_k, m_k|\mathbf{x}_k) \propto\ & \rho_F^2(1 - P_D)^2\big(1 - P_u(\mathbf{x}_k)\big) + \rho_F(1 - P_D^u)P_u(\mathbf{x}_k) \\ & + P_D^u\rho_F P_u(\mathbf{x}_k)\sum_{i=1}^{m_k}\mathcal{N}(\mathbf{z}_k^i;\ \mathbf{H}_k^g\mathbf{x}_k,\ \mathbf{R}_k^g) \\ & + \rho_F P_D(1 - P_D)\big(1 - P_u(\mathbf{x}_k)\big)\sum_{i=1}^{m_k}\big\{\mathcal{N}(\mathbf{z}_k^i;\ \mathbf{H}_k\mathbf{x}_k^1,\ \mathbf{R}_k) + \mathcal{N}(\mathbf{z}_k^i;\ \mathbf{H}_k\mathbf{x}_k^2,\ \mathbf{R}_k)\big\} \\ & + P_D^2\big(1 - P_u(\mathbf{x}_k)\big)\sum_{\substack{i,j=1\\ i\neq j}}^{m_k} p_k^{ij}(\mathbf{x}_k)\ \mathcal{N}(\mathbf{z}_k^i;\ \mathbf{H}_k\mathbf{x}_k^1,\ \mathbf{R}_k)\ \mathcal{N}(\mathbf{z}_k^j;\ \mathbf{H}_k\mathbf{x}_k^2,\ \mathbf{R}_k). \end{aligned} \tag{7.24}$$

7.1.3 A Formation Tracking Example

If the spatial false return density is not too high, JPDA-type approximations [5] can be applied (JPDA: Joint Probabilistic Data Association). According to this philosophy, the joint state mixture density $p(\mathbf{x}_k^1, \mathbf{x}_k^2|Z^k)$ resulting from the likelihood function previously discussed is approximated by a single Gaussian with the same expectation vector and covariance matrix as the mixture $p(\mathbf{x}_k^1, \mathbf{x}_k^2|Z^k)$ (moment matching [5, p. 56 ff]). Objects moving closely-spaced for some time irreversibly lose their identity if no other information is available such as measurements of object attributes [16]. When they dissolve again, a unique track-to-target association is impossible. It is thus reasonable to deal with densities that are symmetric under permutations of the individual targets. Thus, no statistically relevant information is lost and the filter performance remains unchanged, while the mean number of hypotheses involved may be significantly reduced. Within the framework of JPDA-type approximations, this has the following effect: Before combining two components of the mixture via moment matching, we check if the components are more 'similar' to each other when the target indices are switched. If this is the case, we combine them instead. These considerations are also a useful and simple means to avoid track coalescence.

Figure 7.3 shows a set of data from a typical medium-range radar. The scan interval is 5 s and the detection probability about 80 %. The example clearly shows that resolution must be taken into account as soon as the targets begin to move

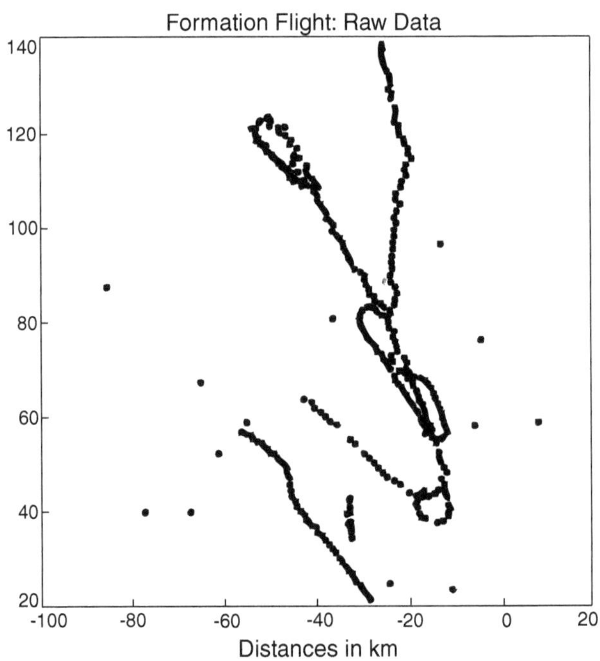

Fig. 7.3 Partly irresolved aircraft formation: accumulated raw data of a mid-range radar

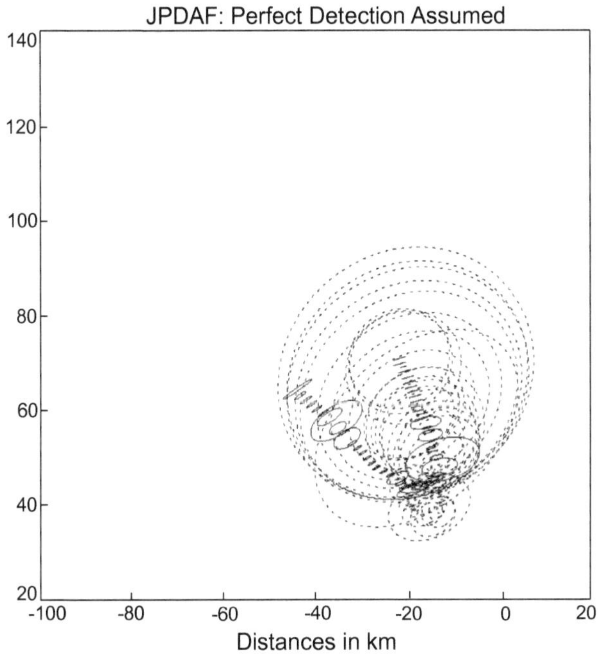

Fig. 7.4 Tracking of an aircraft formation: filtering results (JPDA, no resolution model)

closely-spaced. Figures 7.4 and 7.5 show the estimation error ellipses for two targets solid and dashed lines resulting from JPDA filtering. While in Fig. 7.4 perfect sensor resolution was assumed (wrongly!), in Fig. 7.5 the previously discussed resolution model was used. JPDA filtering without considering resolution phenomena evidently fails after a few frames, as indicated by diverging tracking error ellipses. This has a simple explanation: without modeling the limited sensor resolution, an actually produced irresolved plot can only be treated as a single target measurement along with a missed detection. In consequence, the related covariance matrices increase in size. This effect is further intensified by subsequent irresolved returns. If hypotheses related to resolution conflicts are taken into account, however, the tracking remains stable. The error ellipses in Figs. 7.4 and 7.5 have been enlarged to make their data-driven adaptivity more visible. The ellipses shrink, for instance, if both targets are actually resolved in a particular scan. The transient enlargement halfway during the formation flight is caused by a crossing target situation.

7.1.4 Resolution: Summary of Results

MHT filtering with explicit handling of resolution conflicts can successfully be applied to real radar data [6]. The main conclusions of extensive simulations based on exemplary scenarios and typical radar parameters are [7]:

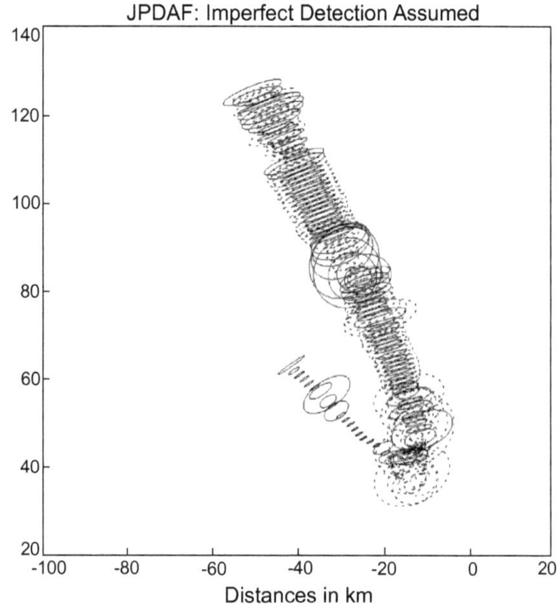

Fig. 7.5 Tracking of an aircraft formation: filtering results (with resolution model)

1. For objects with overlapping expectation gates and potentially irresolved measurements, MHT filters that handle data association conflicts in combination with resolution phenomena by far outperform more conventional trackers (monohypothesis approximations or filters ignoring imperfect resolution). Much higher false return densities and significantly lower detection probabilities can be tolerated, the tracks are more accurate, the correlation gates are reduced in size, and the critical phases of joining and splitting-off are supported.
2. Provided only primary radar data are available, information on the object identity rapidly fades out while the objects move closely-spaced and produce potentially irresolved plots. After splitting off again, a unique track-to-target correlation is no longer possible. We may thus drop the notion of identity and deal with indistinguishable targets. By this, no statistically relevant information is lost, i.e. the number of hypotheses involved can significantly be reduced without affecting the track accuracy.
3. Whether an object group is resolvable or not is highly dependent on the specific sensor-to-object geometry considered and on the position of the objects relative to each other. This phenomenon is adaptively taken into account by the resolution model used. As the correct association hypotheses can reliably be reconstructed by retrodiction techniques at the expense of some delay, the resolution model may in a retrospective view provide information on the relative position of the targets within the formation.
4. Besides the ambiguity due to irresolved or missed detections, overlapping correlation gates, and false returns, scenarios with highly maneuvering targets are also

ambiguous with respect to the object evolution model assumed to be in effect. Hypotheses related to resolution conflicts fit well into the more complex framework of IMM-MHT and provide performance improvements over more simplified dynamics models.

Further Reading

A detailed discussion of this approach has been published in:

- W. Koch, G. van Keuk
  Multiple hypothesis track maintenance with possibly unresolved measurements. IEEE Trans. Aerosp. Electron. Syst. 33(3), 883–892, 1997.
  An extended version with results from various related conference papers of the author has been published as a handbook chapter in W. Koch. Target tracking, Chap. 8, in *Advanced Signal Processing: Theory and Implementation for Sonar, Radar, and Non-Invasive Medical Diagnostic Systems*, ed. by S. Stergiopoulos. CRC Press, Boca Raton (2001).

  Abstract
  In surveillance problems, dense clutter/dense target situations call for refined data association and tracking techniques. In addition, closely-spaced targets may exist which are not resolved. This phenomenon has to be considered explicitly in the tracking algorithm. We concentrate on two targets that temporarily move in close formation and derive a generalization of MHT methods on the basis of a simple resolution model.

  Key words: Sensor resolution, Bayesian multiple target tracking, Multiple hypothesis tracking, Target formations

7.2 GMTI Radar: Doppler Blindness

Ground surveillance comprises track extraction and maintenance of single ground-moving vehicles and convoys, as well as low-flying objects such as helicopters or Unmanned Aerial Vehicles. As ground object tracking is a challenging problem, all available information sources must be exploited, i.e. the sensor data themselves, as well as context knowledge about the sensor performance and the underlying scenario.

7.2.1 Air-to-Ground Surveillance

For long-range, wide-area, all-weather, and all-day surveillance operating at high data update rates, GMTI radar proves to be the sensor system of choice (GMTI:

Ground Moving Target Indication). By using airborne sensor platforms in stand-off ground surveillance applications, the effect of topographical screening is alleviated, thus extending the sensors' field of view. In Ref. [8] characteristic problems of signal processing related to GMTI tracking with STAP radar are discussed. In this context, the following topics are of particular interest:

- *Doppler-Blindness.* Ground moving vehicles can well be masked by the clutter notch of the sensor. This physical phenomenon directly results from the low-Doppler characteristics of ground-moving vehicles and causes interfering fading effects that seriously affect track accuracy and track continuity. The problems are even more challenging in the presence of Doppler ambiguities.
- *Collectively Moving Targets.* Collectively moving convoys consisting of individual vehicles are typical of certain applications and have to be treated as aggregated entities. In some cases, the kinematic states of the individual vehicles can be treated as internal degrees of freedom. In addition, the convoy extension can become part of the object state (see Sects. 2.51 and 8.2).
- *Road-Map Information.* Even military targets usually move on road networks, whose topographical coordinates are known in many cases. Digitized topographical road-maps such as provided by Geographical Information Systems (GIS) should therefore enter into the target tracking and sensor data fusion process (see Sect. 9.1).
- *Multisensor Data.* Since a single GMTI sensor on a moving airborne platform can record a situation of interest only over short periods of time, sensor data fusion proves to be of particular importance. The data processing and fusion algorithms used for ground surveillance are closely related to the statistical, logical, and combinatorial methods applied to air surveillance.

7.2.2 A Model for Doppler Blindness

For physical and technical reasons, the detection of ground-moving targets by airborne radar, typically on a moving platform, is limited by strong ground clutter returns. This can be much alleviated by STAP techniques [8]. The characteristics of STAP processing, however, directly influence the GMTI tracking performance. Even after platform motion compensation by STAP filtering low-Doppler targets can be masked by the clutter notch of the GMTI radar. Let $\mathbf{e}_k^p = (\mathbf{r}_k - \mathbf{p}_k)/|\mathbf{r}_k - \mathbf{p}_k|$ denote the unit vector pointing from the platform position $\mathbf{p}_k$ at time t_k to the target at the position $\mathbf{r}_k$ moving with the velocity $\dot{\mathbf{r}}_k$. The kinematic object state is given by $\mathbf{x}_k = (\mathbf{r}_k^\top, \dot{\mathbf{r}}_k^\top)^\top$. Doppler blindness occurs if the radial velocities of the object as well as of the surrounding main-lobe clutter return are nearly identical, i.e. if the function

$$h_n(\mathbf{r}_k, \dot{\mathbf{r}}_k; \mathbf{p}_k) = \frac{(\mathbf{r}_k - \mathbf{p}_k)^\top \dot{\mathbf{r}}_k}{|\mathbf{r}_k - \mathbf{p}_k|} \tag{7.25}$$

is close to zero. In other words, $h_c(\mathbf{x}_k; \mathbf{p}_k) \approx 0$ holds if the target's velocity vector is nearly perpendicular to the sensor-to-target line-of-sight. For this reason, the equation $h_c(\mathbf{x}_k; \mathbf{p}_k) = 0$ defines the location of the GMTI clutter notch in the state space of a ground target and as such reflects a fundamental physical/technical fact without implying any further modeling assumptions.

Qualitative Discussion

Any GMTI detection model for air-to-ground radar must thus reflect the following phenomena:

1. The detection probability P_D depends on the target state and the sensor/target geometry.
2. P_D is small in a certain region around the clutter notch characterized by the Minimum Detectable Velocity (MDV), an important sensor parameter that must enter into the tracking process.
3. Far from the clutter notch, the detection probability depends only on the directivity pattern of the sensor and the target range.
4. There exists a narrow transient region between these two domains.

GMTI models are adapted to STAP techniques in that the detection probability assumed in the tracking process is described as a function of the GMTI-specific clutter notch. While the current location of the notch is determined by the kinematical state of the target and the current sensor-to-target geometry, its width is given by a characteristic sensor parameter (MDV). In this way, more detailed information on the sensor performance can be incorporated into the tracking process. This in particular permits a more appropriate treatment of missing detections. In other words, information on the potential reasons that might have caused the missing detections enters into the tracking filter. We observed that by this measure, the number of lost tracks can significantly be reduced, while the track continuity is improved, finally leading to a more reliable ground picture. This qualitative discussion of the observed detection phenomena related to the GMTI clutter notch is similar in nature to that of resolution effects.

Quantitative Discussion

In a generic description of the detection performance of GMTI sensors it seems plausible to write $P_D = P_D(\mathbf{x}_k)$ as a product with one factor reflecting the directivity pattern and propagation effects due to the radar equation [9], $p_D = p_D(r_k, \varphi_k)$, the other factor being related to the clutter notch. To this end, let us consider functions of the following form:

$$P_D(r_k, \varphi_k, \dot{r}_k) = p_d(r_k, \varphi_k) \left(1 - \mathrm{e}^{-\frac{1}{2}\left(\frac{h_n(r_k, \varphi_k, \dot{r}_k)}{\mathrm{MDV}}\right)^2}\right). \tag{7.26}$$

In this expression the sensor parameter MDV has a clear and intuitive meaning: In the region defined by $|n_c(\mathbf{x}_k)| < \mathrm{MDV}$ we have $P_D < \frac{1}{2}\, p_d$. The parameter MDV is thus a quantitative measure of the minimum radial velocity with respect to the

Table 7.1 Simplified GMTI tracking scenario: selected sensor and platform parameters.

Sensor	h_p (km)	v_p (m/sec)	Δr (km)	$\Delta\varphi$ (deg)	ΔT (sec)	MDV (m/sec)
1	10	200	[232, 292]	[−128, −67]	15	2
2	1	40	[22, 54]	[77, 172]	10	2

sensor platform that a ground-moving target must at least have to be detectable by the sensor. The actual size of MDV depends on the particular signal processor used.

For SWERLING I targets p_d is given by: $p_d(r,\varphi) = p_F{}^{1/[1+\mathrm{snr}(r,\varphi)]}$ with the false alarm probability p_F and the signal-to-noise ratio $\mathrm{snr}(r,\varphi) = \mathrm{snr}_0\ D(\varphi)\ (\sigma/\sigma_0)\ (r/r_0)^{-4}$ according to Ref. [9] and the discussion in Sect. 2.3.4. Let the sensor's directivity pattern be described by $D(\varphi) = \sin^2(\varphi - \varphi_a)$.

After rearranging the terms in Eq. 7.26, we can formally introduce Gaußian likelihood functions, where $h_n(\mathbf{x}_k)$ appears as a fictitious nonlinear measurement function:

$$P_D(\mathbf{x}_k; \mathbf{p}_k) = P_D - P_D^n\, \mathcal{N}(0; h_n(\mathbf{x}_k; \mathbf{p}_k), R_n)\big), \tag{7.27}$$

with a detection parameter P_D^n and a related 'variance' R_n that may be a function of MDV.

Impact of Sensor-to-Object Geometry

Assuming a flat earth, Fig. 7.6 shows an idealized scenario with two airborne GMTI sensors observing a ground vehicle moving at a constant speed (15 m/s = 54 km/h) parallel to the x-axis for most of the time. This situation is typical of stand-off or gap-filling ground surveillance missions. In the second half of the observation period over $\Delta t_{\max} = 25$ min the target stops for 7 min. Then it speeds up again reaching its initial velocity. Finally, the target leaves the field of view of sensor 2. In Table 7.1 selected sensor and platform parameters are summarized. h_p, v_p denote the constant height and speed of the sensor platforms over ground. Δr, $\Delta\varphi$ are the range and azimuth regions covered by each sensor during observation. The revisit intervals are given by ΔT. Unless appropriately handled, two phenomena in particular can cause problems in GMTI tracking:

1. Sensor-to-target geometries can occur where targets to be tracked are masked by the clutter notch of the sensor. This results in a series of missing detections until the geometry changes again.
2. As stopping targets are indistinguishable from ground clutter, the early detection of a stopping event itself as well as tracking of 'stop & go' targets can be important to certain applications.

The impact of these effects on the detection probability (see definitions in Sect. 2.3.2) is shown in Fig. 7.7 for the scenario previously introduced. For both sensors we observe deep notches (dashed line: platform 1, dotted line: platform 2). In the center of these notches the radial velocities of the target and the surrounding ground patch

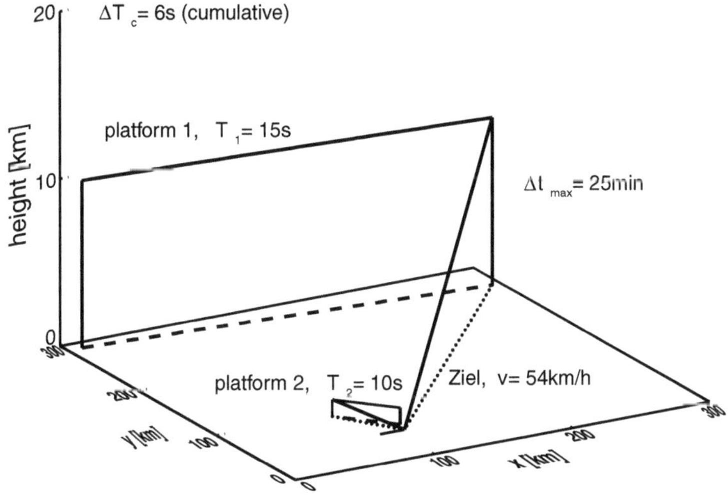

Fig. 7.6 Simplified ground target tracking scenario: two moving airborne GMTI radar platforms and a single ground moving target

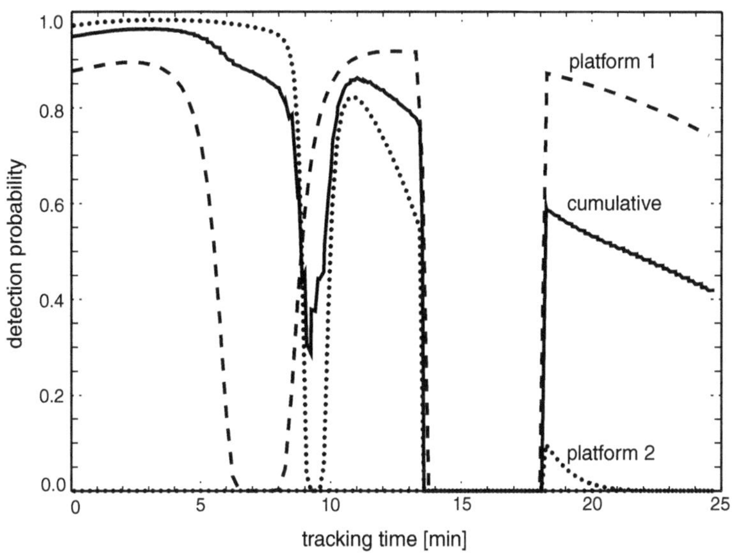

Fig. 7.7 GMTI tracking: detection probability of the individual sensors and the mean accumulated detection probability as a function of the tracking time

are very close to each other, thus making target discrimination by Doppler processing (STAP [8]) impossible. This is particularly true if the target stops.

The dashed and solid lines in Fig. 7.8 denote the radial velocities of ground patches around the target and target returns, respectively. The area shaded in gray reflects

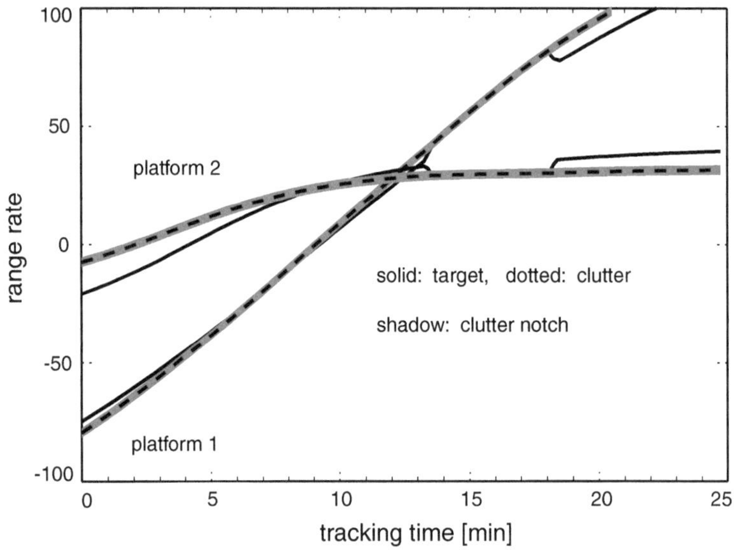

Fig. 7.8 GMTI tracking: range rate of the ground target and the surrounding ground patch relative to the moving GMTI sensors

the width of the clutter notches of the sensors, which is determined by the individual Minimum Detectable Velocities (MDVs). For each sensor, both curves are closely adjacent to each other, indicating that the target is moving at a much lower speed than the sensor platforms. We notice sliding intersections between the curves. They are responsible for the relatively long duration of Doppler-blind phases.

Assuming an idealized processing architecture (measurement fusion), the *mean cumulative revisit interval* ΔT_c results from the individual revisit intervals $\Delta T_1 =$ 15 s, $\Delta T_2 = 10$ s, yielding $\Delta T_c = 6$ s. The *mean cumulative detection probability* P_D^c is shown in Fig. 7.7 (solid line). The impact of the clutter notches is more or less compensated for. Due to the fact that P_D^c is related to the mean cumulative revisit interval $\Delta T_c = 6$ s, being shorter than those of the individual sensors ($\Delta T_1 = 10$ s, $\Delta T_2 = 15$ s), P_D^c is smaller than the detection probability of the sensor dominating at that time.

On Convoy Resolution

Since in certain applications, ground traffic vehicles often move in convoys, at first view resolution phenomena seem to be typical of long-range ground surveillance. Due to the asymmetric effect of range and angle resolution, however, Doppler-blindness in many cases superimposes resolution effects. As soon as convoy targets cease to be resolvable, they are at the same time buried in the clutter notch and thus escape detection. Vice versa, resolvable convoy targets are rarely Doppler-screened. A separate modeling of the sensor resolution might therefore be omitted.

As an example we assume two targets moving in a row along a straight road with 30 km/h as typical of military applications. Their mutual distance is 50 m.

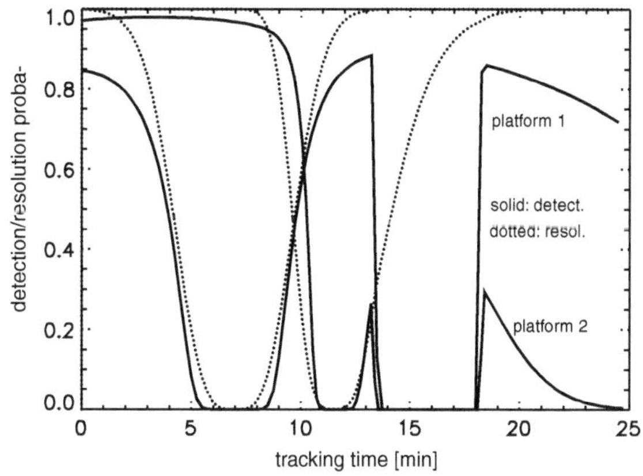

Fig. 7.9 Detection and resolution probability

The target/sensor geometry is as depicted in Fig. 7.6. Let the sensor resolution be given by: $\alpha_r = 10$ m (range), $\alpha_\varphi = 0.1°$ (azimuth), $\alpha_{\dot{r}} = 0.5$ m/s (range-rate). Figure 7.9 shows the detection probabilities of both sensors (solid lines). The width of the notches is larger than in Fig. 7.7 due to the smaller convoy speed. The dotted lines denote the resolution probabilities P_r of the sensors modeled according to the discussion in Sect. 7.1:

$$P_r = 1 - e^{-\log 2(\Delta r/\alpha_r)^2} e^{-\log 2(\Delta\varphi/\alpha_\varphi)^2} e^{-\log 2(\Delta\dot{r}/\alpha_{\dot{r}})^2}. \tag{7.28}$$

Δr, $\Delta\varphi$, $\Delta\dot{r}$ are the distances between the targets in sensor coordinates. If P_r is dominated by the angular resolution (i.e. Δr and $\Delta\dot{r}$ are small), Doppler-blindness occurs. Outside of the notch the high range/range-rate resolution guarantees resolved returns.

7.2.3 *Essentials of GMTI Tracking*

The choice of a suitable coordinate system for describing the underlying sensor/target geometry, the sensor platform trajectory, and the available a priori information on the dynamical behavior of ground-moving targets are prerequisites to target tracking. In wide-area applications a flat earth model is often not admissible. We consider three coordinate systems in which the underlying physical phenomena become transparent:

1. Appropriate *ground* coordinates, typically WGS84, where the description of the target and platform kinematics is of a particularly simple form,
2. the moving Cartesian *antenna* coordinate system, whose x-axis is oriented along the array antenna of the GMTI radar mounted on the airborne sensor platform,
3. the *sensor* coordinate system, in which the measurements of the kinematical target parameters are described (target range, azimuth, and range-rate).

Under the assumptions made in Sect. 2.3.3, the likelihood is given by the following expression (single vehicle, mild residual clutter density ρ_F, m_k plots in each sensor scan $Z_k = \{\mathbf{z}_k^j\}_{j=1}^{m_k}$):

$$p(Z_k, m_k|\mathbf{x}_k) = \big(1 - P_D(\mathbf{x}_k; \mathbf{p}_k)\big)\rho_F \; + \; P_D(\mathbf{x}_k; \mathbf{p}_k) \textstyle\sum_{j=1}^{m_k} \mathcal{N}\big(\mathbf{z}_k^j;\; \mathbf{h}(\mathbf{x}_k),\; \mathbf{R}_k\big)$$
$$= p_0(Z_k, m_k|\mathbf{x}_k) + p_n(Z_k, m_k|\mathbf{x}_k) \qquad (7.29)$$

where $p_0 = p_0(Z_k, m_k|\mathbf{x}_k)$ denotes the standard likelihood without considering clutter notches:

$$p_0 = (1 - P_d)\rho_F + P_d \textstyle\sum_{j=1}^{m_k} \mathcal{N}\big(\mathbf{Z}_K^J;\; \mathbf{h}(\mathbf{x}_k),\; \mathbf{R}_k\big), \qquad (7.30)$$

$p_n = p_n(Z_k, m_k|\mathbf{x}_k)$ is the part of the overall likelihood function characteristic of the GMTI problem. For a generalization in case of Doppler-unambiguous measurements see [10, 11].

If the GMTI detection model is inserted into this expression, we immediately see that the effect of the GMTI-specific clutter notch on the likelihood function can formally be described by a fictitious measurement of a quantity defined by pseudo measurement function $\mathbf{h}_k^n$, where the minimum detectable velocity plays the role of a fictitious measurements error standard deviation.

According to Bayes' rule, the processing of the new sensor data Z_k received at revisit time t_k is based on the predicted density $p(\mathbf{x}_k|Z^{k-1})$ and the likelihood function $p(Z_k, m_k|\mathbf{x}_k)$. Assuming a Gaussian sum representation for $p(\mathbf{x}_k|Z^{k-1})$, the Gaußian sum structure of the likelihood function guarantees that also $p(\mathbf{x}_k|Z^k)$ belongs to this family. According to Bayes Theorem we obtain up to a normalizing constant:

$$p(\mathbf{x}_k|Z^k) \propto p(Z_k, m_k|\mathbf{x}_k)\; p(\mathbf{x}_k|Z^{k-1}) \qquad (7.31)$$
$$\propto \sum_i p_k^i\; \mathcal{N}\big(\mathbf{x}_k;\; \mathbf{x}_{k|k}^i,\; \mathbf{P}_{k|k}^i\big). \qquad (7.32)$$

The same type of mixture reduction techniques can be applied as discussed in Sect. 3.3.3 (pruning, local combining) in order to keep the number of mixture components under control. Simulations showed that even a representation by only two mixture components is sufficient in many practical cases and seems to mirror the underlying physics of the detection process quite well.

7.2.4 Effect of GMTI-Modeling

Figures 7.10, 7.11 and 7.12 provide a qualitative insight into the effect of the refined sensor model on target tracking/data fusion. While a high adaptivity is evident near the clutter notch, far from the notch no difference to standard filters is observed.

Figure 7.10 displays the probability density functions resulting from processing the event that a missing detection occurred near the notch. To show the most

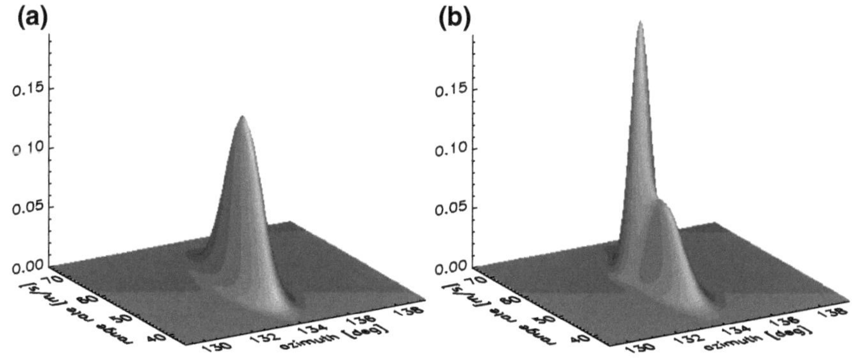

Fig. 7.10 Effect of GMTI modeling (missing detection near the clutter notch): **a** standard filter, **b** GMTI filter)

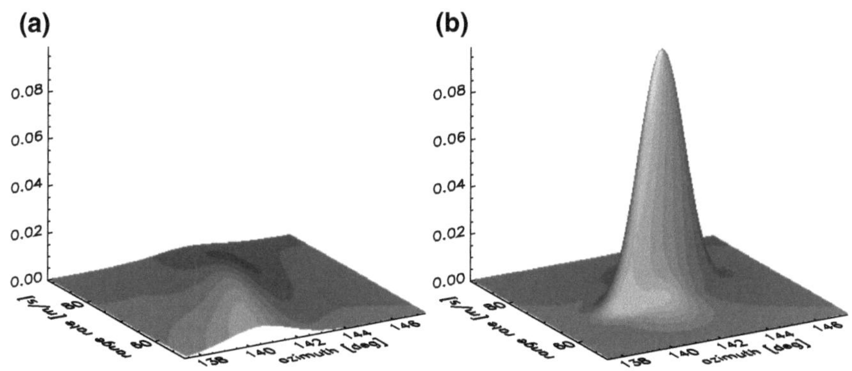

Fig. 7.11 Effect of GMTI modeling (target buried in the notch for several revisits): **a** standard filter. **b** GMTI filter

interesting features, the densities are projected on the azimuth/range-rate plane. While the probability density of the standard tracker (Fig. 7.10a) is identical with the corresponding predicted density, the refined sensor model leads to a bimodal structure (Fig. 7.10b). The broader peak refers to the possible event that the missing detection has purely statistical reasons as in the case of standard filtering, while the sharper peak behind it reflects the hypothesis that the target was not detected because it is masked by the clutter notch.

The situation where the target is buried in the clutter notch for several revisits is represented in Fig. 7.11. Obviously, the probability density of the standard filter totally faded away permitting no reasonable state estimation (Fig. 7.11a). The refined filter, however, preserved a definite shape (Fig. 7.11b). This can be explained as follows. Instead of actual sensor data, the very information that several successively missing detections occurred was processed. This event provides a hint to the filter that the kinematical target state probably obeys a certain relation determined by

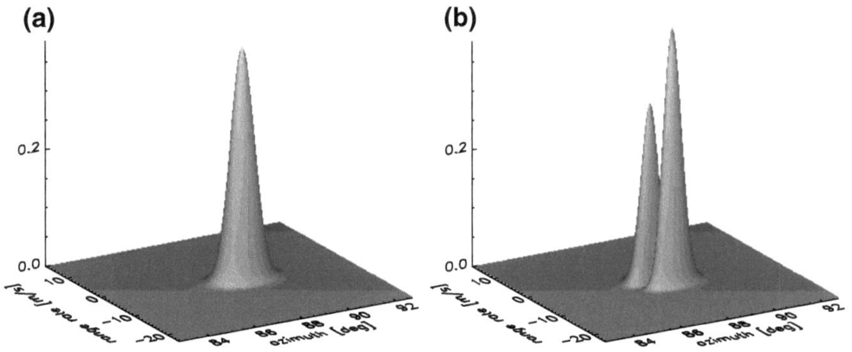

Fig. 7.12 Effect of GMTI modeling (detection occurs near the clutter notch): **a** standard filter. **b** GMTI filter

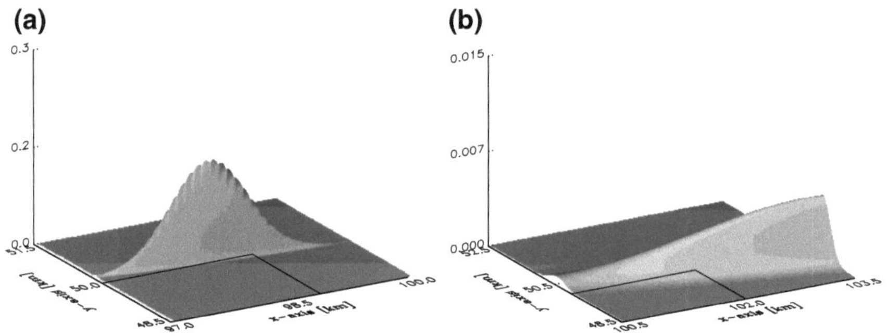

Fig. 7.13 Gain by processing GMTI data from sensor 1 only: **a** during tracking. **b** target stop

the clutter notch. Apparently, this piece of evidence proves to be as valuable as a measurement of one of the components of the target state.

Figure 7.12 refers to the event that a detection occurred near the clutter notch. While the standard filter produced a simple Gaußian, the refined filter shows a more complex structure. In fact, the probability density is a two-component mixture whose weighting factors differ in their sign (but sum up to one). The resulting shape permits an intuitive interpretation. The sensor model inherently takes into account the fact that the target state $\mathbf{x}_k$ does not lead to a small value of $n_c(\mathbf{x}_k)$; otherwise the target would not have been detected at all. For this reason, the sharp cut in the probability density simply indicates the location of the clutter notch.

Gain by Sensor Data Fusion

Figures 7.13, 7.14 and 7.15 show the probability densities of the target position in Cartesian ground coordinates after filtering. The prolated structure of the probability densities mirrors the predominant impact of cross-range errors. Their shape is rotated with respect to each other due to the different sensor-to-target geometries. This effect can be much more pronounced in other situations. We indicated the true target

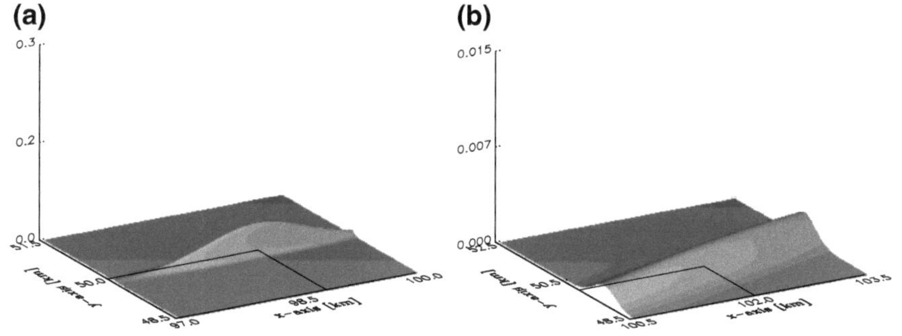

Fig. 7.14 Gain by processing GMTI data from sensor 2 only: **a** during tracking. **b** target stop

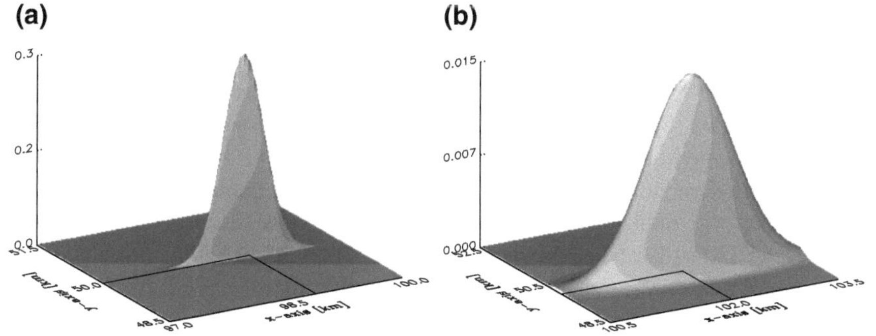

Fig. 7.15 Gain by fusing GMTI data from sensor 1 and 2: **a** during tracking. **b** target stop

position. Figures 7.13a, 7.14a and 7.15a refer to a regular tracking situation (after 10 min, see Figs. 7.6 and 7.7). Doppler-blindness occurred for sensor 2 during the previous revisits. The probability densities shown in Figs. 7.13b, 7.14b and 7.15b have been calculated at a time when the target has stopped for 3 min. Evidently in Figs. 7.13b and 7.14b the dissipation of the density functions is confined to a particular direction according to the GMTI sensor model.

Gain by Sensor Data Fusion

Figures 7.13, 7.14 and 7.15 show the probability densities of the target position in Cartesian ground coordinates after filtering. The prolated structure of the probability densities mirrors the predominant impact of cross-range errors. Their shape is rotated with respect to each other due to the different sensor-to-target geometries. This effect can be much more pronounced in other situations. We indicated the true target position. Figures 7.13a, 7.14a and 7.15a refer to a regular tracking situation (after 10 min, see Figs. 7.1, 7.7). Doppler-blindness occurred for sensor 2 during the previous revisits. The probability densities shown in Figs. 7.13b, 7.14b and 7.15b have been calculated at a time when the target has stopped for 3 min. Evidently in Figs. 7.13b and 7.14b the dissipation of the density functions is confined to a particular direction according to the GMTI sensor model.

Figure 7.15 shows the probability densities obtained by sensor data fusion. In both cases we observe a significant fusion gain. It is a consequence of the different orientation of the density functions and leads to improved state estimates. The result for the stopping targets is particularly remarkable. Though no sensor data are available from both sensors, the very fusion of the sensor output 'target under track is no longer detected' implies an improved target localization. This is a consequence of the different target/sensor geometries.

Further Reading

A detailed discussion of this approach has been published in:

- W. Koch, R. Klemm
  Ground target tracking with STAP radar. IEE Proc. Radar Sonar Navig. **148**(3), 173–185, 2001 (Special Issue on: *Modeling and Simulation of Radar Systems*, ed. by S. Watts, invited paper).
  An extended version with results from various related conference papers of the author has been published as a handbook chapter in W. Koch, Ground target tracking with STAP radar: Selected tracking aspects, Chap. 14, in *Applications of Space-time Adaptive Processing. Institution of Electrical Engineers* ed. by R. Klemm. IEE Press, London, 41 pages (2004).

Abstract
The problem of tracking ground-moving targets with a moving radar (airborne, spaceborne) is addressed. Tracking of low Doppler targets within a strong clutter background is of special interest. The motion of the radar platform induces a spreading of the clutter Doppler spectrum so that low Doppler target echoes may be buried in the clutter band. Detection of such targets can be much alleviated by space-time adaptive processing (STAP) which implicitly compensates for the Doppler spread effect caused by the platform motion. Even if STAP is applied, low Doppler targets can be masked by the clutter notch. This physical phenomenon is frequently observed and results in a series of missing detections, which may seriously degrade the tracking performance. We propose a new sensor model adapted to STAP and discuss its benefits to tracking well-separated targets. By exploiting a priori information on the sensor specific clutter notch, the model in particular provides a more appropriate treatment of missing detections. In this context the Minimum Detectable Velocity (MDV) proves to be an important sensor parameter explicitly entering into ground-moving target indication (GMTI) tracking.

Key words: Air-/spaceborne radar, STAP, GMTI radar, GMTI tracking, Minimum detectable velocity (MDV), Bayesian target tracking, Probabilistic data association (PDA)

7.3 Main-Lobe Jamming

The degrees of freedom available in applications with airborne phased-array radar enable suppression of so called main-lobe jammers that try to blind the radar by transmitting specially designed radiation directly into the main beam of the radar, by using adaptive array signal processing techniques [12]. Following the spirit of the discussions in the previous sections, the current position of the resulting jammer notch as well as information on the distribution of the related monopulse measurements [17] can be incorporated into a more sophisticated sensor performance model of airborne phased-array radar. The proposed model does not only improve object tracking in the vicinity of a jammer notch in terms of a shorter extraction delay, improved track accuracy/continuity. It also has strong impact on strategies for adaptive sensor control.

7.3.1 Modeling the Jammer Notch

Tracking of an approaching missile under mainlobe jamming conditions is among the most challenging data fusion tasks [13]. Advanced sensor models can contribute to their efficient and robust solution. An example is the simulated situation in Fig. 7.16, which shows the trajectories of a sensor (AESA: Active Electronically Scanned Array) on a moving platform, of an object to be tracked, and the jammer.

By using adaptive digital beamforming techniques, AESA radars of modern interceptor aircraft are able to electronically produce a sector of vanishing susceptibility in their receive beam pattern. Excepting this "blind spot", also called jammer notch, the radar is operating more or less normally. A non-cooperative missile, however, is expected to approach the interceptor aircraft as long as possible in the shadow of the jammer notch. The dashed lines in Fig. 7.16 characterize the spatial region of the blind spot depending on the current sensor-to-jammer geometry object.

The effect of the jammer is directly visible in the signal-to-noise-plus-jammer ratio (SNJR) of the target, which is shown in Fig. 7.17 for the scenario discussed as a function of time. Only in the beginning can the missile be detected for a short time. Then it is masked for a long time by the radar's blind spot, until it becomes visible again in close vicinity of the sensor, where the reflected signal is very strong (Burn Through). Sophisticated signal processing provides estimates of the missile direction by using adaptive monopulse techniques [12] as well as the corresponding estimation error covariance matrix $\mathbf{R}(\mathbf{b}_k, \mathbf{j}_k)$ as an additional sensor output. $\mathbf{R}(\mathbf{b}_k, \mathbf{j}_k)$ depends on the current beam direction $\mathbf{b}_k$ of the AESA radar and the jammer direction $\mathbf{j}_k$ and describes in particular the mutual correlation of the estimated direction cosines in the vicinity of the jammer notch. It thus provides valuable context information on the sensor performance.

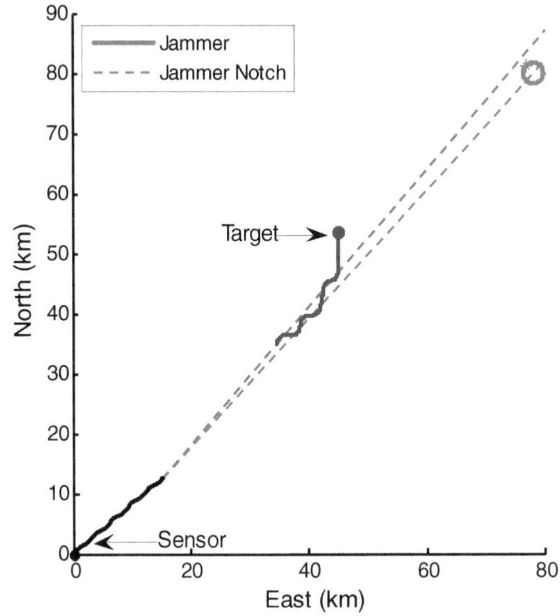

Fig. 7.16 Moving aircraft under mainlobe jamming conditions: approaching missile near the shadow of the jammer notch

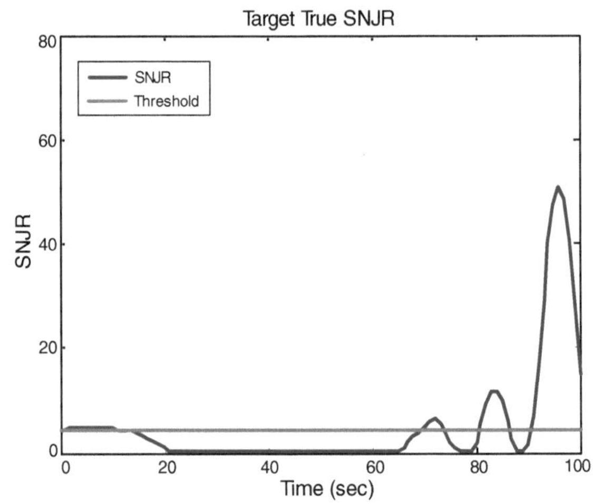

Fig. 7.17 Temporal variation of the signal-to-noise ratio of an approaching missile main-lobe jamming

Following the spirit of the discussion in Sect. 2.3.4, the sensor model is based on an expression for the signal-to-noise+jammer ratio (SNJR) after completing the signal processing chain. The following simple formula mirrors all relevant phenomena observed:

$$\mathrm{SNJR}(\mathbf{d}_k, r_k; \mathbf{b}_k, \mathbf{j}_k) = \mathrm{SNR}_0 \left(\frac{r_k}{r_0}\right)^{-4} D(\mathbf{d}_k) \times \mathrm{e}^{-\log 2|\mathbf{d}_k - \mathbf{b}_k|^2/b^2} \left(1 - \mathrm{e}^{-\log 2|\mathbf{d}_k - \mathbf{j}_k|^2/j^2}\right).$$

The vectors $\mathbf{b}_k$ and $\mathbf{j}_k$ denote the angular position of the current beam and the jammer, respectively (assumed to be known). b is a measure of the beam width, while j indicates the width of the jammer notch produced by adaptive nulling, and r_0 is the radar's instrumented range. $\mathbf{d}_k$ is the object's direction vector and r_k its range from the sensor. $D(\mathbf{d}_d)$ reflects the antenna's directivity pattern. In the case of Swerling I fluctuations of the objects' radar cross section and for a simple detection model (see the discussion in Sect. 2.3.4, the detection probability is a function of $\mathbf{d}_k$, r_k, $\mathbf{b}_k$, and $\mathbf{j}_k$:

$$P_D(\mathbf{d}_k, r_k; \mathbf{b}_k, \mathbf{j}_k) = P_F^{\frac{1}{1+\mathrm{SNJR}(\mathbf{x}_k; \mathbf{b}_k, \mathbf{j}_k)}}. \tag{7.33}$$

P_D can be approximated by using Gaußians linearly depending on the object state. Essentially, we enter this expression of the detection probability into the likelihood function in Eq. 2.40, yielding a Gaußian sum type expression for it.

7.3.2 Tracking Filter Alternatives

According to the previous discussion, the signal-to-noise-plus-jammer is essential in the modeling of the detection probability and thus enters into the likelihood function ratio. After some approximations, the likelihood function can be represented by a Gaußian mixture, finally leading to a version of the Gaußian sum filter. Since the number of mixture components grows in each update step, adaptive approximation schemes must be applied. By using Monte-Carlo simulations five competing approaches have been evaluated and compared with each other:

1. *Method 1 (Fixed EKF).* This tracking filter serves as a reference and uses no sophisticated sensor model. The impact of the jammer notch on P_D and the measurement error covariance matrix $\mathbf{R}$ are not taken into account.
2. *Method 2 (Variable EKF).* Here, only the monopulse error covariance $\mathbf{R}(\mathbf{b}_k, \mathbf{j}_k)$ is used as an improvement of the sensor model. The detection probability P_D is is assumed to be constant.
3. *Method 3 (Fixed Pseudo-bearing EKF).* This approach assumes a constant error covariance matrix $\mathbf{R}$, but uses the correct likelihood function, i.e. the jammer notch, in a second-order approximation.
4. *Method 4 (Variable Pseudo-bearing EKF).* In addition to the previous realization, here also the covariance matrix $\mathbf{R}(\mathbf{b}_k, \mathbf{j}_k)$ is part of the sensor model.

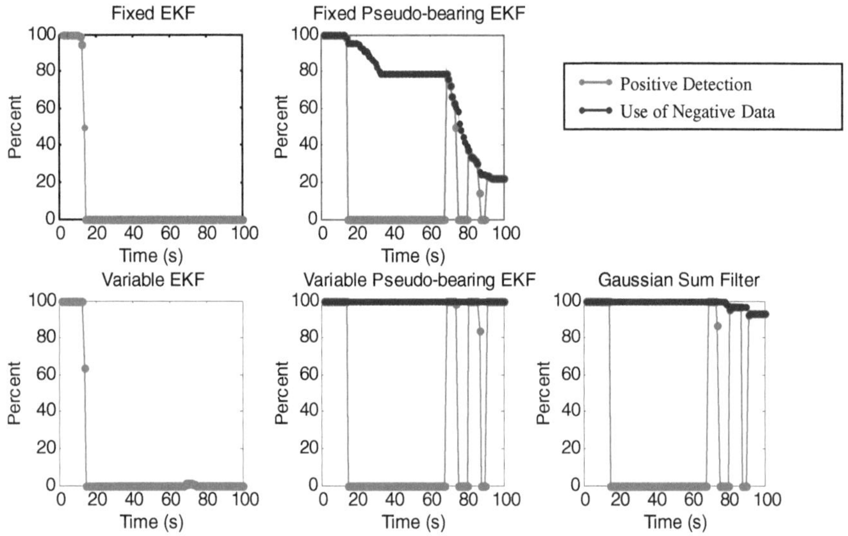

Fig. 7.18 Simulation results (250 runs) characterizing track continuity for different tracking filters

5. *Method 5 (Gaussian Sum Filter).* In this tracker the complete likelihood function and the monopulse covariance $\mathbf{R}(\mathbf{b}_k, \mathbf{j}_k)$ are used. The number of the mixture components involved to represent $p(\mathbf{x}_k|Z^k)$ is confined by three.

For the methods 3–5 the following is true: If the radar beam points to the vicinity of the blind spot and no detection occurs, a local search is performed similar to the ideas discussed in Sect. 10.4.3. By this, probability mass is concentrated near the blind spot provided the target is actually there.

7.3.3 Selected Simulation Results

Figure 7.18 shows the mean track continuity averaged over 250 Monte-Carlo runs. The superiority of tracking methods that use context information on the spatial position of the blind spot is obvious. The use of the monopulse covariance matrix is necessary, but not sufficient for avoiding track loss. The methods 3, 4, and 5 can, using "negative" sensor evidence, bridge over the missing data in the jammer notch. In spite of the fact that method 5 is more computationally intensive than method 4, it shows deficiencies if compared with method 4. This is an indication for the fact that further performance improvements are possible by more advanced approximation methods.

Further Reading

A detailed discussion of this approach has been published in:

- W. Blanding, W. Koch, U. Nickel
  Adaptive phased-array tracking in ECM using negative information. IEEE Trans. Aerosp. Electron. Syst. **45**(1), 152–166, 2009.

  Abstract
  Advances in characterizing the angle measurement covariance for phased array monopulse radar systems that use adaptive beamforming to null a jammer source allow for the use of improved sensor models in tracking algorithms. Using a detection probability likelihood function consisting of a Gaussian sum that incorporates negative contact measurement information, four tracking systems are compared when used to track a maneuvering target passing into and through standoff jammer interference. Each tracker differs in how closely it replicates sensor performance in terms of accuracy of measurement covariance and the use of negative information. Only the tracker that uses both the negative contact information and corrected angle measurement covariance is able to consistently reacquire the target when it exits the jammer interference.

  Keywords: Target tracking, Adaptive beamforming, Standoff jamming, Gaußian sum filter.

7.4 Negative Sensor Information

More advanced sensor models especially enable the exploitation of 'negative' sensor evidence. By this we mean the rigorous drawing of conclusions from expected but actually missing sensor measurements. These conclusions aim at an improvement of the position or velocity estimates for objects currently kept under track. Even a failed attempt to detect an object in the field of view of a sensor is to be considered as a useful sensor output, which can be processed by using appropriate sensor models, i.e. by background information on the sensors, with benefits for target tracking, sensor management, and sensor data fusion. The technical term chosen here for denoting such pieces of evidence, i.e. 'negative' information, seems to be accepted in the data fusion community (see, e.g. [14, 15]).

7.4.1 A Ubiquitous Notion

A very simple example illustrates that negative sensor information is an ubiquitous phenomenon, which often appears in disguise. The notion fits well into the Bayesian formalism. Assume a sensor producing at discrete time instants t_k mutually independent measurements $\mathbf{z}_k$ of a single object with Gaußian likelihood $\mathcal{N}(\mathbf{z}_k;\ \mathbf{H}\mathbf{x}_k,\ \mathbf{R})$. Absence of clutter is assumed ($\rho_F = 0$). The objects are detected with a constant detection probability $P_\text{D} < 1$. We thus have classical Kalman filtering under the constraint that there exists not at each time a measurement. The likelihood function is thus given by Eq. 2.41 and yields:

1. In the case of a positive sensor output ($m_k = 1$), $\mathbf{z}_k$ is processed by Kalman filtering leading to $p(\mathbf{x}_k|\mathcal{Z}^k) = \mathcal{N}\big(\mathbf{x}_k;\ \mathbf{x}_{k|k},\ \mathbf{P}_{k|k}\big)$ with $\mathbf{x}_{k|k}$ and $\mathbf{P}_{k|k}$ given by:

$$\mathbf{P}_{k|k} = (\mathbf{P}_{k|k-1}^{-1} + \mathbf{H}^\top\mathbf{R}^{-1}\mathbf{H})^{-1} \tag{7.34}$$

$$\mathbf{x}_{k|k} = \mathbf{P}_{k|k}(\mathbf{P}_{k|k-1}^{-1}\mathbf{x}_{k|k-1} + \mathbf{H}^\top\mathbf{R}^{-1}\mathbf{z}_k). \tag{7.35}$$

2. For a negative sensor output ($m_k = 0$), the likelihood function is a constant $1 - P_\text{D}$. By filtering the prediction density is not modified: $\mathbf{x}_{k|k} = \mathbf{x}_{k|k-1}$, $\mathbf{P}_{k|k} = \mathbf{P}_{k|k-1}$. According to 7.34 and 7.35 this result could formally be interpreted as the processing of a fictitious measurement with an infinite measurement error covariance $\mathbf{R}$, since $\mathbf{R}^{-1} = 0$.

7.4.2 Lessons Learned from Examples

The Bayes formalism and the likelihood function thus precisely indicate, in which way a negative sensor output, i.e. a missing detection has to be processed. This observation notion can be generalized and leads to the following conclusions:

1. Missing but expected (i.e. negative) sensor data can convey information on the current target position or a more abstract function of the kinematic object state. This type of negative evidence can be included in data fusion within the rigorous Bayesian structure. There is no need for recourse to ad hoc or empirical schemes.
2. The prerequisite for processing negative evidence is a refined sensor model, which provides additional background information for explaining its data. As a consequence, negative evidence often appears as an artificial sensor measurement, characterized by a corresponding measurement matrix and a measurement error covariance.
3. The particular form of the fictitious measurement equation involved is determined by the underlying model of the sensor performance, while the fictitious measurement error covariance is characterized by sensor parameters such as sensor resolution, radar beam width, or minimum detectable velocity.
4. Negative evidence implies well-defined probability densities of the object states that prove to be Gaussian mixtures with potentially negative coefficients summing up to one. Intuitively speaking, these components reflect that the targets keep a certain distance from each other, from the last beam position, or the clutter/jammer notch.

5. If the fictitious measurement depends on the underlying sensor-to-target geometry, we can even introduce the fusion of negative evidence.

Further Reading

A detailed discussion of this approach has been published in:

- W. Koch, On exploiting 'negative' sensor evidence for target tracking and sensor data fusion. Int. J. Inf. Fusion **8**(1), 28–39, 2007 (Elsevier) (Special Issue: *Best Papers of FUSION 2004* ed. by P. Svensson, J. Schubert, invited paper).

  Abstract
  In various applications of target tracking and sensor data fusion all available information related to the sensor systems used and the underlying scenario should be exploited for improving the tracking/fusion results. Besides the individual sensor measurements themselves, this especially includes the use of more refined models for describing the sensor performance. By incorporating this type of background information into the processing chain, it is possible to exploit 'negative' sensor evidence. The notion of 'negative' sensor evidence covers the conclusions to be drawn from expected but actually missing sensor measurements for improving the position or velocity estimates of targets under track. Even a failed attempt to detect a target is a useful sensor output, which can be exploited by appropriate sensor models providing background information. The basic idea is illustrated by selected examples taken from more advanced tracking and sensor data fusion applications such as group target tracking, tracking with agile beam radar, ground-moving target tracking, or tracking under jamming conditions.

  Keywords:
  Negative information/evidence, Target tracking, Sensor resolution, Local search, Adaptive beam positioning, GMTI sensor fusion

References

1. M. Wunder, J. Grosche (eds.), *Verteilte Führungsinformationssysteme* (Springer, Berlin, 2009)
2. F.E. Daum, R.J. Fitzgerald, The importance of resolution in multiple target tracking. SPIE Signal Data Process. Small Targets **2235**, 329 (1994)
3. W. Koch, G. van Keuk, Multiple hypothesis track maintenance with possibly unresolved measurements. IEEE Trans. Aerosp. Electron. Syst. **33**(3), 883–892 (1997)
4. W. Koch, On exploiting 'negative' sensor evidence for object tracking and sensor data fusion. Inf. Fusion J. **8**(1), 28–39 (2007)
5. Y. Bar-Shalom, X.-R. Li, T. Kirubarajan, *Estimation with Applications to Tracking and Navigation* (Wiley, New York, 2001)

6. W. Koch, Experimental results on Bayesian MHT for maneuvering closely-spaced objects in a densely cluttered environment, in *IEE International Radar Conference RADAR'97*, Edinburgh, UK, 1999 , p. 729.
7. W. Koch, On Bayesian MHT for Formations with possibly unresolved measurements—Quantitative results. SPIE Signal Data Process. Small Targets **3163**, 417 (1997)
8. R. Klemm, *Principles of Space–Time Adaptive Processing*, 3rd edn. (IEE Publishers, London, 2006)
9. P.L. Bogler, *Radar Principles with Applications to Tracking Systems* (Wiley, New York, 1990)
10. W. Koch, Ground target tracking with STAP radar: Selected tracking aspects, Chap. 15, in *The Applications of Space–Time Adaptive Processing*, ed. by R. Klemm (IEE Publishers, London, 2004)
11. W. Koch, On the effect of Doppler ambiguities on GMTI tracking, in *Proceedings of IEE International Radar Conference RADAR'02*, Edinburgh, UK, October 2002, , pp. 153–157
12. U. Nickel, Performance measure of corrected adaptive monopulse estimation. IEE Proc. Radar Sonar Navig. **146**, 17–24 (1999)
13. W. Blanding, W. Koch, U. Nickel, Adaptive phased-array tracking in ECM using negative information. IEEE Trans. Aerosp. Electron. Syst. **45**(1), 152–166 (2009). (Preliminary version appeared in Proceedings of FUSION 2006, Florence 2006)
14. K.J.S. Agate, Utilizing negative information to track ground vehicles through move-stop-move cycles, in *Proceedings of Signal Processing, Sensor Fusion, and Target Recognition XIII*, SPIE vol 5429, Orlando, FL, April 2004
15. H. Sidenbladh, Multi-target particle filtering for the probability hypothesis density, in *Proceedings of The 7th International Conference on Information Fusion (FUSION 2003)*, Cairns, Australia, 2003, , pp. 800–806
16. M. Mertens, M. Ulmke, Ground Target Tracking with RCS Estimation Utilizing Probability Hypothesis Density Filters, in *Proceedings of 16th International Conference on Information Fusion*, Istanbul, July (2013)
17. U. Nickel, Overview of Generalized Monopulse Estimation, IEEE Aerospace and Electronic Systems Magazine. **21**(6) II, pp. 27–56 (2006)

Chapter 8
Integration of Advanced Object Properties

In several applications, it is necessary to learn more from the sensor data received than the time-varying geolocation of moving objects of interest. Rather, we wish to understand *what* the objects we observe are, i.e. we aim to learn as much as possible about their attributes in order to be able to classify or even identify them. Many relevant object attributes can be derived even from their purely kinematic properties such as speed, heading vector, and normal acceleration as well as from mutual interrelations inferable from multiple object tracks, as has been extensively discussed in the introductory chapter, Sect. 1.3.5.

This chapter illustrates this concept with selected application examples and shows how object attributes such as the maneuvering class an air target currently belongs to, the spatial extension of an extended object or a collectively moving group, including size, shape, and orientation, or even the "anomalous" behavior of an individual in a person stream can be considered as state quantities that can be included into a more general notion of an object state and be tracked by fusing imperfect data within the Bayesian framework.

In particular, the notion of an 'object extension' is introduced by symmetrical and positively definite random matrices serving as state quantities that complement the kinematic state vectors. In this way, matrix-variate analysis is brought into play, by which is made possible to deal with collectively moving object groups and extended objects in a unified approach. This point of view is all the more appropriate, the smaller the mutual distances between the individual objects within a group are, or the larger an extended object is.

In another example, chemical sensors are discussed that make it possible to classify objects with respect to characteristic chemical signatures. Due to their fundamental lack of space-time resolution, chemical sensors develop their full potential for the classification of individuals only if the output of multiple chemical sensors distributed in space is fused with kinematic person or object tracks. The fusion result enables to identify which individual person in a person stream, for example, is actually carrying a hazardous carry-on item. Obviously, this type of behavior is a fairly well defined pattern of "anomalous behavior" that can easily be recognized using methodologies of multiple sensor data fusion.

W. Koch, *Tracking and Sensor Data Fusion*,
Mathematical Engineering, DOI: 10.1007/978-3-642-39271-9_8,

8.1 Model History MHT Tracking

As discussed in the introductory remarks of Sect. 7.1, air surveillance in a dense object/dense clutter environment is a difficult task that requires refined data association and tracking techniques that make use of multiple model and multiple hypothesis tracking methods as discussed in Part I. While the multiple hypothesis character mirrors the uncertain origin of the data, the multiple models refer to the different maneuvering phases. For this reason, adaptive IMM modeling describing the object evolution and more sophisticated sensor models for ambiguous data including resolution phenomena (Sect. 7.1, Eq. 7.24) have to be combined in a unified approach (IMM-MHT [1]).

Moreover, since the standard IMM approach with the approximations discussed in Sect. 3.36 does not lead to satisfying results in certain applications, the consideration of evolution model histories is inevitable. Simulation studies indicate that a worst/best case modeling (i.e. 2 evolution models) and model histories up to a length of 4 processing steps into the past provide enough internal degrees of freedom to make an adequate representation of the actual object behavior possible. In a worst-/best-case analysis, reasonable and practical assumptions for the switching probabilities are easily obtainable. IMM-MHT thus enables a more simplified dynamics modeling at a nearly optimal level of performance.

In many applications, IMM-MHT methods result in trees of hypothetical state estimates that are simply structured and provide satisfying target trajectories. In a more difficult dense target/dense clutter environment, however, the current sensor data may be highly ambiguous, thus leading to a filtering output that is hard to interpret (and to display) and may lead to unsatisfying estimates of the current target states. Nevertheless, even under those conditions pruning methods do remain applicable that preserve the statistically relevant information on the targets and avoid memory explosions: The filtering results are represented by potentially large and fluctuating but *limited* hypothesis trees.

As this phenomenon is a direct consequence of the current lack of information, hard pruning measures guaranteeing a fixed amount of computational load are not an answer to obtain more easily interpretable results. At the expense of a certain time delay, however, a retrodictive analysis of the MHT output may provide significantly improved estimates of the trajectories. This might well alleviate the surveillance task in dense target/dense clutter situations (see the discussion in Sect. 3.4.4).

8.1.1 IMM-MHT Tracking

By making use of the Total Probability Theorem, the multiple model approach can easily be adopted to multiple hypothesis filtering. In the probability density $p(\mathbf{x}_k|Z^k) = \sum_{\mathbf{j}_k} p(\mathbf{x}_k, \mathbf{j}_k|Z^k)$ at each step k of the filtering loop (Eq. 3.38), the individual terms of the sum become mixture densities themselves,

$$p(\mathbf{x}_k, \mathbf{j}_k|Z^k) = \sum_{i_k,\dots i_1} p(\mathbf{x}_k, i_k, \dots i_1, \mathbf{j}_k|Z^k) \tag{8.1}$$

$$= \sum_{i_k,\dots i_1} p(i_k, \dots i_1, \mathbf{j}_k|Z^k)\, p(\mathbf{x}_k|i_k, \dots i_1, \mathbf{j}_k, Z^k). \tag{8.2}$$

Hence, in the optimal approach to IMM-MHT filtering the conditional densities $p(\mathbf{x}_k, \mathbf{j}_k|Z^k)$ of the joint state $\mathbf{x}_k$ of the targets are sums over every possible sequence of dynamics models $i_k, \dots i_1$ from the initial observation through the most recent measurement at scan k ("Dynamics Histories"). As the number of terms in the sum (8.2) increases exponentially with increasing k, various techniques have been developed that represent the densities (8.2) approximately by mixtures with a *constant* number of components at each scan k.

Let us denote the dynamics histories "n scans back" by $\mathbf{i}_k$, an n-tuple of indices, $\mathbf{i}_k = (i_k, i_{k-1}, \dots i_{k-n+1})$. In particular, we are looking for approximations by Gaußian mixtures,

$$p(\mathbf{x}_k, \mathbf{j}_k|Z^k) \approx \sum_{\mathbf{i}_k} p(\mathbf{i}_k, \mathbf{j}_k|Z^k)\, p(\mathbf{x}_k|\mathbf{i}_k, \mathbf{j}_k, Z^k) \tag{8.3}$$

$$= \sum_{\mathbf{i}_k} p_{k|k}^{\mathbf{i}_k,\mathbf{j}_k}\, \mathcal{N}\big(\mathbf{x}_k;\, \mathbf{x}_{k|k}^{\mathbf{i}_k,\mathbf{j}_k},\, \mathbf{P}_{k|k}^{\mathbf{i}_k,\mathbf{j}_k}\big). \tag{8.4}$$

The weighting factors of the data association history $\mathbf{j}_k$ (Sect. 3.3.3), $p_{k|k}^{\mathbf{j}_k} = p(\mathbf{j}_k|Z^k)$, are given by $p_{k|k}^{\mathbf{j}_k} = \sum_{\mathbf{i}_k} p_{k|k}^{\mathbf{i}_k,\mathbf{j}_k}$. Due to Bayes' Theorem, the expectation vectors $\mathbf{x}_{k|k}^{\mathbf{i}_k,\mathbf{j}_k}$ and covariance matrices $\mathbf{P}_{k|k}^{\mathbf{i}_k,\mathbf{j}_k}$ of the mixtures are iteratively obtained by formulae that are essentially based on Kalman filtering. Also the weighting factors $p_{k|k}^{\mathbf{i}_k,\mathbf{j}_k}$ obey simple update formulae. In case of a single dynamics model ($r = 1$), the conditional densities $p(\mathbf{x}_k|\mathbf{j}_k, Z^k)$ are strictly given by Gaussians; i.e. up to measures controlling the growth of the tree of association hypotheses, the filtering loop may be performed in an approximately optimal way.

For $n = 1$ and assuming well-separated targets, the density $p(\mathbf{x}_k, \mathbf{j}_k|Z^k)$ is approximated by a mixture with r components according to the r dynamics models used. GPB and IMM algorithms are possible realizations of this scheme. More generally speaking, the basic idea of the original IMM algorithm (moment matching *directly after* the prediction step) may easily be adopted to $n > 1$ providing a more accurate approximation of the densities $p(\mathbf{x}_k, \mathbf{j}_k|Z^k)$ and essentially leads to a disguised version of GPBn [2]. The book chapter [3] is a reference going beyond the standard IMM approach with $n = 1$.

The IMM approach can easily be adopted to fixed-interval retrodiction. In a direct analogy to Eq. 8.4, the densities $p(\mathbf{x}_l|Z^k) = \sum_{\mathbf{j}_k} p(\mathbf{x}_l, \mathbf{j}_k|Z^k)$ in the retrodiction loop (Eq. 3.65) are approximately represented by the same class of functions previously used in the filtering loop:

$$p(\mathbf{x}_l, \mathbf{j}_k|Z^k) \approx \sum_{\mathbf{i}_l} p_{l|k}^{\mathbf{i}_l\mathbf{j}_k} \, \mathcal{N}\big(\mathbf{x}_l;\, \mathbf{x}_{l|k}^{\mathbf{i}_l\mathbf{j}_k},\, \mathbf{P}_{l|k}^{\mathbf{i}_l\mathbf{j}_k}\big) \tag{8.5}$$

with $l < k$. The backward iteration is initialized by the filtering result at the present scan k: $p_{k|k}^{\mathbf{i}_k\mathbf{j}_k}$, $\mathbf{x}_{k|k}^{\mathbf{i}_k\mathbf{j}_k}$, $\mathbf{P}_{k|k}^{\mathbf{i}_k\mathbf{j}_k}$. In analogy to the reasoning in Sect. 3.4.3, approximate update formulae for the parameters defining $p(\mathbf{x}_l, \mathbf{j}_k|Z^k)$ can be derived.

8.1.2 Performance Evaluation

The performance of IMM-MHT techniques is demonstrated by a simulation example. The operational conditions chosen (spatial false return density, detection probability) are challenging, but not extreme in so far as tracks are seldom lost, and the algorithms produce useful results even under significantly worse conditions. We consider a scenario with two maneuvering closely-spaced targets characterized by data association and resolution conflicts. Since the resulting MHT output is hard to interpret and leads to unsatisfactory tracks, the benefits of retrodiction become obvious. The treatment of potentially irresolved measurements is based on the resolution model in Sect. 7.1.

Pruning methods have been used but were not optimized with respect to this particular example. We observed that retrodiction-based cut-off (Sect. 3.4.4) may well stabilize tracking in that the number of track losses is reduced while the mean number of hypotheses and the track accuracy are kept roughly constant. To achieve the same probability of track loss by using pruning methods based only on the current filtering results, more hypotheses are required on average. This observation indicates that retrodiction-based approximations might be advantageous for realizing MHT trackers. A more detailed quantitative study of the impact of various pruning methods on the filter performance, however, is not within the scope of this section.

Sensor Model and Scenario

Let us consider a typical 2D medium range radar characterized by the following parameters: sensor resolution in range and azimuth $\alpha_r = 150$ m, $\alpha_\varphi = 1.5°$, measurement error $\sigma_r = 30$ m, $\sigma_\varphi = 0.2°$, measurement error for unresolved returns $\sigma_r^u = 75$ m, $\sigma_\varphi^u = 0.75°$, scan period of the radar $T = 5$ s, maximum range $r_{\max} = 100$ km, detection probability $P_D = 0.8$ and spatial false return density $\rho_F = 0.002/\text{km}^2$.

In Fig. 8.1, the trajectories of two targets moving with constant speed ($v = 300$ m/s) are displayed over a period of 100 scans (8'20"). The radar is located in (0,0). At the beginning, the targets are well-separated. After a crossing and a join-maneuver, they fly in echelon formation (separation $d = 150$ m) approaching the radar radially. At a distance of 17 km from the radar the formation changes its direction of flight. Finally, a split-off maneuver is followed by a second crossing. The crossings differ in their target-to-sensor geometry. Both targets maneuver with constant normal acceleration ($q = 30\,\text{m/s}^2$) for 45 s (joining/splitting-off) and 22 s (in formation). During the radial flight, the range difference between the targets is constant ($\Delta r = d/\sqrt{2}$). After the change of heading, Δr is increasing.

The resolution and measurement accuracy depend on the target-to-sensor geometry (Fig. 8.2). First, the probability of receiving resolved returns (solid line) is equal to one (except the crossing). During the radial flight it is determined by the range resolution mainly and thus constant. After the change of heading the resolution is improved as Δr is increasing. The peak occurs during the split-off maneuver. Due to the particular target-to-sensor geometry at that time, both targets differ in azimuth only before they are separated enough to be perfectly resolvable. See [4, 5] for details of the resolution model used. The mean major semi axes (dashed line) of the measurement error ellipses reflect the azimuth error, the minor semi axes (dotted line) the error in range. Unresolved returns have greater measurement errors than resolved returns. The different flight phases may easily be identified.

We apply two Van-Keuk-evolution models ($r = 2$, Eq. 2.2.1) characterized by two parameters (maneuver correlation time $\theta_i = 60\,\text{s}, 30\,\text{s}$ and acceleration bandwidth $\Sigma_i = 3, 30\,\text{m/s}^2$, i $= 1, 2$). The model transition probabilities are given by $p_{11} = 0.85$, $p_{22} = 0.7$, $p_{12} = 1 - p_{11}$, $p_{21} = 1 - p_{22}$. In our simplified example we further assume that the two targets obey the same dynamical model during a particular scan and consider "standard" IMM-MHT (i.e. $n = 1$).

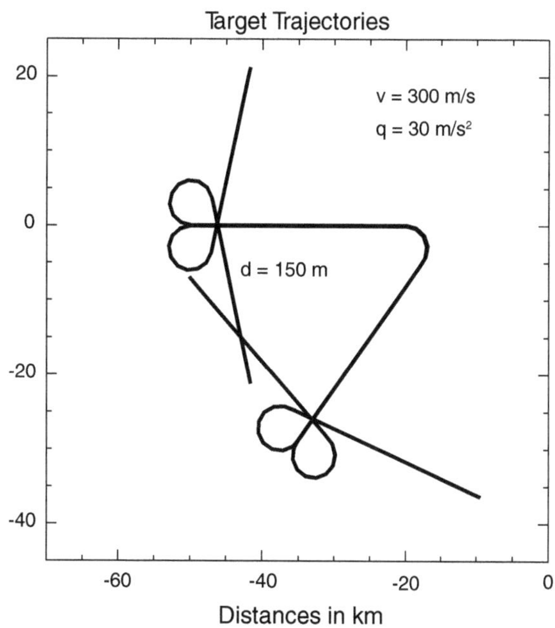

Fig. 8.1 Trajectories of *two* maneuvering targets moving partly closely spaced

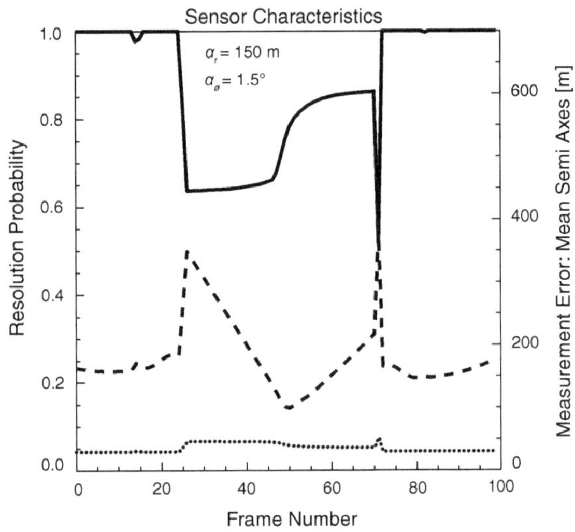

Fig. 8.2 Sensor characteristics depending on the sensor-to-target geometry

Measures of Performance

Performance evaluation for surveillance systems employing MHT methods is a difficult task that involves various conflicting aspects and user-dependent limitations [6, 7]. To illustrate the potential gain by retrodiction, we have here confined the discussion to intuitively clear and simple performance measures obtained by simulation (1000 runs; 43 lost tracks). The filter was initiated in the first 5 scans by using the true measurements only. Monohypothesis approximations to MHT (e.g. JPDAF) proved to be inapplicable under the conditions considered here.

As a function of the scan number k, we evaluated the *mean number of hypotheses* $n^l(k)$ and the *mean model probabilities* $\mu^l(k)$ provided a delay of l scans is tolerated. The *mean reconstruction time* $\tau_r(k)$ denotes the average time delay required to uniquely represent the targets at scan k by "reconstruction of histories" (Sect. 4.4). A simple quality measure for multiple hypothesis tracks may be obtained by summing up the distances between the hypothetical estimates and the true states weighted by the coefficients $p_{k|k}^{\mathbf{j}_k} = p(\mathbf{j}_k|Z^k)$ (Sects. 3.3.3, 3.4.4). As this quantity reflects the scattering around the true trajectories in some sense, it seems to be more meaningful and has better statistical properties than the mean distance between the MMSE estimate (obtained by global combining) and the true states. Following this definition let $p^l(k)$, $v^l(k)$, $h^l(k)$, $q^l(k)$ denote the mean position, speed, heading, and acceleration error at scan k with l scans delay.

Discussion of Results

In Fig. 8.3, n^l is displayed for $l = 0, 1, 3, 6$ (solid lines). Figure 8.3a refers to single model MHT (SM-MHT, $r = 1$), while in Fig. 8.3b IMM-MHT ($r = 2$) is used. n^0 is particularly large during the formation flight. The peaks at scan 15 and 82 correspond with the crossings. In the maneuvering phases, the hypothesis numbers

Fig. 8.3 **a** Single Model MHT, **b** IMM MHT ($r = 2$), Performance of Logical Reconstruction

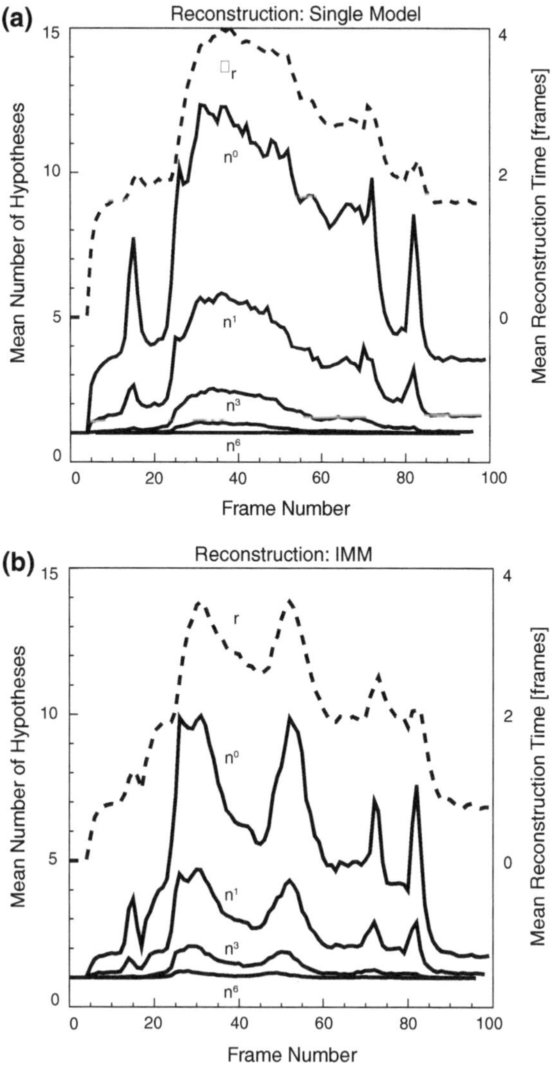

for SM- and IMM-MHT differ less than in between the maneuvers. This is due to the adaptability of IMM-MHT. In both cases a delay of *one* scan already significantly reduces the number of hypotheses required to represent the target states. The dashed curves represent the mean reconstruction time and show an analogous behavior.

In Fig. 8.4 we compared the normal acceleration of one target (solid line) and the x-component of the velocity (dashed line) with the mean model probabilities μ^l for delays $l = 0, 3, 6$. While at present (no delay, dashed line) no clear structure is visible, by a delay of only three scans the structure of the underlying target dynamics can clearly be identified.

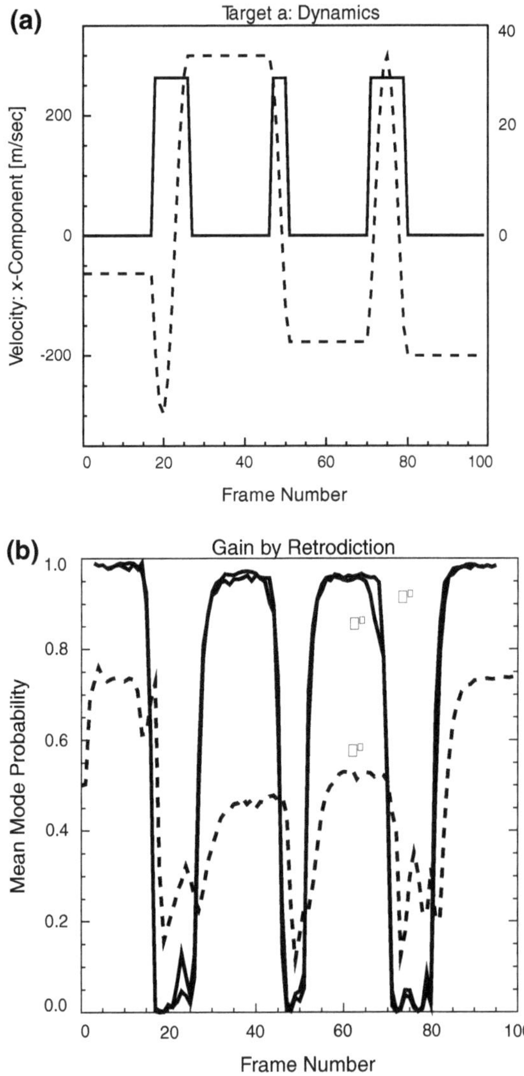

Fig. 8.4 **a** True normal acceleration (*solid*), x-component of the velocity (*dashed*), **b** mean model probabilities μ^l for delays $l = 0, 3, 6$, Identification of the Underlying Target Dynamics

The gain in track accuracy by tolerating delay is displayed in Fig. 8.5. While p^l, h^l, q^l, $l = 1, 3, 6$ denote the track accuracy in position, heading, and acceleration with l scans delay, p^*, h^*, q^* denote the maximum track accuracy that is achievable by applying the backward-time recursion over the previously obtained forward-time filtered estimates. Even by a delay of one scan (i.e. $l = 1$) a significant improvement may be obtained on average. These quantities are close to the results obtained for $\rho_F = 0$ (dotted lines in Fig. 8.5, Table 8.1). Again, the pronounced adaptability of IMM-MHT is evident between the maneuvers where after a small delay accurate estimates are achieved. Shortly after a maneuver has begun the mean heading error for

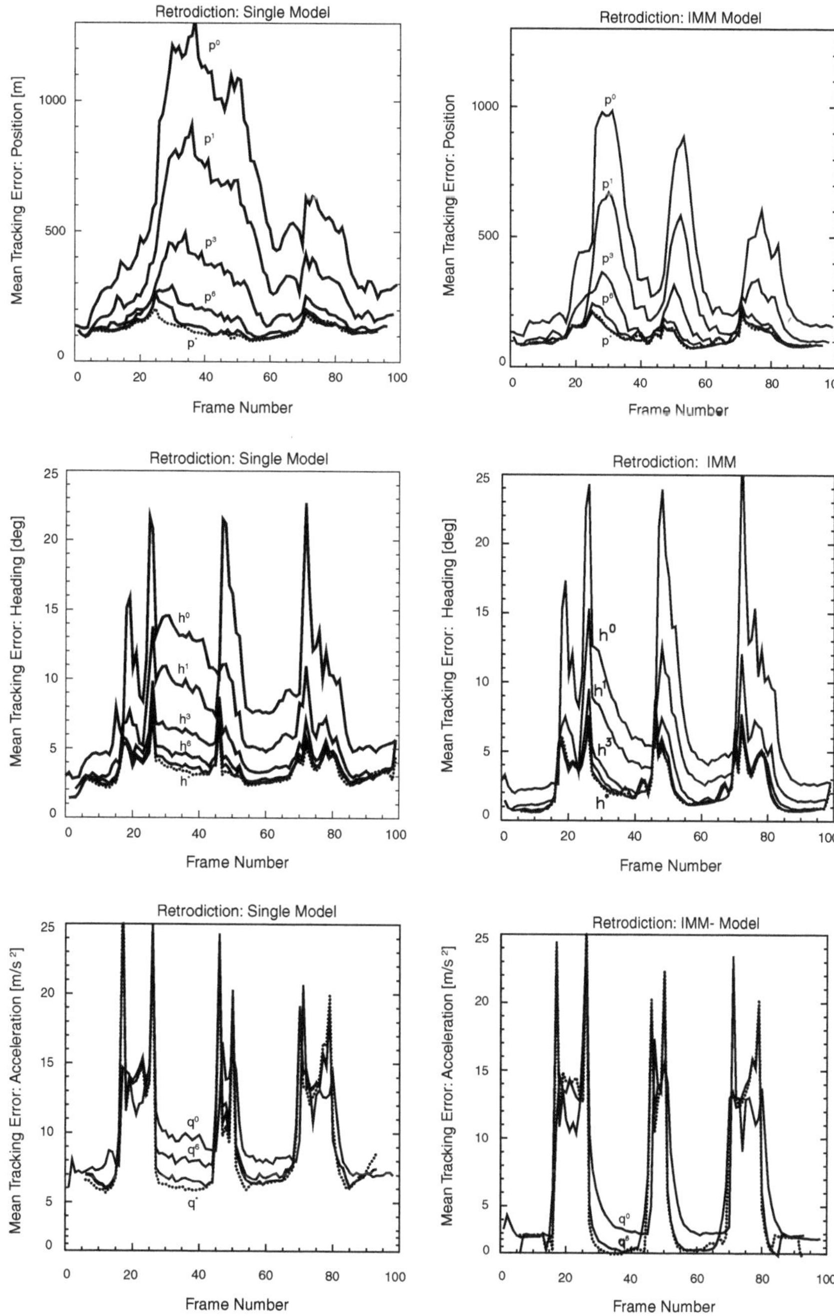

Fig. 8.5 Achievable advantage by tolerating delay

Table 8.1 Quantitative results for $P_D = 0.8$, $\rho_F = 0.002/\text{km}^2$

model	delay [scans]	$\bar{n}^l$	$\bar{p}^l$ [m]	$\bar{v}^l$ [m/s]	$\bar{h}^l$ [deg]	rgp^l	rgv^l	rgh^l
SM-MHT	$l = 0$	9.1	806	105.7	12.1	0.00	0.00	0.00
	$l = 1$	4.1	522	70.2	7.8	0.42	0.43	0.48
$\bar{\tau}_r = 3.0$	$l = 3$	1.8	283	45.7	5.2	0.77	0.73	0.77
$TL = 74$	$l = 6$	1.2	173	34.5	3.9	0.94	0.87	0.88
	*	1.0	141	30.4	3.6	0.98	0.92	0.93
$\rho_F = 0$	*	1	125	29.0	3.5	–	–	–
IMM-MHT	$l = 0$	6.5	506	79.9	9.9	0.44	0.31	0.24
	$l = 1$	3.0	322	47.2	5.9	0.72	0.71	0.68
$\bar{\tau}_r = 2.5$	$l = 3$	1.6	186	29.4	3.6	0.92	0.93	0.94
$TL = 43$	$l = 6$	1.1	141	24.7	3.1	0.98	0.99	0.99
	*	1.0	130	23.7	3.1	1.0	1.0	1.0
$\rho_F = 0$	*	1	130	23.4	3.0	–	–	–

IMM-MHT is larger than for SM-MHT reflecting the switching mechanism between the models.

In Table 8.1 we collected time averages of the above performance measures ($\bar{n}^l = \frac{1}{60}\sum_{k=20}^{79} n^l(k)$, etc.) that enable a more quantitative comparison between SM- and IMM-MHT. TL is the number of lost tracks in 1000 runs. rgp^l, rgv^l, rgh^l denote the relative gain in position, speed, heading, and acceleration by tolerating l scans delay with reference to the maximum accuracy (IMM-MHT, p^*, v^*, h^*). We observed:

1. In general, IMM-MHT filtering is more stable than SM-MHT in that track loss occurs less often and the tracks are more accurate.
2. There is a significant gain by retrodiction for both, SM- and IMM-MHT.
3. The "steady state" accuracy (p^*, v^*, h^*) is comparable for SM- and IMM-MHT, and approximately equal to the results obtained in absence of false returns. Hence in this example, the deterioration of track quality due to clutter and maneuvers may be removed by tolerating delay.
4. Roughly speaking, IMM-MHT without delay provides a track quality comparable to SM-MHT with 1 scan delay, while 1 scan delay (IMM-MHT) refers to 3 scans for SM-MHT.
5. For SM-MHT we obtain approximately 45 % of the maximum gain after a delay of one scan, 75 % after three scans, and 90 % after six scans delay. By using IMM-MHT, however, more than 90 % of the relative gain are achievable after 3 scans delay.

8.1.3 IMM-MHT: Conclusions

1. We have proposed a fixed-interval retrodiction method for Bayesian multiple target tracking that is a straightforward generalization of standard Rauch-Tung-Striebel (RTS) fixed-interval smoothing [8] to applications employing MHT and IMM methods. If the approximations involved preserve the relevant information (pruning/local combining, shortened dynamics histories), retrodiction provides

the best possible knowledge about the targets' history. The quality of the approximations is at the users disposal and depends on his particular needs and computational resources.
2. Besides using the underlying dynamics model, retrodiction only acts on the densities provided by filtering/prediction. It is thus decoupled from the data association/processing task and may be switched off in overload situations without affecting the filter performance. In particular, there is no explicit processing of the sensor returns from multiple scans [9]. Retrodiction might be used for improved pruning (Multiple Frame Data Association [10]).
3. The potential benefits of the method have been demonstrated by an example involving two maneuvering targets in a cluttered environment. At the expense of a relatively small delay (a few scans of the radar), retrodiction may significantly improve the ambiguous MHT output under more serious conditions. The maximum gain achievable by retrodiction is roughly the same for both, single model (SM) and IMM-MHT, and comparable to the results if false returns are absent. Hence, the deterioration of track quality due to clutter and maneuvers may be removed by tolerating some delay. The combined concept, Bayesian MHT supplemented by retrodiction, may thus open a fresh look at target identification/classification or threat assessment in a dense target/dense clutter environment (C^3I applications).
4. Retrodiction algorithms employing multiple dynamics models are superior to those with a single model in that the time delays involved are shorter. IMM without delay provides a track quality approximately comparable to SM with 1 scan delay, while 1 scan delay (IMM) refers to 3 scans for SM. In case of SM, roughly 45 % of the maximum gain is obtained after a delay of one scan, 75 % after three scans, and 90 % after six scans delay. By using IMM, however, more than 90 % of the relative gain is achievable after 3 scans delay and the structure of the underlying target dynamics can be clearly identified.

As a final remark: improvement by retrodiction comes at the price of some (small) delay. In discussing whether delay is admissible in air surveillance, however, it should be noted that for a conventional radar with a scan period T and detection probability $P_D < 1$, an average delay of $\frac{1}{2}T/P_D$ is inevitable.

Further Reading

A detailed discussion of this approach has been published in:

- W. Koch
  Fixed-Interval Retrodiction Approach to Bayesian IMM-MHT for Maneuvering Multiple Targets
  IEEE Transactions on Aerospace and Electronic Systems, Vol. 36, No. 1, p.2-14, January 2000

Abstract
In a Bayesian framework, we propose a hierarchy of suboptimal retrodiction algorithms that generalize Rauch-Tung-Striebel fixed-interval smoothing to MHT applications employing IMM methods (IMM-MHT). As a limiting case we obtain new simple formulae for suboptimal fixed-interval smoothing applied to Markovian switching systems.
Retrodiction techniques provide uniquely interpretable and accurate trajectories from ambiguous MHT output if a certain (small) time delay is tolerated. By a simulated example with two maneuvering targets that operate closely-spaced under relatively hard conditions we demonstrate the potential gain by fixed-interval retrodiction and provide a quantitative idea of the achievable track accuracy and mean time delay involved.

Keywords: Fixed-interval retrodiction, Rauch-Tung-Striebel (RTS) formulae, dense target/dense clutter environment, Probabilistic Data Association (PDA), Multiple Hypothesis Tracking (MHT), Markovian switching systems, Interacting Multiple Model (IMM) algorithms, Generalized Pseudo-Bayesian (GPB) algorithms

8.2 Extended Object Tracking

Due to the increasing resolution capabilities of modern sensors, there is an increasing need for recognizing extended objects as individual units, for initiating extended object tracks, and for extended object track maintenance. Extended objects typically involve a relatively large and often strongly fluctuating number of sensor reports originated by the individual scattering centers that are part of one and the same object. In this context, we usually cannot assume that in subsequent target illuminations the same scattering centers are always responsible for the measurements. The individual sensor reports can therefore no longer be treated in analogy to point object measurements produced by a group of well-separated targets.

Related problems arise if a group of closely-spaced objects is to be tracked. For sensors such as radar, the resolution capability in range is usually much better than in cross-range. As a consequence, two or more targets within the group can be irresolvable, depending on the current sensor-to-target geometry [4, 11, 12]. In addition, little is known about the measurement error of irresolved measurements produced by an unknown number of targets involved. Practically important examples are aircraft formations or ground moving convoys. Under these circumstances, it seems to be reasonable to treat the group as an individual object and to estimate and track its current extension from the sensor data.

In view of the discussion in Sect. 2.1 and 2.3.5, the object extension should be considered as an additional 'internal degree of freedom' characterizing an extended object or a collectively moving object group (cluster) to be tracked. The object extension is thus a part of the object state and has to be estimated jointly with the kinematic properties involved. This paper section discusses a realization of this concept within a Bayesian framework. Temporally changing object extensions are tractable within the proposed framework. An extension increasing along a certain direction, e.g., can indicate that an object is beginning to separate into individual subgroups or parts, which then have to be tracked individually.

8.2.1 Generalized Formalism

In a Bayesian view, a tracking algorithm for an extended object or a collectively moving object group is an updating scheme for $p(\mathbf{x}_k, \mathbf{X}_k|Z^k)$ at each time t_k given the accumulated sensor data $Z^k = \{Z_l, m_l\}_{l=1}^k$ and underlying models describing the object's temporal evolution and the sensor performance. Evidently the joint density

$$p(\mathbf{x}_k, \mathbf{X}_k|Z^k) = p(\mathbf{x}_k|\mathbf{X}_k, Z^k)\; p(\mathbf{X}_k|Z^k) \tag{8.6}$$

can be written as a product of a vector-variate probability density $p(\mathbf{x}_k|\mathbf{X}_k, Z^k)$ and a matrix-variate density $p(\mathbf{X}_k|Z^k)$ [13]. Furthermore, the probabilistic formalism indicates that the density $p(\mathbf{x}_k|\mathbf{X}_k, Z^k)$, describing the kinematical object properties in the product representation in Eq. 8.6, should show an explicit dependency on the current object extension $\mathbf{X}_k$. To the author's knowledge, random matrices were first introduced for describing physical phenomena by Eugene Wigner [14].

Extended target tracking, i.e. the iterative calculation of the joint density $p(\mathbf{x}_k, \mathbf{X}_k|Z^k)$, basically consists of two steps: prediction and filtering. This scheme is completed by the notion of retrodiction.

Prediction

Each update of the joint probability density $p(\mathbf{x}_k, \mathbf{X}_k|Z^k)$ of the extended target state $(\mathbf{x}_k, \mathbf{X}_k)$ is preceded by a *prediction* step,

$$p(\mathbf{x}_{k-1}, \mathbf{X}_{k-1}|Z^{k-1}) \xrightarrow[\text{models}]{\text{evolution}} p(\mathbf{x}_k, \mathbf{X}_k|Z^{k-1}), \tag{8.7}$$

based on the underlying evolution models. More precisely, we interpret the prediction density $p(\mathbf{x}_k, \mathbf{X}_k|Z^{k-1})$ as a marginal density to be calculated by integration:

$$\begin{aligned} p(\mathbf{x}_k, \mathbf{X}_k|Z^{k-1}) = &\int d\mathbf{x}_{k-1} d\mathbf{X}_{k-1} \\ &\times\; p(\mathbf{x}_k, \mathbf{X}_k|\mathbf{x}_{k-1}, \mathbf{X}_{k-1}, Z^{k-1})\; p(\mathbf{x}_{k-1}, \mathbf{X}_{k-1}|Z^{k-1}). \end{aligned} \tag{8.8}$$

For the (joint) transition density in the previous representation,

$$p(\mathbf{x}_k, \mathbf{X}_k|\mathbf{x}_{k-1}, \mathbf{X}_{k-1}, Z^{k-1}) = p(\mathbf{x}_k|\mathbf{X}_k, \mathbf{x}_{k-1}, \mathbf{X}_{k-1}, Z^{k-1})\, p(\mathbf{X}_k|\mathbf{x}_{k-1}, \mathbf{X}_{k-1}, Z^{k-1}), \tag{8.9}$$

we make use of natural Markov-type assumptions for its kinematical part, i.e. $p(\mathbf{x}_k|\mathbf{X}_k, \mathbf{x}_{k-1}, \mathbf{X}_{k-1}, Z^{k-1}) = p(\mathbf{x}_k|\mathbf{X}_k, \mathbf{x}_{k-1})$, and assume that the object's kinematical properties have no impact on the temporal evolution of the object extension and previous measurements if $\mathbf{X}_{k-1}$ is given, i.e.:

$$p(\mathbf{X}_k|\mathbf{x}_{k-1}, \mathbf{X}_{k-1}, Z^{k-1}) = p(\mathbf{X}_k|\mathbf{X}_{k-1}). \tag{8.10}$$

This restriction can be justified in many practical cases. We thus have:

$$p(\mathbf{x}_k, \mathbf{X}_k|\mathbf{x}_{k-1}, \mathbf{X}_{k-1}, Z^{k-1}) = p(\mathbf{x}_k|\mathbf{X}_k, \mathbf{x}_{k-1})\, p(\mathbf{X}_k|\mathbf{X}_{k-1}). \tag{8.11}$$

The probabilistic formalism clearly indicates that the evolution of the object kinematics, described by $p(\mathbf{x}_k|\mathbf{X}_k, \mathbf{x}_{k-1})$, is affected by the current object extension $\mathbf{X}_k$ as well. This dependence cannot be ignored.

With the previous filtering update $p(\mathbf{x}_{k-1}, \mathbf{X}_{k-1}|Z^{k-1})$ we obtain the following prediction formula:

$$p(\mathbf{x}_k, \mathbf{X}_k|Z^{k-1}) = \int d\mathbf{x}_{k-1} d\mathbf{X}_{k-1} \times \underbrace{p(\mathbf{x}_k|\mathbf{X}_k, \mathbf{x}_{k-1})\, p(\mathbf{X}_k|\mathbf{X}_{k-1})}_{\text{evolution model}}\ \underbrace{p(\mathbf{x}_{k-1}|\mathbf{X}_{k-1}, Z^{k-1})\, p(\mathbf{X}_{k-1}|Z^{k-1})}_{\text{previous update}}. \tag{8.12}$$

The transition densities $p(\mathbf{x}_k|\mathbf{X}_k, \mathbf{x}_{k-1})$ and $p(\mathbf{X}_k|\mathbf{X}_{k-1})$ will be specified below using suitable models that describe the temporal evolution of extended or group targets.

Further discussion is much simplified if we additionally assume that the *temporal change* of the object extension has no impact on the prediction of the *kinematical* object properties, i.e. if we are allowed to assume $p(\mathbf{x}_{k-1}|\mathbf{X}_{k-1}, Z^{k-1}) \approx p(\mathbf{x}_{k-1}|\mathbf{X}_k, Z^{k-1})$ or, in other words, to replace $\mathbf{X}_{k-1}$ by $\mathbf{X}_k$. Such an assumption seems to be justified in many practical cases. By this approximation, the predicted density

$$p(\mathbf{x}_k, \mathbf{X}_k|Z^{k-1}) = p(\mathbf{x}_k|\mathbf{X}_k, Z^{k-1})\, p(\mathbf{X}_k|Z^{k-1}) \tag{8.13}$$

is given by two factors to be obtained by independent integrations:

$$p(\mathbf{x}_k|\mathbf{X}_k Z^{k-1}) = \int p(\mathbf{x}_k|\mathbf{X}_k, \mathbf{x}_{k-1})\, p(\mathbf{x}_{k-1}|\mathbf{X}_k, Z^{k-1})\, d\mathbf{x}_{k-1} \tag{8.14}$$

$$p(\mathbf{X}_k|Z^{k-1}) = \int p(\mathbf{X}_k|\mathbf{X}_{k-1})\, p(\mathbf{X}_{k-1}|Z^{k-1})\, d\mathbf{X}_{k-1}. \tag{8.15}$$

Filtering

The prediction is followed by a *filtering* step, in which the current sensor information Z_k at time t_k is to be processed:

$$p(\mathbf{x}_k, \mathbf{X}_k|Z^{k-1}) \xrightarrow[\text{sensor model}]{\text{data:} Z_k, m_k} p(\mathbf{x}_k, \mathbf{X}_k|Z^k). \tag{8.16}$$

More precisely, in the filtering step the sensor-specific likelihood function $p(Z_k, m_k| \mathbf{x}_k, \mathbf{X}_k)$, defined by the current data and the underlying sensor model, is combined with the predicted density by exploiting Bayes' formula [15, 16]:

$$p(\mathbf{x}_k, \mathbf{X}_k|Z^k) = \frac{p(Z_k, m_k|\mathbf{x}_k, \mathbf{X}_k)\, p(\mathbf{x}_k, \mathbf{X}_k|Z^{k-1})}{\int p(Z_k, m_k|\mathbf{x}_k, \mathbf{X}_k)\, p(\mathbf{x}_k, \mathbf{X}_k|Z^{k-1})\, d\mathbf{x}_k d\mathbf{X}_k}. \tag{8.17}$$

8.2.2 Extended Object Prediction

The probability density $p(\mathbf{x}_k, \mathbf{X}_k|Z^k)$ of an extended or group target state is given by Eq. 8.17. The joint densities in this equation can be written as products:

$$\begin{aligned} p(\mathbf{x}_k, \mathbf{X}_k|Z^k) &= p(\mathbf{x}_k|\mathbf{X}_k, Z^k)\, p(\mathbf{X}_k|Z^k) \\ p(\mathbf{x}_k, \mathbf{X}_k|Z^{k-1}) &= p(\mathbf{x}_k|\mathbf{X}_k, Z^{k-1})\, p(\mathbf{X}_k|Z^{k-1}) \\ p(Z_k, m_k|\mathbf{x}_k, \mathbf{X}_k) &= p(Z_k|m_k, \mathbf{x}_k, \mathbf{X}_k)\, p(m_k|\mathbf{x}_k, \mathbf{X}_k). \end{aligned} \tag{8.18}$$

The kinematical state variable $\mathbf{x}_k$ at time t_k is given by $\mathbf{x}_k = (\mathbf{r}_k^\top, \dot{\mathbf{r}}_k^\top, \ddot{\mathbf{r}}_k^\top)^\top$ with the spatial state components $\mathbf{r}_k$. Let the dimension d of the vector $\mathbf{r}_k$ be also the dimension of the $d \times d$ SPD matrix $\mathbf{X}_k$ that describe the current ellipsoidal object extension (SPD: symmetrical and positively definite). $\dot{\mathbf{r}}_k, \ddot{\mathbf{r}}_k$ denote the corresponding velocity and acceleration. The dimension of the kinematical state vector $\mathbf{x}_k$ is thus $s \times d$, where $s - 1$ describes up to which derivative the object kinematics is modeled. Here we have $s = 3$.

Extended Object Evolution

The temporal evolution of an extended or collective object is modeled as usual in Kalman filtering theory: $\mathbf{x}_k = \boldsymbol{\Phi}_{k|k-1}\mathbf{x}_{k-1} + \mathbf{v}_k$, $p(\mathbf{v}_k) = \mathcal{N}(\mathbf{v}_k;\, \mathbf{0},\, \boldsymbol{\Delta}_{k|k-1})$. Using the Kronecker product [13], the evolution matrix $\boldsymbol{\Phi}_{k|k-1}$ can be written as:

$$\boldsymbol{\Phi}_{k|k-1} = \mathbf{F}_{k|k-1} \otimes \mathbf{1}_d, \tag{8.19}$$

where the $s \times s$ matrix $\mathbf{F}_{k|k-1}$ is given for example by van Keuk's model (Sect. 2.2.1). The use of Kronecker products will prove to be very convenient in the subsequent calculations. For the dynamics noise covariance $\boldsymbol{\Delta}_{k|k-1}$, we postulate the following structure:

$$\boldsymbol{\Delta}_{k|k-1} = \mathbf{D}_{k|k-1} \otimes \mathbf{X}_k. \tag{8.20}$$

Model parameters describing the underlying dynamics are part of a $s \times s$ matrix $\mathbf{D}_{k|k-1}$, as given by van Keuk's model, for example (Sect. 2.2.1). The $s \times s$ matrices $\mathbf{F}_{k|k-1}, \mathbf{D}_{k|k-1}$ also appear in this form in the 1D tracking problem. The system noise is thus a band limited Gaussian acceleration noise process with a covariance proportional to the extension matrix $\mathbf{X}_k$. This has the effect of directing the acceleration of the group (or object) centroid along the direction of the major axis of the ellipse.

The assumption of a dynamics covariance matrix $\boldsymbol{\Delta}_{k|k-1}$ depending on the current object extension $\mathbf{X}_k$, which is a consequence of the probability formalism, needs a discussion with more physical arguments:

1. The collective character of a group motion is the more pronounced the smaller the group is. The dynamical behavior of a smaller group is thus to a larger extent deterministic in nature ('maneuvering becomes dangerous').
2. For a group dissolving into subgroups, i.e. if its extension is increasing, the knowledge of its dynamical behavior decreases, and the motion of the group is not easily predictable, being expressed by the increasing dynamics noise covariance.
3. In addition, large extended or group objects will produce so many sensor measurements that the prediction part of the tracking process, i.e. exploitation of information on the object evolution, seems to be negligible if compared to the gain obtained in the filtering step.
4. In case of extended objects like submarines or ground moving convoys, which show a clear orientation, the proposed dynamics model provides a natural description of their actual movement along the major axes of the extension ellipse.

Besides these more or less physically motivated reasons, an important formal argument exists in favor of the model: A dynamics model of the proposed form implies a formal structure of the densities $p(\mathbf{x}_k, \mathbf{X}_k|Z^k)$, which enables a rigorous application of the Bayesian formalism under certain assumptions.

Structure of the Predicted Density

According to Eq. 8.18, the kinematics can be discussed separately from the extension estimation in the tracking process. Let us assume that the density of the kinematical state variable $p(\mathbf{x}_{k-1}|\mathbf{X}_k, Z^{k-1})$ after filtering at time t_{k-1} is a Gaussian with the following special structure:

$$p(\mathbf{x}_{k-1}|\mathbf{X}_k, Z^{k-1}) = \mathcal{N}(\mathbf{x}_{k-1};\ \mathbf{x}_{k-1|k-1},\ \mathbf{P}_{k-1|k-1} \otimes \mathbf{X}_k). \tag{8.21}$$

Then the previous evolution model guarantees that this structure is preserved by the prediction process (Eq. 8.14):

$$p(\mathbf{x}_k|\mathbf{X}_k, Z^{k-1}) = \int \mathcal{N}(\mathbf{x}_k;\ (\mathbf{F}_{k|k-1} \otimes \mathbf{1}_d)\mathbf{x}_{k-1},\ \mathbf{D}_{k|k-1} \otimes \mathbf{X}_k)$$
$$\times\ \mathcal{N}(\mathbf{x}_{k-1};\ \mathbf{x}_{k-1|k-1},\ \mathbf{P}_{k-1|k-1} \otimes \mathbf{X}_k)\, d\mathbf{x}_{k-1} \tag{8.22}$$
$$= \mathcal{N}(\mathbf{x}_k;\ \mathbf{x}_{k|k-1},\ \mathbf{P}_{k|k-1} \otimes \mathbf{X}_k) \tag{8.23}$$

according to the usual rules for Kronecker products (see Appendix A. 4) with $\mathbf{x}_{k|k-1}$ and $\mathbf{P}_{k|k-1}$ given by:

$$\mathbf{x}_{k|k-1} = (\mathbf{F}_{k|k-1} \otimes \mathbf{1}_d)\mathbf{x}_{k-1|k-1} \tag{8.24}$$
$$\mathbf{P}_{k|k-1} = \mathbf{F}_{k|k-1}\mathbf{P}_{k-1|k-1}\mathbf{F}_{k|k-1}^{\top} + \mathbf{D}_{k|k-1} \tag{8.25}$$

in close analogy to standard Kalman filtering.

Moreover, let us assume that the densities of the extension state variable $p(\mathbf{X}_{k-1}| Z^{k-1})$ are given by Inverted Wishart densities [13] defined up to a factor independent of $\mathbf{X}_{k-1}$ by (see Appendix A. 11):

$$p(\mathbf{X}_{k-1}|Z^{k-1}) = \mathcal{IW}\big(\mathbf{X}_{k-1};\ \nu_{k-1|k-1},\ \mathbf{X}_{k-1|k-1}\big) \tag{8.26}$$
$$\propto |\mathbf{X}_{k-1}|^{-\frac{1}{2}\nu_{k-1|k-1}}\ \mathrm{etr}\Big[-\tfrac{1}{2}\mathbf{X}_{k-1|k-1}\mathbf{X}_{k-1}^{-1}\Big]. \tag{8.27}$$

d is the dimension of the measurement vectors $\mathbf{z}_k^j$ and etr[$\mathbf{A}$] an abbreviation for exp[tr$\mathbf{A}$] with tr$\mathbf{A}$ denoting the trace of a matrix $\mathbf{A}$. The expectation of $\mathbf{X}_{k-1}$ is given by $\mathbb{E}[\mathbf{X}_{k-1}|Z^{k-1}] = \frac{\mathbf{X}_{k-1|k-1}}{\nu_{k-1|k-1}-2d-2}$.

In the prediction step, the parameters $\nu_{k|k-1}, \mathbf{X}_{k|k-1}$ defining $p(\mathbf{X}_k|Z^{k-1})$ have to be calculated from $\nu_{k-1|k-1}, \mathbf{X}_{k-1|k-1}$ available after the previous filtering step according to appropriate modeling assumptions. In a first heuristic approach, we postulate that the expectation of the predicted density shall be equal to the expectation of the previous filtering step; i.e.: $\frac{\mathbf{X}_{k|k-1}}{\nu_{k|k-1}-2d-2} = \frac{\mathbf{X}_{k-1|k-1}}{\nu_{k-1|k-1}-2d-2}$. The degrees of freedom of an inverse Wishart density are related to the 'precision' of the corresponding expectation. The 'precision' of predictions, however, will decrease with increasing update intervals $\Delta t_k = t_k - t_{k-1}$. With a temporal decay constant τ as an additional modeling parameter, the following prediction update equations seem to be plausible:

$$\nu_{k|k-1} = \mathrm{e}^{-\Delta t_k/\tau}\ \nu_{k-1|k-1} \tag{8.28}$$
$$\mathbf{X}_{k|k-1} = \frac{\mathrm{e}^{-\Delta t_k/\tau}\,\nu_{k-1|k-1}-d-1}{\nu_{k-1|k-1}-d-1}\ \mathbf{X}_{k-1|k-1}. \tag{8.29}$$

$\tau = \infty$ represents a static object or group extension.

8.2.3 Extended Object Filtering

In the case of extended or group targets, the significance of a single measurement is obviously dominated by the underlying object extension. The sensor-specific measurement error that describe the precision by which a given scattering center is currently measured is the more unimportant, the larger the actual extension of the object is compared to the measurement error. The individual measurements must therefore rather be interpreted as measurements of the centroid of the extended or collective object, since it is unimportant for the extended object tracking task which of the varying scattering centers was actually responsible for the measurement.

We thus interpret each individual measurement produced by an extended object as a measurement of the object centroid with a corresponding 'measurement error' that is proportional to the object extension $\mathbf{X}_k$ to be estimated. By means of this 'measurement error', however, the object extension $\mathbf{X}_k$ becomes explicitly part of the likelihood function $p(Z_k, m_k|\mathbf{x}_k, \mathbf{X}_k)$, which describes what the measured quantities Z_k, m_k can say about the state variables $\mathbf{x}_k$ and $\mathbf{X}_k$. As a consequence of this interpretation, the object extension $\mathbf{X}_k$ can also be estimated by exploiting the sensor data (besides the kinematical state vector $\mathbf{x}_k$).

By using the Kronecker product, we also assume that the measurement matrix has the following special structure:

$$(h_k^1 \mathbf{1}_d, h_k^2 \mathbf{1}_d, h_k^3 \mathbf{1}_d) = \mathbf{H}_k \otimes \mathbf{1}_d. \tag{8.30}$$

With $\mathbf{H}_k = (1, 0, 0)$, e.g., scenarios with range and azimuth measurements are accessible after transforming them into Cartesian coordinates. According to the previous considerations, the corresponding measurement error covariance is given by the extension matrix $\mathbf{X}_k$ to be estimated.

Likelihood Function

In order to exploit Bayes' formula Eq. 8.17, a likelihood function factorized according to Eq. 8.18 needs to be defined. For the sake of simplicity, let us exclude false or unwanted measurements at present. In a first approximation, the number m_k of measurements in Z_k is assumed to be independent of the state variables $\mathbf{x}_k$, $\mathbf{X}_k$; i.e. $p(m_k|\mathbf{x}_k, \mathbf{X}_k)$ is assumed to be a constant. According to the discussion in Sect. 2.3.5, the joint density $p(Z_k|m_k, \mathbf{x}_k, \mathbf{X}_k)$ can be factorized:

$$p(Z_k|m_k, \mathbf{x}_k, \mathbf{X}_k) \propto \mathcal{N}\big(\mathbf{z}_k;\ (\mathbf{H}_k \otimes \mathbf{1}_d)\mathbf{x}_k,\ \tfrac{\mathbf{X}_k}{m_k}\big)\ \mathcal{LW}\big(\mathbf{Z}_k;\ m_k - 1,\ \mathbf{X}_k\big). \tag{8.31}$$

with a centroid measurement $\mathbf{z}_k$, a corresponding scattering matrix $\mathbf{Z}_k$, and a Wishart density in $\mathbf{Z}_k$ with $m_k - 1$ degrees of freedom.

Structure After Filtering

With these preliminaries, it is possible to exploit the Bayes formula Eq. 8.17. To this end, we have to calculate the product:

$$\begin{aligned} p(Z_k|m_k, \mathbf{x}_k, \mathbf{X}_k)\ p(\mathbf{x}_k, \mathbf{X}_k|\mathbf{Z}^{k-1}) \propto\ & \mathcal{N}\big(\mathbf{z}_k;\ (\mathbf{H}_k \otimes \mathbf{1}_d)\mathbf{x}_k,\ \tfrac{\mathbf{X}_k}{m_k}\big) \\ & \times\ \mathcal{N}\big(\mathbf{x}_k;\ \mathbf{x}_{k|k-1},\ \mathbf{P}_{k|k-1} \otimes \mathbf{X}_k\big) \\ & \times\ \mathcal{LW}\big(\mathbf{Z}_k;\ m_k - 1,\ \mathbf{X}_k\big)\ \mathcal{IW}\big(\mathbf{X}_k;\ \nu_{k|k-1},\ \mathbf{X}_{k|k-1}\big). \end{aligned} \tag{8.32}$$

By standard calculations (product formula for Gaussians and properties of Kronecker products, see Appendix A. 5, A. 4), the product of the two Gaussians in the previous equation yields:

$$\begin{aligned}&\mathcal{N}\big(\mathbf{z}_k;\ (\mathbf{H}_k \otimes \mathbf{1}_d)\mathbf{x}_k,\ \tfrac{\mathbf{X}_k}{n_k}\big)\ \mathcal{N}\big(\mathbf{x}_k;\ \mathbf{x}_{k|k-1},\ \mathbf{P}_{k|k-1} \otimes \mathbf{X}_k\big)\\ &= \mathcal{N}\big(\mathbf{z}_k;\ (\mathbf{H}_k \otimes \mathbf{1}_d)\mathbf{x}_{k|k-1},\ S_{k|k-1}\mathbf{X}_k\big)\ \mathcal{N}\big(\mathbf{x}_k;\ \mathbf{x}_{k|k},\ \mathbf{P}_{k|k} \otimes \mathbf{X}_k\big)\end{aligned} \tag{8.33}$$

where the quantities $\mathbf{x}_{k|k}$ and $\mathbf{P}_{k|k}$ are given by

$$\mathbf{x}_{k|k} = \mathbf{x}_{k|k-1} + (\mathbf{W}_{k|k-1} \otimes \mathbf{1}_d)(\mathbf{z}_k - (\mathbf{H}_k \otimes \mathbf{1}_d)\mathbf{x}_{k|k-1}) \tag{8.34}$$

$$\mathbf{P}_{k|k} = \mathbf{P}_{k|k-1} - \mathbf{W}_{k|k-1} S_{k|k-1} \mathbf{W}_{k|k-1}^\top \tag{8.35}$$

with a scalar *innovation factor* and a gain vector defined by

$$S_{k|k-1} = \mathbf{H}_k \mathbf{P}_{k|k-1} \mathbf{H}_k^\top + \tfrac{1}{n_k} \tag{8.36}$$

$$\mathbf{W}_{k|k-1} = \mathbf{P}_{k|k-1} \mathbf{H}_k^\top S_{k|k-1}^{-1}. \tag{8.37}$$

The first factor on the right side in Eq. 8.31 does not depend on the kinematical state variable $\mathbf{x}_k$. It can be rewritten as

$$\mathcal{N}\big(\mathbf{z}_k;\ (\mathbf{H}_k \otimes \mathbf{1}_d)\mathbf{x}_{k|k-1},\ S_{k|k-1}\mathbf{X}_k\big) \propto |\mathbf{X}_k|^{-\frac{1}{2}}\ \mathrm{etr}\Big[-\tfrac{1}{2}\mathbf{N}_{k|k-1}\mathbf{X}_k^{-1}\Big] \tag{8.38}$$

up to a factor independent of the state variables and with an *innovation matrix* $\mathbf{N}_{k|k-1}$ defined by

$$\mathbf{N}_{k|k-1} = S_{k|k-1}^{-1}\big(\mathbf{z}_k - (\mathbf{H}_k \otimes \mathbf{1}_d)\mathbf{x}_{k|k-1}\big)\ \big(\mathbf{z}_k - (\mathbf{H}_k \otimes \mathbf{1}_d)\mathbf{x}_{k|k-1}\big)^\top. \tag{8.39}$$

The remaining two factors on the right side of Eq. 8.32 yield:

$$\begin{aligned}&\mathcal{LW}\big(\mathbf{Z}_k;\ n_k - 1,\ \mathbf{X}_k\big)\,\mathcal{IW}\big(\mathbf{X}_k;\ \nu_{k|k-1},\ \mathbf{X}_{k|k-1}\big)\,|\mathbf{X}_k|^{-\frac{1}{2}}\ \mathrm{etr}\Big[-\tfrac{1}{2}\mathbf{N}_{k|k-1}\mathbf{X}_k^{-1}\Big]\\ &\qquad \propto \mathcal{IW}\big(\mathbf{X}_k;\ \nu_{k|k},\ \mathbf{X}_{k|k}\big)\end{aligned} \tag{8.40}$$

with the simple update equations:

$$\mathbf{X}_{k|k} = \mathbf{X}_{k|k-1} + \mathbf{N}_{k|k-1} + \mathbf{Z}_k \tag{8.41}$$

$$\nu_{k|k} = \nu_{k|k-1} + n_k. \tag{8.42}$$

Joint Density after Filtering

The probability density function of the joint state $(\mathbf{x}_k, \mathbf{X}_k)$ after processing the current sensor data Z_k at time t_k is thus given by:

$$p(\mathbf{x}_k, \mathbf{X}_k|Z^k) = \mathcal{N}(\mathbf{x}_k;\ \mathbf{x}_{k|k}, \mathbf{P}_{k|k} \otimes \mathbf{X}_k)\ \mathcal{IW}\big(\mathbf{X}_k;\ \nu_{k|k}, \mathbf{X}_{k|k}\big). \tag{8.43}$$

Important Remark: By means of the *innovation matrix* $\mathbf{N}_{k|k-1}$, it is possible to estimate an unknown measurement error covariance even in the case of point source targets or the extension of a completely irresolved target group, i.e. for $m_k = 1$.

8.2.4 Extended Object Kinematics

In many practical applications, we are interested in estimates of the kinematic state variables only, i.e. on the marginal density $p(\mathbf{x}_k|Z^k)$ obtained by integrating the joint density $p(\mathbf{x}_k, \mathbf{X}_k|Z^k)$ over the random matrices $\mathbf{X}_k$:

$$p(\mathbf{x}_k|Z^k) = \int p(\mathbf{x}_k, \mathbf{X}_k|Z^k)\, d\mathbf{X}_k \tag{8.44}$$

$$= \int \mathcal{N}(\mathbf{x}_k;\ \mathbf{x}_{k|k}, \mathbf{P}_{k|k} \otimes \mathbf{X}_k)\ \mathcal{IW}\big(\mathbf{X}_k;\ \nu_{k|k}, \mathbf{X}_{k|k}\big)\, d\mathbf{X}_k. \tag{8.45}$$

By lengthy but elementary algebraic calculations using the facts listed in Appendix (see also [17]) the integrand can be transformed into the following product:

$$\mathcal{N}\big(\mathbf{x}_k;\ \mathbf{x}_{k|k},\big)\mathbf{P}_{k|k} \otimes \mathbf{X}_k\ \mathcal{IW}\big(\mathbf{X}_k;\ \nu_{k|k}, \mathbf{X}_{k|k}\big) \propto$$

$$|\mathbf{Y}_k(\mathbf{x}_k)|^{-\frac{(\nu_{k|k}+s-sd)+sd}{2}}\ \mathcal{IW}\big(\mathbf{X}_k;\ \nu_{k|k} + s,\ \mathbf{Y}_k(\mathbf{x}_k)\mathbf{X}_{k|k}\big) \tag{8.46}$$

with a matrix $\mathbf{Y}_k = \mathbf{Y}_k(\mathbf{x}_k)$ depending on the kinematical state variable $\mathbf{x}_k$ whose determinant is given by

$$|\mathbf{Y}_k| = 1 + (\mathbf{x}_k - \mathbf{x}_{k|k})^\top\ (\mathbf{P}_{k|k}^{-1} \otimes \mathbf{X}_{k|k}^{-1})(\mathbf{x}_k - \mathbf{x}_{k|k}). \tag{8.47}$$

With this representation of the integrand, integration over the random matrix $\mathbf{X}_k$ is trivial. We ultimately find that the marginal density with respect to the kinematical state variable $\mathbf{x}_k$ is given by a multivariate version of the Student density with $\nu_{k|k}$ degrees of freedom (see Appendix A. 11):

$$p(\mathbf{x}_k|Z^k) = \mathcal{T}\big(\mathbf{x}_k;\ \nu_{k|k} + s - sd,\ \mathbf{x}_{k|k}, \mathbf{P}_{k|k} \otimes \mathbf{X}_{k|k}\big). \tag{8.48}$$

By exploiting the multivariate t-density a 'gating' can be constructed that is simply a version of the Hotelling-t^2-test.

It is immediately clear that the marginalized prediction and retrodiction densities are also given by Student densities: $p(\mathbf{x}_l|Z^{l-1}) = \mathcal{T}\big(\mathbf{x}_l;\ \nu_{l|l-1} + s - sd,\ \mathbf{x}_{l|l-1}, \mathbf{P}_{l|l-1} \otimes \mathbf{X}_{l|l-1}\big)$, $p(\mathbf{x}_l|Z^k) = \mathcal{T}\big(\mathbf{x}_l;\ \nu_{l|k} + s - sd,\ \mathbf{x}_{l|k}, \mathbf{P}_{l|k} \otimes \mathbf{X}_{l|k}\big)$.

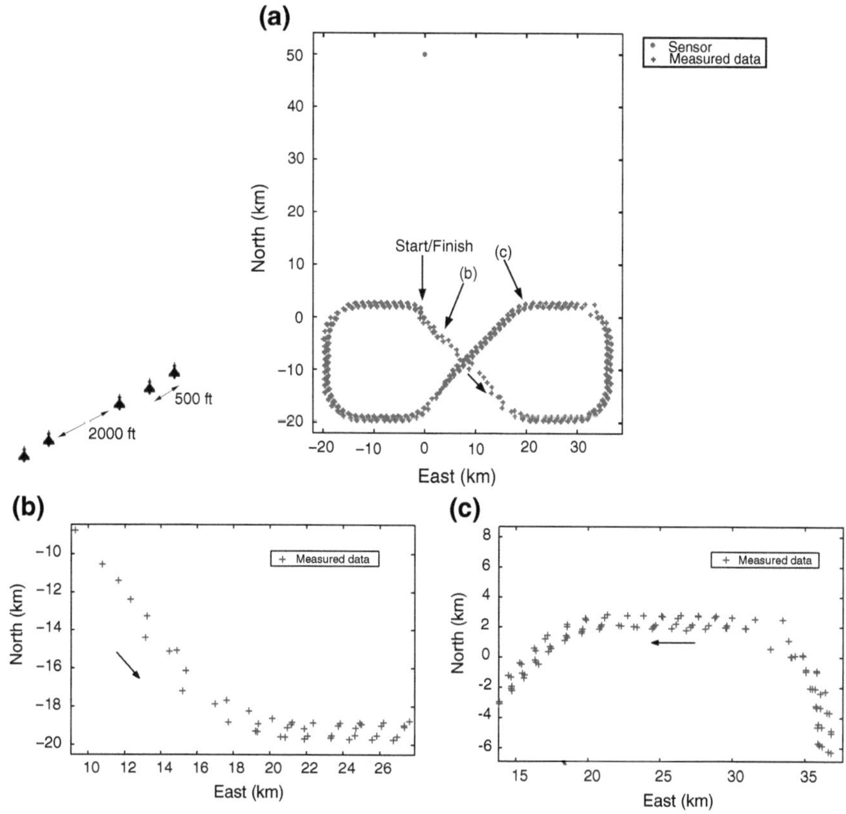

Fig. 8.6 Measurements of a Partly Unresolved Formation (resolution: 50 m, 1.0°, measurement error: 10 m, 0.1°): Accumulated radar data (**a**), Details (**b, c**)

8.2.5 Selected Simulation Results

For the sake of simplicity, aircraft trajectories are simulated in a plane and partitioned into straight and circular segments where each aircraft is moving with a constant tangential speed as shown in Fig. 8.6. In an echelon formation consisting of five aircraft, the leading aircraft is responsible for navigating, while the other aircraft try to preserve their relative position to the leading aircraft. The underlying radar sensor has a finite resolution capability (range resolution: 50 m, azimuth resolution: 1.0°). The corresponding measurement error standard deviations for resolvable objects are 10 m and 0.1°, respectively. The orientation of the aircraft formation varies as it moves around the trajectory. The update interval is 5 s. For the parameters of the Van-Keuk-evolution model, we chose $\Sigma = 1$g, $\theta = 40$s. The normal acceleration during the maneuvers is $1g$, the speed is 250 m/s. The formation starts at the origin of the coordinate system.

Simulating a Partly Resolvable Formation

For the simulation of radar measurements, the corresponding measurements errors and the sensor resolution have to be taken into account. The generation of false returns is not considered here. For a group of two targets at positions (r_1, φ_1), (r_2, φ_2) in polar coordinates with respect to the sensor position, the probability of being unresolved, P_u, can be modeled as in Sect. 7.1:

$$P_u(\Delta r, \Delta\varphi) = \mathrm{e}^{-\frac{1}{2}(\Delta r/\alpha_r)^2}\, \mathrm{e}^{-\frac{1}{2}(\Delta\varphi/\alpha_\varphi)^2} \tag{8.49}$$

with $\Delta r = r_2 - r_1$, $\Delta\varphi = \varphi_2 - \varphi_1$, where the sensor parameters α_r, α_φ characterize the radar's resolution capability in range and azimuth, respectively. According to this probability and for pairs of aircraft, it can be simulated whether an unresolved measurement occurs or not. In case of a resolution conflict the pair is replaced by a single unresolvable object at the centroid position. For large formations with more than two targets, a list is created containing all possible pairs of aircraft. A pair of this list is selected at random according to P_u and merged. In this case, one of the aircraft is to be removed the list, which thus has to be recalculated. If no resolution conflict occurs according to the probability $1 - P_u$, the pair is removed from the list. The previous reasoning is repeated for the remaining pairs. If the list is empty, the algorithm terminates.

We finally have to consider the effect of successive mergings on the simulated measurement errors of unresolvable objects. To this end, we assume that an unresolved measurement error resulting from m aircraft is to be simulated according to $\sigma^u_{r,\varphi} = m\sigma_{r,\varphi}$ where $\sigma_{r,\varphi}$ denote the standard deviations of resolvable range and azimuth measurements, respectively. It is reasonable to delimit the growth of the measurement error by the sensor resolution: $\sigma^u_{r,\varphi} \leq \alpha_{r,\varphi}$. In the same manner, missing detections can be simulated. We here assumed a detection probability $P^u_D = 1$ for unresolvable aircraft and $P^r_D = 0.9$ otherwise.

Impact of the Resolution Parameters

Figure 8.6 displays the radar data simulated according to these assumptions. The details in Fig. 8.6b, c clearly reveal the impact of resolution phenomena and make it obvious that they depend heavily on the current sensor-to-target geometry. The discussed phenomena make it clearly evident that even a very regular target formation is very similar in appearance to an extended object producing a highly fluctuating number of measurements. There is no reasonable hope to be able to track the single components of such a formation individually.

Discussion of Results

As before in the case of a totally irresolvable formation, in Fig. 8.7 the root mean squared errors of the position estimates of the extended target filter are compared with the corresponding results produced by standard Kalman filtering. As the measurement error in the Kalman filter, we used the scattering matrix calculated from the true target positions within the formation and processed averaged measurements. The extended target filter shows significantly smaller estimation errors.

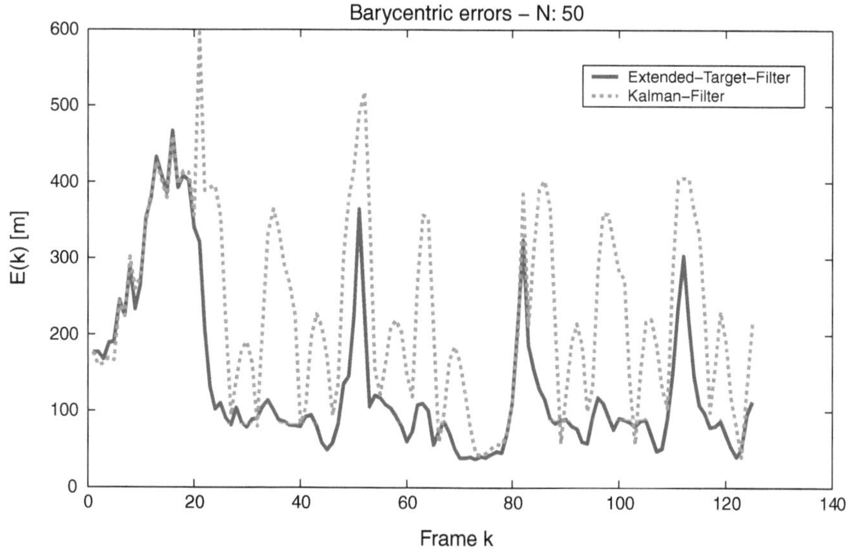

Fig. 8.7 Position error: extended target filter versus standard Kalman filter

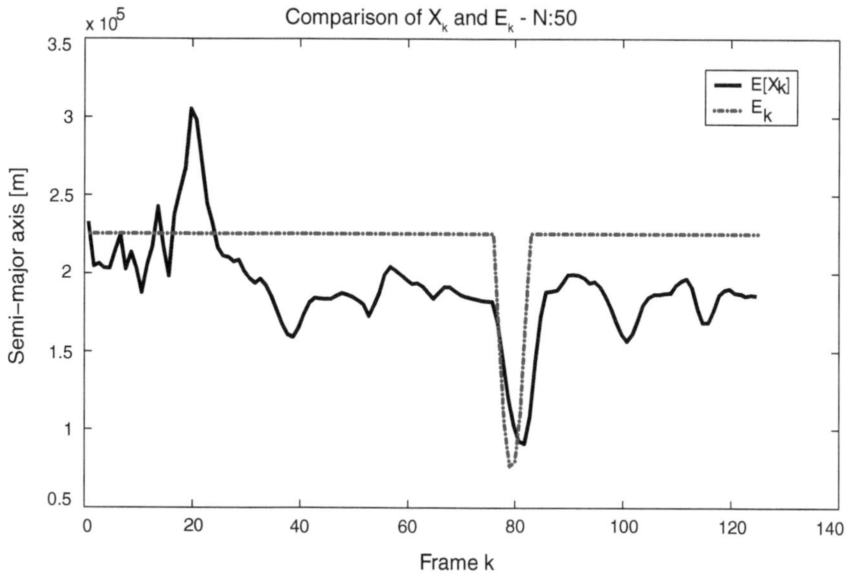

Fig. 8.8 Mean major semi-axes of $\mathbb{E}[\mathbf{X}_k|\mathcal{Z}^k]$versus$\mathbf{X}_k$

In Fig. 8.8, the estimated major semi-axes are compared with the major semi-axes of the scattering matrix of the true target positions. The concordance seems to be

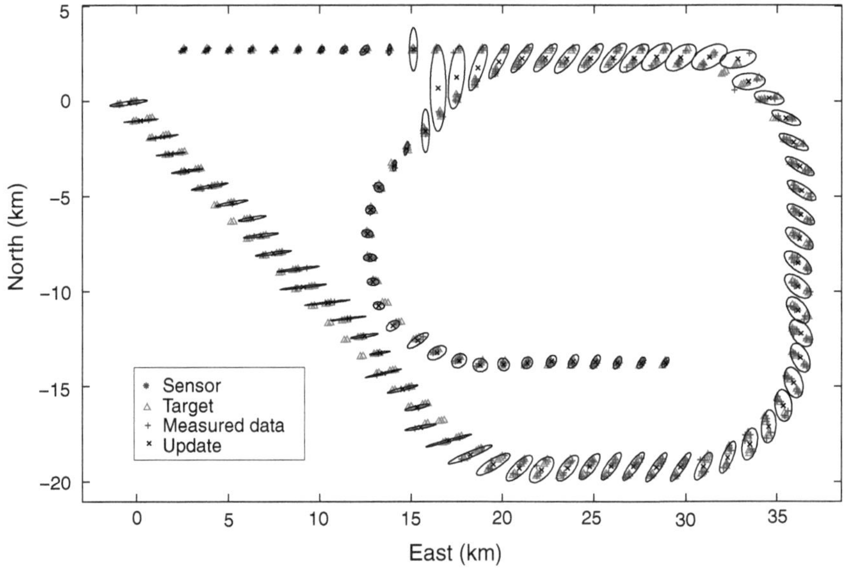

Fig. 8.9 Echelon formation: split-off maneuver

fairly good. The peak in the middle of the time axis is due to the reorganization of the formation.

Figures 8.9 and 8.10 a 'split-off' maneuver is clearly indicated by the increasing eccentricity of the estimated extension ellipse. As soon as the eccentricity exceeds a certain threshold, two extended object tracks are initiated and the sub-groups are tracked separately. The proposed filter thus provides a criterion of when a single extended object track has to be split into two extended object tracks. An analogous mechanism is possible in the case of a larger formation being created by merging two or more converging sub-groups.

8.2.6 Summary of Results

The essential theoretical result of this paper seems to be the insight that the Bayesian formalism can be applied to extended objects or collectively moving target clusters with approximations to be justified in many applications. Basically, the application of the Bayesian formalism relies on closure properties of matrix-variate Wishart and Inverted Wishart densities under multiplication.

In view of practical applications, the following aspects seem to be of particular relevance:

1. There exists a natural extension of the standard Kalman filter equations to objects whose spatial extension is approximately described by ellipsoids.

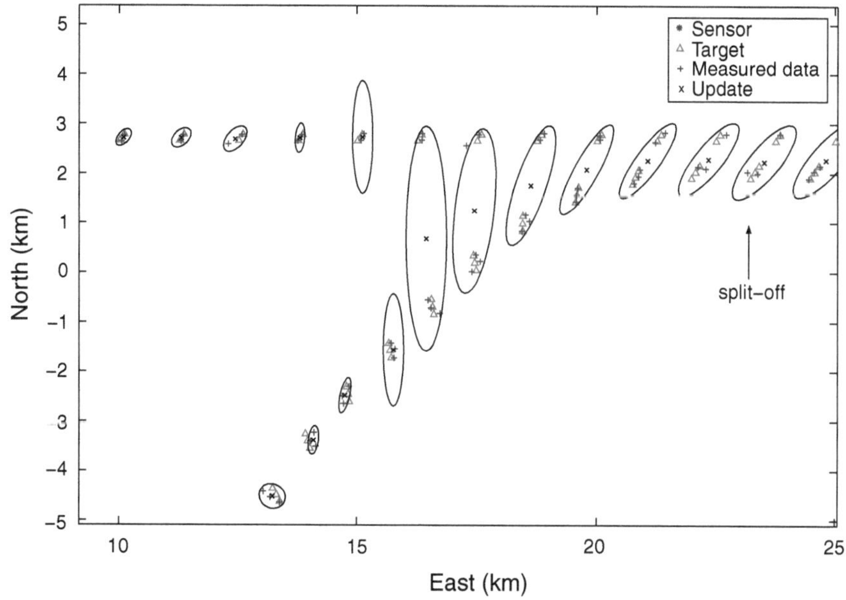

Fig. 8.10 Split-off maneuver (detail)

2. The object extension can be modeled by symmetrical, positive definite random matrices, whose statistical properties are described by well-known matrix-variate probability densities [13].
3. Due to the mild character of the approximations used, a representation of the probability densities involved by particle filtering techniques such as proposed in [15, 18], does not seem to be necessary. The densities are characterized by a finite parameter set.
4. Information on the objects' kinematic properties is represented by vector-variate Student densities. 'Gating', i.e. exclusion of unwanted measirements, is provided by a Hotelling test.
5. Tracking of point source targets with an unknown measurement error is a limiting case of the proposed method (e.g. tracking of an irresolvable formation).
6. With respect to the kinematical properties, the achievable filter performance is only slightly different from Kalman filtering with a known measurement error covariance matrix.
7. The estimated measurement error covariance matrix corresponds to the true measurement error covariance matrix (simulated) relatively well. This is an interesting side result, considering the small number of data in the case of a totally unresolved group.
8. The proposed filter can successfully be applied to target formations, which are only partly resolvable depending on the underlying sensor-to-target geometry.

9. "Split-off" maneuvers, indicating that an object is beginning to separate into individual subgroups or parts, can be detected by analyzing the extension ellipsoid (e.g. by designing a test based on its eccentricity).

In principle, the proposed approximate Bayesian method for dealing with extended objects or collectively moving target clusters can be embedded into multiple-object, multiple-hypothesis tracking techniques and can also be combined with context information (e.g. road-map assisted convoy tracking). This opens an interesting field for further research.

Key Publication

A detailed discussion of this approach has been published in:

- W. Koch Bayesian Approach to Extended Object and Cluster Tracking using Random Matrices. *IEEE Transactions on Aerospace and Electronic Systems, Vol. 44, Nr. 3, p. 1042-1059, July 2008.*

  Abstract In algorithms for tracking and sensor data fusion, the targets to be observed are usually considered as point source objects; i.e. compared to the sensor resolution, their extension is neglected. Due to the increasing resolution capabilities of modern sensors, however, this assumption is often no longer valid, since different scattering centers of an object can cause distinct detections when passing the signal processing chain. Examples of extended targets are found in short-range applications (littoral surveillance, autonomous weapons, or robotics). A collectively moving target group can also be considered as an extended target. This point of view is the more appropriate, the smaller the mutual distances between the individual targets are. Due to the resulting data association and resolution conflicts, any attempt to track the individual objects within the group seems to be no longer reasonable.
  With simulated sensor data produced by a partly unresolvable aircraft formation, the addressed phenomena are illustrated, and an approximate Bayesian solution to the resulting tracking problem is proposed. Ellipsoidal object extensions are modeled by random matrices, which are treated as additional state variables to be estimated or tracked. We expect that the resulting tracking algorithms are also relevant for tracking large, collectively moving target swarms.

  Keywords: Target tracking, Extended targets, Group targets, Target clusters, Sensor resolution, Random matrices, Matrix-variate analysis

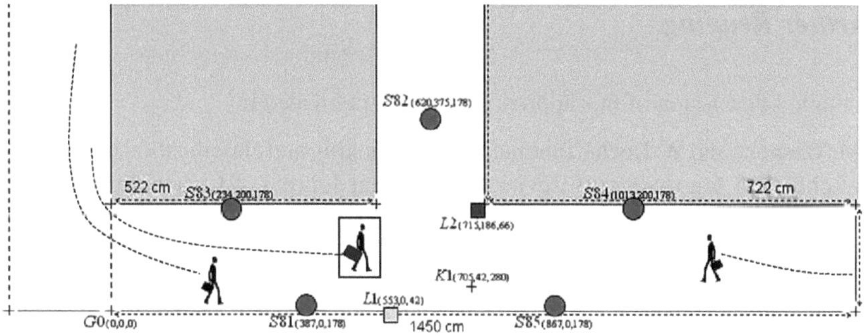

Fig. 8.11 Experimental corridor with 5 chemical and 2 tracking sensors

8.3 Tracking and Classification

Emerging applications in the domain of homeland security require non-standard information sources such as chemical sensors for detecting explosives or other hazardous materials. While in the biosphere "noses" belong to the oldest of senses, their technical equivalents are still in the beginning of their development.

Nevertheless, there exists a fundamental commonality between natural "noses" and chemical sensors in view of their limited space-time resolution: we may be able to detect a specific smell or chemical signature, but this typicaUy occurs with a time delay.

Moreover, due to the non-directional spread of chemical signatures we are usually unable to localize its source, to associate it with an individual, or to track it. This deficiency, however, can be compensated by fusing the output of multiple chemical sensors distributed in space with kinematic data produced by laser-range-scanners or video cameras. In other words, tracking spans an additional temporal dimension for processing chemical sensor attributes. Multiple person tracking and chemical attribute fusion are thus to be performed within a single framework (see Fig. 8.11, [19, 20]).

In designing a multiple sensor system for decision support in security applications we wish to know which person going through an access area in an airport, e.g., may be carrying explosives. With reference to the experimental corridor sketched in Fig. 8.11, five chemical sensors measure at each scan the chemical signatures with respect to the position of each of the chemical sensors symbolized (green filled circles). Furthermore, there two laser-range-scanners (cyan and blue filled rectangles) are used as tracking sensors. In this example, three persons are walking along the corridor, one of them carrying hazardous material. Space-time processing for multiple person tracking and classification obviously plays a key role: Only in an integrated framework can the potential of chemical sensors or other attribute sensors of this type be exploited entirely.

Further Reading

A detailed discussion of this approach has been published in:

- M. Wieneke and W. Koch Combined person tracking and classification in a network of chemical sensors textitElsevier International Journal of Critical Infrastructure Protection, vol. 2, nr. 1–2, p. 51–67, May 2009..

Abstract Transportation infrastructures play a crucial role in the way of life and the economic vitality of a modern society. Access points like stations, airports or harbors are among the most critical elements in these infrastructures because they offer a possibility to bring in hazardous materials that can be used for attacks against people and against the transportation network itself. A timely recognition of such threats is essential and can be significantly supported by systems that monitor critical areas continuously and call the security personnel in case of anomalies. We are describing the concept and the realization of an indoor security assistance system for real time decision support. The system is specifically designed for the surveillance of entrance areas in transportation access facilities and consists of multiple heterogeneous sensors: Chemical sensors detecting hazardous materials provide data for the classification of persons. But due to their limited spatial?temporal resolution, a single chemical sensor cannot localize a substance and assign it to a person. We compensate for this deficiency by fusing the output of multiple, distributed chemical sensors with kinematical data from laser range?scanners. Both tracking and fusion of tracks with chemical attributes can be processed within a single framework called Probabilistic Multi-Hypothesis Tracking (PMHT). An extension of PMHT for dealing with classification measurements (PMHT-c) already exists.We show how PMHT- c can be applied to assign chemical attributes to person tracks. This affords the localization of threats and a timely notification of the security personnel.

Keywords: Person tracking, Probabilistic Multiple Hypothesis Tracking (PMHT), Classification, Attributes, Data fusion, Security assistance systems

References

1. W. Koch, Fixed-interval retrodiction approach to bayesian IMM-MHT for maneuvering multiple targets. in *IEEE Transactions on Aerospace and Electronic Systems*, AES-36, No. 1 (2000)
2. H.A.P. Blom, Y. Bar-Shalom, The interacting multiple model algorithm for systems with Markovian switching coefficients. IEEE Trans. Autom. Control **33**(8), 780–783 (1988)
3. X.R. Li, in *Control and Dynamic Systems*, ed. by C.T. Leondes. Hybrid Estimation Techniques, vol 76 (Academic Press, San Diego, 1996)
4. W. Koch, G. van Keuk, Multiple hypothesis track maintenance with possibly unresolved measurements. IEEE Trans. Aerosp. Electron. Syst. **33**(3), 883–892 (1997)
5. W. Koch, On Bayesian MHT for formations with possibly unresolved measurements—quantitative results. SPIE Signal Data Process. Small Targets **3163**, 417 (1997)

6. S.S. Blackman, R. Popoli, *Design and Analysis of Modern Tracking Systems* (Artech House, Norwood, 1999)
7. Y. Bar-Shalom, X.-R. Li, T. Kirubarajan, *Estimation with Applications to Tracking and Navigation* (Wiley & Sons, New York, 2001)
8. A. Gelb (ed.), *Applied Optimal Estimation* (M.I.T Press, Cambridge, 1974)
9. J.A. Roecker, Multiple scan joint probabilistic data association. IEEE Trans. Aerosp. Electron. Syst. **31**(3), 1204–1210 (1995)
10. O.E. Drummond, Target tracking with retrodicted discrete probabilities. in *Proceedings of SPIE 3163, Signal and Data Processing of Small Targets*, (1997), p. 249
11. F.E. Daum, R.J. Fitzgerald, The importance of resolution in multiple target tracking. SPIE Signal Data Process. Small Targets **2235**, 329 (1994)
12. H.A.P. Blom, E.A. Bloem, Approximate Bayesian tracking of two targets that maneuver in and out formation amidst unresolved and false measurements. in *IEEE Transactions on Transaction in Aerospace and Electronic Systems* (to appear) (2008)
13. A.K. Gupta, D.K. Nagar, *Matrix Variate Distributions* (Chapman & Hall/CRC, Boca Raton, 1999)
14. E.P. Wigner, Random matrices in physics. SIAM Rev. **9**, 1–13 (1967)
15. B. Ristic, S. Arulampalam, N. Gordon, *Beyond the Kalman Filter:Particle Filters for Tracking Applications* (Artech House Radar Library, Boston, 2004)
16. W. Koch, in *Advanced Signal Processing: Theory and Implementation for Sonar, Radar, and Non-Invasive Medical Diagnostic Systems*, ed. by S. Stergiopoulos. Target Tracking (CRC Press, Boca Raton, 2001)
17. M. Feldmann, W. Koch, Comments on Bayesian Approach to Extended Object and Cluster Tracking using Random Matrices. IEEE Trans. Aerosp. Electron. Syst. **48**(2), 1687–1693 (2012)
18. A. Doucet, N. de Freitas, N. Gordon (eds.) *Sequential Monte Carlo Methods in Practice* (Springer, New York, 2001)
19. M. Wieneke, W. Koch, Combined person tracking and classification in a network of chemical sensors. Int. J. Crit. Infrastruct. Prot. (In Press, Accepted Manuscript, Available online 24 November 2008, Elsevier)
20. C. Becher, G.L. Foresti, P. Kaula, W. Koch, F.P. Lorenz, D. Lubczyk, C. Micheloni, C. Piciarelli, K. Safenreiter, C. Siering, M. Varela, S. Waldvogel, M. Wieneke, A Security Assistance System combining Person Tracking with Chemical Attributes and Video Event Analysis. in *Proceedings of 11th ISIF International Conference on Information Fusion*, Cologne, July 2008

Chapter 9
Integration of Topographical Information

Complex sensing environments, such as in airborne ground surveillance or reconnaissance in an urban terrain, are increasingly present in modern applications of sensor data fusion. In order to fulfill user requirements even under those challenging conditions, it is not enough to use a broad spectrum of heterogeneous sensor systems and to integrate all available context information on the sensor performance and the object characteristics in the sense of the previous chapters. We also have to take advantage of context information on the sensing environment in itself insofar as it is the stationary or only slowly changing "stage" where a dynamic scenario evolves.

Typical examples of such environmental information are digital topographical maps and map-related information, which is made available by Geographical Information Systems. In the application examples of this chapter, road-maps in undeveloped areas are considered as well as city street maps and cadastre information on the urban canyons. Another category of topographical context information is provided by digital elevation models, which enable the identification of occluded areas, or vegetation maps indicating regions, where a high clutter background is to be taken into account, for example. This category of topographical information is especially important in mission planing as well as in sensor deployment and sensor management (see the discussion in Sect. 1.1.3, [4, Chap. 20]).

The discussion below shows for a simplified example how road-map information can be used to improve the track accuracy of road-moving vehicles and, most importantly, significantly enhance their track continuity. By using road-maps it is also possible to design statistical tests to decide whether an object is moving on a road or not (see the discussion in Sect. 1.3.5). Moreover, tracks produced by observing road-moving vehicles are well-suited for extracting road-map information, which is highly up-to-date and fairly precise. Sensor data fusion is thus even able to produce scenario-related context information, which can be stored as a data exploitation product in a Geographical Information System.

Another example discussed here deals with emitter localization and tracking in a city street using a receiver dislocated from the emitter, an application, which is dominated by multipath propagation phenomena. In an urban scenario, however,

W. Koch, *Tracking and Sensor Data Fusion*,
Mathematical Engineering, DOI: 10.1007/978-3-642-39271-9_9,

cadastre information on the location of the buildings is available, which can serve as input for a ray-tracing algorithm, which evaluates the most likely propagation paths. For each randomly chosen candidate of emitter positions, we can thus algorithmically calculate the likelihood function related to the measurements produced by the receiver.

9.1 Road-map Assisted Tracking

In many practical cases, ground vehicles move on roads, whose topographical coordinates are available up to a certain error (digitized vector map information). Figure 9.1 shows a schematic representation of a road-map by a graph, whose nodes are given by sources or sinks of roads, or by 'traffic signs', indicating that a road changes its quality. The edges represent individual road segments. For the sake of simplicity, we here confine the discussion to road segments. For a solution of the road-intersection problem see [2].

In this context, it seems reasonable to describe the kinematic state vector $\mathbf{x}_k^r$ of a road-moving vehicle at time t_k by its position on the road l_k (i.e. the arc length of the curve) and its scalar speed $\dot{l}_k$: $\mathbf{x}_k^r = (l_k, \dot{l}_k)^\top$. The model for describing the dynamic behavior of road vehicles is therefore a 2D version of Eq. 3.10:

$$p(\mathbf{x}_{k-1}^r|\mathcal{Z}^{k-1}) \xrightarrow[\text{model}]{\text{evolution}} p(\mathbf{x}_k^r|\mathcal{Z}^{k-1}). \tag{9.1}$$

A given road through a road network is mathematically described by a continuous curve in Cartesian ground coordinates parameterized by the corresponding arc length l. Digitized vector road-maps provide a piecewise linear approximation of the road by a polygonal curve $\mathcal{R}$. For a further discussion within the Bayesian framework, it is essential to introduce road-map and discretization errors.

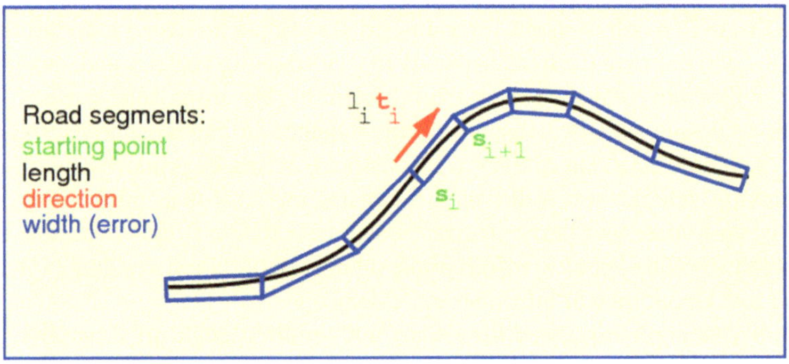

Fig. 9.1 Representation of a road

The processing of the sensor data itself has to be performed in ground or sensor-oriented coordinates depending on the underlying application:

$$p(\mathbf{x}_k^s|\mathcal{Z}^{k-1}) \xrightarrow[\text{model, data}]{\text{sensor}} p(\mathbf{x}_k^s|\mathcal{Z}^k). \tag{9.2}$$

The key to incorporate road-map information into the Bayesian tracking paradigm is therefore given by establishing a transformation of the probability densities of the vehicle states in the different frames of reference:

$$\underbrace{p(\mathbf{x}_k^r|\mathcal{Z}^{k-1})}_{\text{road coordinates}} \xrightarrow[\text{mapping/discretization errors}]{\text{road network}} \underbrace{p(\mathbf{x}_k^s|\mathcal{Z}^{k-1})}_{\text{sensor coordinates}} \tag{9.3}$$

Some calculations and mild approximations [2, 3] yield that the probability densities in sensor coordinates are approximately given by Gaußian mixtures referring to the linear road segments involved, thus enabling a seamless embedding of this type of context information into the Bayesian formalism. Quite naturally, crossing roads imply an inherent multiple hypothesis structure of the tracking algorithm.

9.1.1 Modeling of Road Segments

A given road through a real road network is mathematically described by a continuous 3D curve $\mathcal{R}^*$ in Cartesian ground coordinates. For the sake of simplicity the effect of crossroads is not considered here. See [2–4] for a more detailed discussion. Let $\mathcal{R}^*$ be parameterized by the corresponding arc length l. The exploitation of digitized road-maps provides the data base for a piecewise linear approximation of the road curve $\mathcal{R}^* : l \mapsto \mathcal{R}^*(l)$ by a polygonal curve $\mathcal{R}$. Let us furthermore assume that the curve $\mathcal{R}$ is characterized by n_r node vectors

$$\mathbf{s}_m = \mathcal{R}^*(l_m), \quad m = 1, \ldots, n_r. \tag{9.4}$$

From these quantities $n_r - 1$ normalized tangential vectors

$$\mathbf{t}_m = (\mathbf{s}_{m+1} - \mathbf{s}_m)/||\mathbf{s}_{m+1} - \mathbf{s}_m||, \quad m = 1, \ldots, n_r - 1 \tag{9.5}$$

can be derived. The Euclidian distance $||\mathbf{s}_{m+1} - \mathbf{s}_m||$ between two adjacent node vectors, however, is usually not identical with the distance $\lambda_m = l_{m+1} - l_m$ actually covered by a vehicle when it moves from $\mathbf{s}_m$ to $\mathbf{s}_{m+1}$ along the road. Besides the vectors $\mathbf{s}_m$ the scalar quantities $\lambda_m \geq |\mathbf{s}_{m+1} - \mathbf{s}_m|$ should therefore enter into the road model to make it more realistic. The differences $\sigma_d = |\lambda_m - ||\mathbf{s}_{m+1} - \mathbf{s}_m|||$ can obviously serve as a quantitative measure of the discretization errors we have to deal with. Using the characteristic functions defined by

$$\chi_m(l) = \begin{cases} 1 \text{ for } l \in (l_m, l_{m+1}) \\ 0 \text{ else} \end{cases} \tag{9.6}$$

$$m = 0, \ldots, n_r, \quad l_0 = -\infty, \quad l_{n_r+1} = \infty \tag{9.7}$$

$$\text{and} \quad \mathbf{s}_0 = \mathbf{s}_1, \quad \mathbf{t}_0 = \mathbf{t}_1, \quad l_0 = l_1, \quad \mathbf{t}_{n_r} = \mathbf{t}_{n_r-1}, \tag{9.8}$$

we obtain a mathematically simple description of the polygon curve $\mathcal{R}$, by which the road $\mathcal{R}^*$ is approximated:

$$\mathcal{R} : l \mapsto \mathcal{R}(l) = \sum_{m=0}^{n_r} \left[\mathbf{s}_m + (l - l_m)\mathbf{t}_m\right] \chi_m(l) \tag{9.9}$$

$$\text{with:} \quad \mathcal{R}^*(l_m) = \mathcal{R}(l_m) = \mathbf{s}_m, \quad m = 0, \ldots, n_r. \tag{9.10}$$

9.1.2 Road-Constrained Densities

The Bayesian formalism previously discussed can be directly applied to road-moving vehicles, if it is possible to find a transformation operator $\mathcal{T}_{g\leftarrow r}$ by which the predicted density $p(\mathbf{x}_k^r|\mathcal{Z}^{k-1})$ in road coordinates can be transformed into ground coordinates:

$$p(\mathbf{x}_k^g|\mathcal{Z}^{k-1}) = \mathcal{T}_{g\leftarrow r}[p(\mathbf{x}_k^r|\mathcal{Z}^{k-1})]. \tag{9.11}$$

When available in ground coordinates, the linearized versions of the transforms $\mathbf{t}_{s\leftarrow g}$ and $\mathbf{t}_{g\leftarrow s}$ (Sect. 2.1) can be used to represent the densities in sensor coordinates, where the filtering step is performed. For this purpose, we write the density $p(\mathbf{x}_k^g|\mathcal{Z}^{k-1})$ as a sum over the $n_r + 1$ road segments considered:

$$p(\mathbf{x}_k^g|\mathcal{Z}^{k-1}) = \sum_{m=0}^{n_r} p(\mathbf{x}_k^g, m|\mathcal{Z}^{k-1}) \tag{9.12}$$

$$= \sum_{m=0}^{n_r} p(\mathbf{x}_k^g|m, \mathcal{Z}^{k-1})\, p(m|\mathcal{Z}^{k-1}) \tag{9.13}$$

$$= \sum_{m=0}^{n_r} p_{g\leftarrow r}^m\, \mathcal{T}_{g\leftarrow r}^m\left[p(\mathbf{x}_k^r|\mathcal{Z}^{k-1})\right]. \tag{9.14}$$

In Eq. 9.13 the probability

$$p(m|\mathcal{Z}^{k-1}) = \int d\mathbf{x}_k^r\, p(m, \mathbf{x}_k^r|\mathcal{Z}^{k-1}) \tag{9.15}$$

$$= \int d\mathbf{x}_k^r\, \chi_m(\mathbf{H}_r\mathbf{x}_k^r)\, p(\mathbf{x}_k^r|\mathcal{Z}^{k-1}) \tag{9.16}$$

$$=: p^m_{g\leftarrow r} \tag{9.17}$$

denotes the probability that the vehicle moves on segment m given the accumulated sensor data $\mathcal{Z}^{k-1}$. The matrix $\mathbf{H}_r$ is defined by $\mathbf{H}_r\mathbf{x}^r_k = l_k$. Later on, it will be intuitively interpreted as a fictitious measurement matrix. Since the density $p(\mathbf{x}^r_k|\mathcal{Z}^{k-1}) = \sum_{j=0}^{1} p^j_{k-1}\ \mathcal{N}\big(\mathbf{x}^r_k;\ \mathbf{x}^{rj}_{k|k-1},\ \mathbf{P}^{rj}_{k|k-1}\big)$ is a Gaußian mixture due to the GMTI sensor model, the probabilities $p^m_{g\leftarrow r}$ can be explicitly expressed by error functions:

$$p^m_{g\leftarrow r} = p^j_{k-1}\ (\Phi[\lambda(l^j_{m+1})] - \Phi[\lambda(l^j_m)]),\quad m = 0,\ldots,n_r \tag{9.18}$$

with:

$$\Phi(\lambda) = 1/\sqrt{2\pi}\int_{-\infty}^{\lambda} dt\ \exp(-t^2/2) \tag{9.19}$$

$$\lambda(l)^j = \frac{l - \mathbf{H}_r\mathbf{x}^{rj}_{k|k-1}}{\sqrt{\mathbf{H}_r\mathbf{P}^{rj}_{k|k-1}\mathbf{H}_r^{\top}}}. \tag{9.20}$$

For the remaining term in Eq. 9.14 standard probability reasoning yields:

$$\mathcal{T}^m_{g\leftarrow r}\big[p(\mathbf{x}^r_k|\mathcal{Z}^{k-1})\big] = p(\mathbf{x}^g_k|m,\mathcal{Z}^{k-1}) \tag{9.21}$$

$$= \int d\mathbf{x}^r_k\ p(\mathbf{x}^g_k, \mathbf{x}^r_k|m,\mathcal{Z}^{k-1}) \tag{9.22}$$

$$= \int d\mathbf{x}^r_k\ p(\mathbf{x}^g_k|\mathbf{x}^r_k, m)\ p(\mathbf{x}^r_k|m,\mathcal{Z}^{k-1}). \tag{9.23}$$

Straight Roads

Let us first consider the simple limiting case of a straight road defined by:

$$\mathcal{R}(l) = \mathbf{s} + l\mathbf{t}. \tag{9.24}$$

Under Gaussian assumptions the transform from road to ground coordinates is defined by the normal transition density: $p(\mathbf{x}^g_{k+1}|\mathbf{x}^r_{k+1}) = \mathcal{N}\big(\mathbf{x}^g_{k+1};\ \mathbf{t}_{g\leftarrow r}[\mathbf{x}^r_{k+1}],\ \sigma^2_m\big)$ with the affine transform

$$\mathbf{t}_{g\leftarrow r}[\mathbf{x}_r] = \begin{pmatrix}\mathbf{t} & 0\\ 0 & \mathbf{t}\end{pmatrix}\mathbf{x}_r + \begin{pmatrix}\mathbf{s} - l\mathbf{t}\\ 0\end{pmatrix} \tag{9.25}$$

and σ_m denoting the standard deviation of the mapping error. The transformation of the density $p(\mathbf{x}^r_k|\mathcal{Z}^{k-1})$ into the ground coordinate system is therefore described by $p(\mathbf{x}^g_k|\mathcal{Z}^{k-1}) = \int d\mathbf{x}^r_k\ p(\mathbf{x}^g_k|\mathbf{x}^r_k)\ p(\mathbf{x}^r_k|\mathcal{Z}^{k-1})$. The integration can be carried out explicitly and preserves the Gaußian character of the density functions (normal mixtures). The corresponding inverse is simply provided by a projection of the density $p(\mathbf{x}_k^g|\mathcal{Z}^{k-1})$ on the road. With these transformations the previous considerations directly apply.

Polygonal Roads

The transition density $p(\mathbf{x}_k^g|\mathbf{x}_k^r, m)$ for the road segment m (Eq. 9.23) is characterized by road-map and discretization errors (σ_m, σ_d), which may vary from segment to segment. Under Gaußian assumptions regarding the possible error sources, with the affine transforms

$$\mathbf{t}_{g\leftarrow r}^m[\mathbf{x}_r] = \begin{pmatrix} \mathbf{t}_m & 0 \\ 0 & \mathbf{t}_m \end{pmatrix} \mathbf{x}_r + \begin{pmatrix} \mathbf{s}_m - l_m \mathbf{t}_m \\ 0 \end{pmatrix} \tag{9.26}$$

for each individual road segment m, and the error standard deviation $\sigma_r^2 = \sigma_m^2 + \sigma_d^2$, we obtain normal transition densities

$$p(\mathbf{x}_g|\mathbf{x}_r, m) = \mathcal{N}\big(\mathbf{x}_g;\ \mathbf{t}_{g\leftarrow r}^m[\mathbf{x}_r],\ \sigma_r^2\big). \tag{9.27}$$

With these preliminaries, an application of Bayes' rule to the remaining density in the integrand of Eq. 9.23 yields:

$$p(\mathbf{x}_k^r|m, \mathcal{Z}^{k-1}) = \frac{p(m|\mathbf{x}_k^r)\ p(\mathbf{x}_k^r|\mathcal{Z}^{k-1})}{\int d\mathbf{x}_k^r\ p(m|\mathbf{x}_k^r)\ p(\mathbf{x}_k^r|\mathcal{Z}^{k-1})} \tag{9.28}$$

with probabilities $p(m|\mathbf{x}_k^r)$ given by:

$$p(m|\mathbf{x}_k^r) = \chi_m(\mathbf{H}_r\mathbf{x}_k^r). \tag{9.29}$$

Up to now the derivation was exact. Due to the normalization constant, however, the characteristic functions violate the Gaussian character of the probability densities. To circumvent this problem we propose the following normal approximation:

$$p(m|\mathbf{x}_k^r) \approx \exp[-\tfrac{1}{2}(z_r^m - \mathbf{H}_r\mathbf{x}_r)^2/\lambda_m^2] \tag{9.30}$$

$$c = \sqrt{2\pi}\lambda_m\ \mathcal{N}\big(z_r^m;\ \mathbf{H}_r\mathbf{x}_r,\ \lambda_m^2\big) \tag{9.31}$$

with z_r^m and λ_m^2 given by:

$$z_r^m = \tfrac{1}{2}(l_{m+1} + l_m) \tag{9.32}$$

$$\lambda_m^2 = \frac{(l_{m+1} - l_m)^2}{12}. \tag{9.33}$$

The quantities z_r^m and λ_m^2 can be interpreted as the mean and variance of a uniform density given by $\chi_m(l)$. From Eq. 9.29 and the product formula (Sect. 4.3) we obtain:

$$p(\mathbf{x}_k^r|m, \mathcal{Z}^{k-1}) = \sum_{j=0}^{1} p_{k-1}^m \; \mathcal{N}\big(\mathbf{x}_k^r;\; \mathbf{x}_{k|k-1}^{rmj},\, \mathbf{P}_{k|k-1}^{rmj}\big) \tag{9.34}$$

with Kalman-type update equations for $\mathbf{x}_{k|k-1}^{rmj}$ and $\mathbf{P}_{k|k-1}^{rmj}$, where z_r^m, λ_m^2 are in analogy to a measurement and a related measurement error variance:

$$\mathbf{x}_{k|k-1}^{rmj} = \mathbf{x}_{k|k-1}^{rj} + \mathbf{W}_r^{mj}\,(z_r^m - \mathbf{H}_r \mathbf{x}_{k|k-1}^{rmj}) \tag{9.35}$$

$$\mathbf{P}_{k|k-1}^{rmj} = \mathbf{P}_{k|k-1}^{rmj} - \mathbf{W}_r^{mj}\mathbf{S}_r^{mj}\mathbf{W}_r^{mj\top}. \tag{9.36}$$

with "innovation" covariance matrices $\mathbf{S}_r^{mj}$ and "Kalman Gain" matrices $\mathbf{W}_r^{mj}$ given by:

$$\mathbf{S}_r^{mj} = \mathbf{H}_r \mathbf{P}_{k|k-1}^{rmj} \mathbf{H}_r^\top + \lambda_m^2 \tag{9.37}$$

$$\mathbf{W}_r^{mj} = \mathbf{P}_{k|k-1}^{rmj} \mathbf{H}_r^\top {\mathbf{S}_r^{mj}}^{-1} \tag{9.38}$$

The notation chosen indicates that the effect of road-map information on the probability density functions can formally be described by a fictitious measurement, a corresponding measurement matrix, and a fictitious measurement error. Now the integration in Eq. 9.23 can be carried out explicitly as in the previously discussed limiting case. The transformation from road to ground coordinates is thus known. In analogy to the limiting case of straight roads, the inverse transform is simply provided by individually projecting the densities $p(\mathbf{x}_k^g|m, \mathcal{Z}^k)$ on the road (i.e. after the filtering step has been performed). Before the subsequent prediction is performed, it seems to be reasonable to apply a second-order approximation to the mixture densities:

$$p(\mathbf{x}_k^r|\mathcal{Z}^k) = \sum_{m=0}^{n_r} p(m|\mathcal{Z}^k)\; p(\mathbf{x}_k^r|m, \mathcal{Z}^k) \tag{9.39}$$

$$\approx \sum_{i=0}^{1} p_k^i \; \mathcal{N}\big(\mathbf{x}_k^r;\; \mathbf{x}_{k|k}^{ri},\, \mathbf{P}_{k|k}^{ri}\big). \tag{9.40}$$

9.1.3 Quantitative Discussion

The idealized sensor-to-object scenario discussed in Fig. 7.7, Sect. 7.2, displays features characteristic of military ground surveillance applications with airborne GMTI radar. Based on this example, we quantitatively illustrate the potential gain by exploiting:

1. GMTI sensor modeling,
2. road-map information,

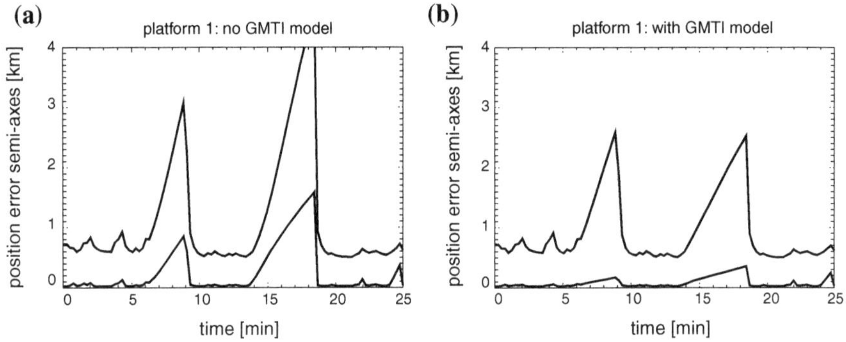

Fig. 9.2 **a** Sensor 1, no GMTI model, no road, **b** Sensor 1, GMTI model, no road

3. and sensor data fusion.

The covariance matrices of the estimates resulting from filtering (Eq. 3.38) provide tracking-inherent measures of performance. For the sake of simplicity we confine the discussion to the semi-axes of the position error ellipses in ground coordinates. The mean squared error with reference to the true trajectory (known in the simulation) is a more direct measure of the tracking performance.

Simulation Parameters

The GMTI sensor reports are randomly generated according to the detection probabilities of the individual sensors as defined in Eq. 7.27. The errors of the range, azimuth, and range-rate measurements are assumed to be bias-free and normally distributed. The corresponding standard deviations σ_φ, $\sigma_{\dot{r}}$ of the azimuth and range-rate measurement errors depend on the underlying signal-to-noise+interference ratio, snir = snir$(r_k, \varphi_k, \dot{r}_k)$, after clutter filtering according to the previous discussion. The standard deviation σ_r of the range errors is assumed to be constant. In order to focus on GMTI-specific aspects, the spatial residual clutter density is assumed to be small. False returns therefore play practically no role in our simulations. In the following list the chosen simulation parameters are summarized:

$$\text{range error: } \sigma_r = 20\,\text{m; azimuth error: } \sigma_\varphi = 0.2\,\text{deg}$$
$$\text{range-rate error: } \sigma_{\dot{r}} = 0.5\,\text{m/s; false alarm probability: } p_f = 10^{-6}$$
$$\text{snir}_0 = 20\,\text{dB; } r_0{}^{1,2} = 200\,\text{km, } 50\,\text{km; dynamics model: } v_t = 18\,\text{m/s,}$$
$$\theta_t = 200\,\text{s; road-map error: } \sigma_m = 20\,\text{m}$$

Numerical Results

Figures 9.2–9.7 show the measures of performance plotted over the time. We consider the sensors on both platforms individually as well as the result of

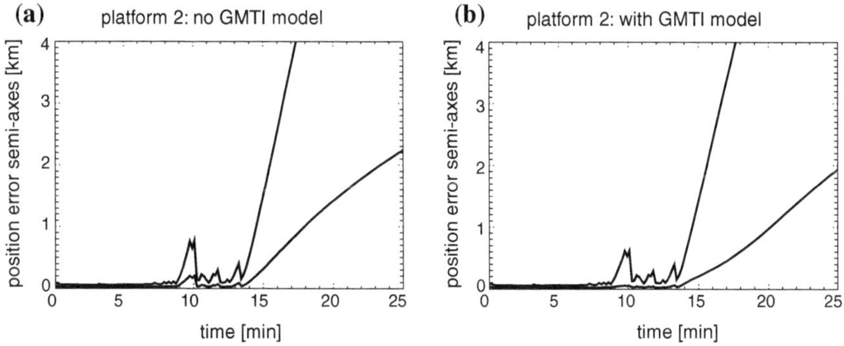

Fig. 9.3 **a** Sensor 2, no GMTI model, no road; **b** Sensor 2, GMTI model, no road

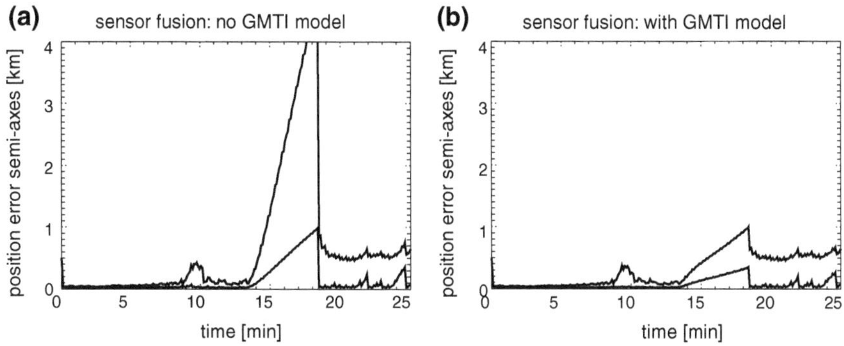

Fig. 9.4 **a** Fusion 2, no GMTI model, no road; **b** Fusion, GMTI model, no road

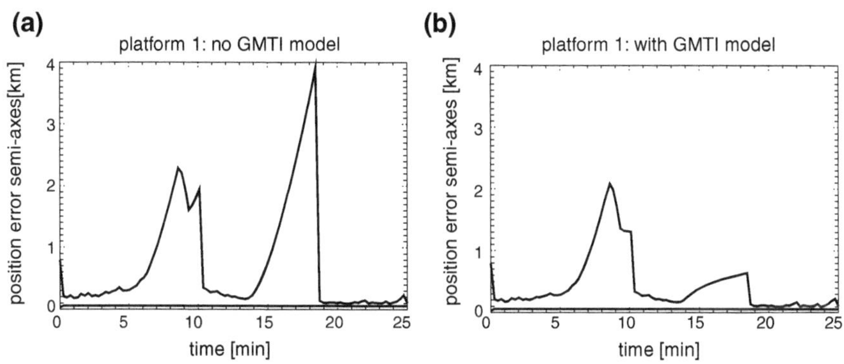

Fig. 9.5 **a** Sensor 1, no GMTI model, road; **b** Sensor 1, GMTI model, road

centralized fusion. How the GMTI model and the fusion with road-map information affect the track quality is examined for each case. The temporal behavior of the performance measures reflects the four different phases of the scenario: (1) high

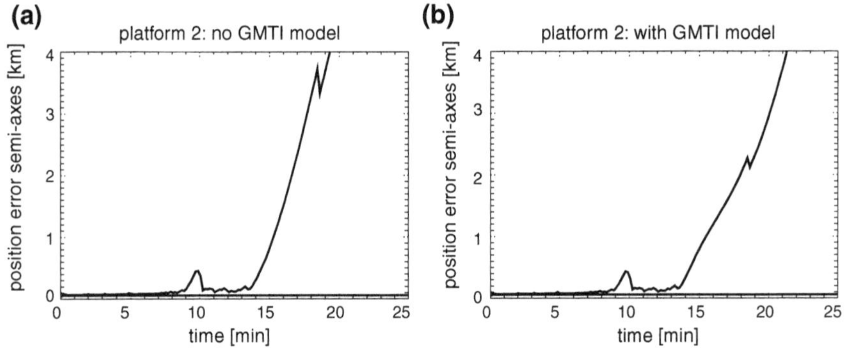

Fig. 9.6 **a** Sensor 2, no GMTI model, road; **b** Sensor 2, GMTI model, no road

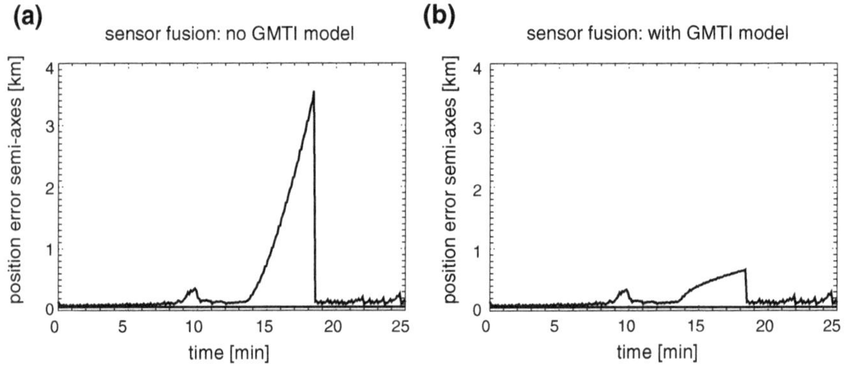

Fig. 9.7 **a** Fusion 2, no GMTI model, road; **b** Fusion, GMTI model, road

P_D far from the notches, (2) missing detections due to unfavorable sensor-to-object geometries, (3) missing detections due to the vehicle stop, (4) decreasing P_D when the vehicle is leaving the field of view (sensor 2).

- *Advantage of a GMTI Sensor Model.* Unless appropriately handled by a GMTI model, the sensor inherent clutter notch can seriously affect the over-all performance of a ground vehicle tracking system. From the figures we can derive the following conclusions:

  1. By missing successive plots due to Doppler-blindness the mean tracking error and filtering covariances strongly increase.
  2. As a consequence the expectation gates also increase, making the discrimination of false or unwanted returns more difficult.
  3. Stopped vehicles escape detection by GMTI radar. An early recognition of this event may be of particular military interest.

  By these observations it is evident that track loss is the probable consequence in ground situations similar to our example. Even if track re-initiation is successful after some delay, the re-establishment of the track continuity may remain a difficult and time consuming task. The simulation results show, however, that the refined sensor model can improve the tracking performance:

1. The minor semi-axes of the position error ellipses are stronger, the corresponding major semi-axes are slightly reduced. This is due to the fact that the GMTI model has a larger effect on the range/range-rate estimates than on the azimuth estimates.
2. For this reason, the bridging over periods characterized by Doppler-blindness and the recognition of stopping vehicles is much alleviated, while the discrimination of false returns can be improved by significantly smaller range/range-rate gates.

- *Advantage by Road-Map Information.* Further improvements are obtained by exploiting information from digitized road-maps:

1. Even without GMTI modeling in the tracking filter, the effect of Doppler-blindness on the track accuracy can be alleviated. In the scenario considered, the road-map information is equivalent to an additional range measurement (minor semi-axis).
2. Depending on the sensor-to-object geometry actually in effect, the azimuth estimates can be much improved. This affects the major semi-axes of the error ellipses. In particular the impact of the object stop on the track accuracy is reduced.
3. If both information on the current location of the GMTI clutter notch and topographical road-maps are exploited, the early recognition of the event 'an object being tracked has stopped' can be assisted, which is important in military applications.

- *Advantage of Sensor Data Fusion.* Even if the GMTI sensor model and road-maps are not exploited in the algorithms, sensor data fusion significantly improves the track accuracy. Additional improvements result if this information is taken into account:

1. The effect of Doppler-blindness on the track accuracy can be reduced even more. We observe a gain by sensor fusion (combined with GMTI-modeling) even in the case of stopping vehicles (i.e. if both sensors do not produce object measurements).
2. The advantage of sensor data fusion is not merely due to the increased data rate, but a consequence of the sensor-to-object geometries considered. Intuitively speaking, the fusion algorithms combine estimation error ellipses rotated with respect to each other.
3. Since the sensor platforms typically move much faster than the observed objects, ground vehicles remain in the field of view for a relatively short time. Evidently, by sensor fusion the total coverage can greatly be increased.

A direct comparison between Figs. 9.5a, 9.6a and 9.7b intuitively illustrates the advantage of using the proposed methods.

Further Reading

A detailed discussion of this approach and the extension to road networks has been published in:

- M. Ulmke and W. Koch
  Road-Map Assisted Ground Moving Target Tracking
  IEEE Transactions on Aerospace and Electronic Systems, Vol. 42, No. 4, p. 1264-1274, October 2006.
  This work is an extension and quantitative evaluation of the methodology published in: W. Koch. Ground Target Tracking with STAP Radar: Selected Tracking Aspects. Chap. 14 *in: Klemm, R. (Ed.): Applications of Space-time Adaptive Processing. Institution of Electrical Engineers, IEE Press, 41 pages, London (2004).*
  Fusion of road-map information in extended convoy tracking based in ideas discussed in Sect. 6.2 has been published in *W. Koch and M. Feldmann. Cluster tracking under kinematical constraints using random matrices. Elsevier Journal on Robotics and Autonomous Systems, vol. 57, nr. 3, p. 296-309, March 2009.*

Abstract
Tracking ground targets with airborne GMTI (ground moving target indicator) sensor measurements is a challenging task due to high target density, high clutter, and low visibility. The exploitation of nonstandard background information such as road-maps and terrain information is therefore highly desirable for the enhancement of track quality and track continuity. The present paper presents a Bayesian approach to incorporate such information consistently. It is particularly suited to deal with winding roads and networks of roads. The target dynamics is modeled in quasi-one-dimensional road coordinates and mapped onto ground coordinates using linear road segments taking road-map errors into account. The case of several intersecting roads with different characteristics, such as mean curvature, slope, or visibility, is treated within an interacting multiple model (IMM) scheme. Targets can be masked both by the clutter notch of the sensor and by terrain obstacles. Both effects are modeled using a sensor-target state-dependent detection probability. The iterative filter equations are formulated within a framework of Gaussian sum approximations on the one hand and a particle filter approach on the other hand. Simulation results for single targets taken from a realistic ground scenario show strongly reduced target location errors compared with the case of neglecting road-map information. By modeling the clutter notch of the GMTI sensor, early detection of stopping targets is demonstrated.

Keywords: Road-maps, Bayesian tracking, Gaußian sums, particle filtering, GMTI tracking

9.2 Track-Based Road-map Extraction

Seen from a different perspective, ground vehicles moving on road networks that are observed by wide-area sensors, such as GMTI radar, produce large data streams that can directly be used for road-map extraction: After a suitable post-processing, GMTI tracks of road targets simply define an approximation of the corresponding road segments currently being used by the ground moving targets.

Tracking-driven road extraction can be beneficial in situations or scenarios where reliable road-maps are not available at all, where the maps provided by geographical information systems are not up-to-date, or where the accuracy of the road-maps is insufficient. In addition, there are fields of applications, in which roads or road-like 'lines of communication' exist only temporarily or may change with time. This can be the case in deserts or in times of a conflict. As practical evidence shows, even in typical off-road scenarios, the existence of structures similar to roads quickly evolve, as a 'second' vehicle usually moves in the tracks of its precursor. This is especially true in an insufficiently explored or dangerous environment (e.g. in a mine field).

9.2.1 Practical Relevance

In the sensor's own coordinate system, the achievable accuracy of road-maps generated by road-target tracking depends on the measurement accuracies of the GMTI sensors, the current sensor-to-target geometry, the scan rate, and the dynamic properties of the ground moving vehicles, i.e. on the accuracy of the produced GMTI tracks. Since usually many vehicles use the same road segments, a significant gain results from fusing several "road tracks". In addition, the underlying sensor-to-target geometry is continuously changing with time as GMTI radar is essentially an airborne, i.e. a moving, sensor system. For this reason, the fusion of "road tracks" produced at different instants of time is expected to improve the achievable accuracy of track-generated road-maps even more, finally leading to high-precision road-maps.

Sensor registration or misalignment errors usually cause serious problems in sensor data fusion. In other words, in a given sensor data fusion application, it cannot always be taken for granted that the data originating from various distributed sensors can be transformed into a common coordinate system. In order to mitigate the corresponding bias errors, the tracking-driven generation of accurate road-maps with reference to the individual sensor coordinate system can well be used. Precisely extracted road-maps with reference to the coordinate system of the individual sensors can easily be matched with each other by using particular road-map features such as characteristic curves or crossings. In this way, a compensation of relative bias errors can easily be achieved. A contribution to remove *absolute* bias errors as well is obtained by matching tracking-generated road-maps with geo-referenced maps stored in a topographical data base.

9.2.2 Road Node Reconstruction

According to the introductory remarks, the track of a road moving vehicle, i.e. the collection of expectation vectors and covariance matrices $\{\mathbf{r}_{l|k}, \mathbf{R}_{l|k}\}_{l=1}^{k}$, provides in itself a first approximation of the road used by the vehicle. Due to low sensor update rates, missing detections, or fading phenomena, Doppler blindness etc., however, the accuracy and sample density of such track generated road-maps may be insufficient.

In applications we therefore wish to produce a suitable interpolation between adjacent 'node vectors' and the related 'mapping error' covariance matrices. This interpolation should take full advantage of the available knowledge of the vehicles' kinematic state vector and the related track accuracy as well as of background information on the vehicle's behavior, i.e. the target dynamics model.

Given two adjacent nodes vectors $\mathbf{r}_{l-1|k}$, $\mathbf{r}_{l|k}$ with their related accuracies $\mathbf{R}_{l-1|k}$, $\mathbf{R}_{l|k}$, we at first have to decide whether it is reasonable to create an additional node at all. Obviously another node is necessary if there are curves or turns to be expected. Vice-versa, for a more or less rectilinear road segment only very few nodes are required.

An intuitively clear indication for the existence of a winding road is given by comparing the direction of the velocity vector estimates $\dot{\mathbf{r}}_{l-1|k}$, $\dot{\mathbf{r}}_{l|k}$ at subsequent instants of time t_{l-1} and t_l, which by definition are proportional to estimates of the tangential vectors to the road at the locations $\mathbf{r}_{l-1|k}$ and $\mathbf{r}_{l|k}$. The decision also depends on the quality of these velocity estimates.

To introduce an additional node vector, let us denote by $\varphi_{l|k}$ the angle between the velocity estimate $\dot{\mathbf{r}}_{l|k}$ and one of the axes of the coordinate system. The corresponding angle for the actual velocity vector is a random variable approximately normally distributed with a variance given by $\Phi_{l|k}$. Let $\psi_{l|k}$ be the corresponding angle of the difference vector $\mathbf{r}_{l|k} - \mathbf{r}_{l-1|k}$. An intuitively plausible decision criterion whether an additional node is to be introduced is thus given by:

$$(\varphi_{l|k} - \psi_{l|k})^2 / \Phi_{l|k} > \kappa^2. \tag{9.41}$$

If the inaccuracy of the heading estimates is large, subsequent headings are allowed to differ more than in case of more precise estimates. It seems to be reasonable to choose the decision parameter around unity. Obviously, the estimate of the complete kinematical state vector enters into this criterion. See Fig. 9.8 for a characteristic example and an intuitive interpretation of this criterion.

Given an additional node is to be introduced, let us consider the probability density $p(\mathbf{x}_{l-\theta}|Z^k)$ with $0 < \theta < 1$, typically $\theta = \frac{1}{2}$ if $t_l = l\Delta t$. This density expresses the available knowledge about the kinematical target state at an intermediate instant of time $t_{l-1} < t_{l-\theta} < t_l$. From this density, an intermediate node vector of the road and a tangential vector can be derived. By considering several $\theta_1, \theta_2, \ldots$ an indication of the arc length λ of the road between the positions $\mathbf{r}_{l-1}$ and $\mathbf{r}_l$ can be obtained.

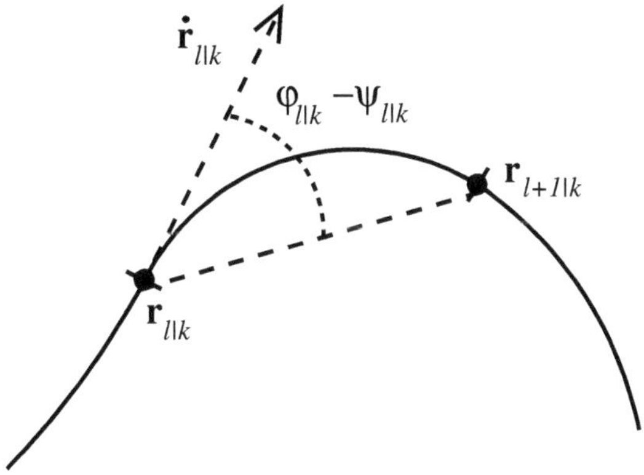

Fig. 9.8 Adding a new node vector

According to the discussion in Sect. 3.4.2, we obtain:

$$p(\mathbf{x}_{l-\theta}|\mathcal{Z}^k) = \mathcal{N}(\mathbf{x}_{l-\theta};\ \mathbf{x}_{l-\theta|k},\ \mathbf{P}_{l-\theta|k}). \tag{9.42}$$

9.2.3 Discussion of an Example

We discuss an example illustrating the iterative process of tracking-driven road-map extraction sketched above.

Simplified Scenario

Figure 9.9 shows a simulated and idealized, but non-trivial GMTI tracking scenario. On a road network a single ground vehicle is moving from 'Start' to 'End'. On its way it passes two regions, where it is not detectable by the radar sensor due to terrain obscurations. The second obscuration hides an intersection. The vehicle stops twice for several minutes (stars). Obviously, during these periods the vehicle is not detectable by a GMTI radar.

Directly before and after the second terrain obscuration the detection probability of the radar is significantly reduced due to the phenomenon of 'Doppler blindness'. In such regions the radial velocity of the moving vehicle relative to the moving sensor platform is equal or close to the corresponding radial velocity of the ground patch surrounding the vehicle. For this reason, the skin echo of the vehicle in most cases can no longer be discriminated from the ground clutter returns by using Pulse-Doppler signal processing (STAP: Space Time Adaptive Processing [5]). The vehicle is thus masked by the 'clutter notch' of the GMTI radar.

The revisit interval of the simulated GMTI radar is 12 s. It is located in a distance of 100 km along the y-axis (stand-off radar). Its measurement accuracy (standard deviation) is 20 m in range (i.e.. along the y-axis) and 400 m in cross-range (i.e. along

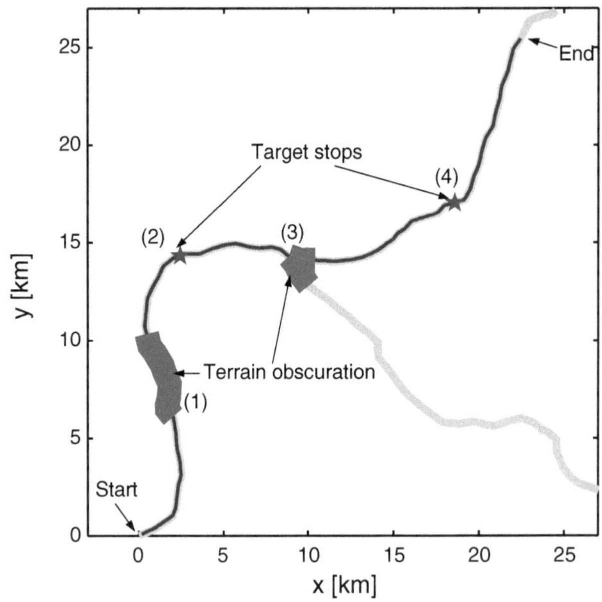

Fig. 9.9 A GMTI tracking scenario

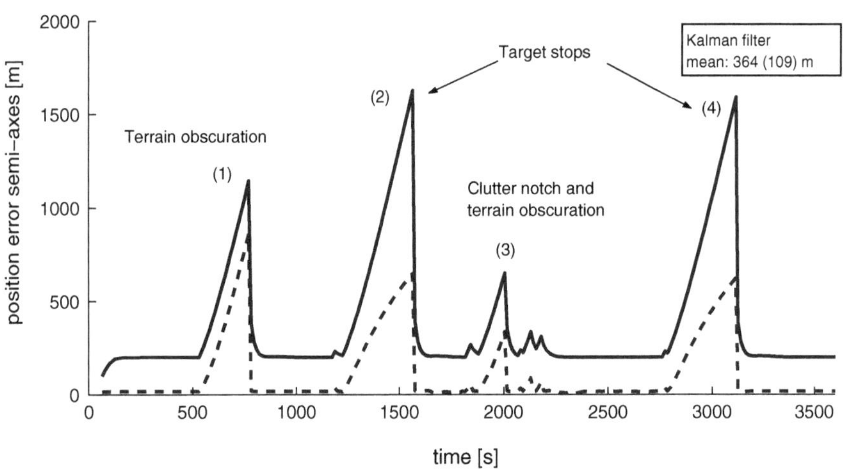

Fig. 9.10 Filtering covariances (major/minor eigenvalues)

the x-axis). The corresponding Minimum Detectable Velocity (MDV) is 2 m/s. In this simplified example, we exclude the treatment of false or unwanted radar returns and consider well-separated ground-moving vehicles only. The total observation time is one hour (300 scans).

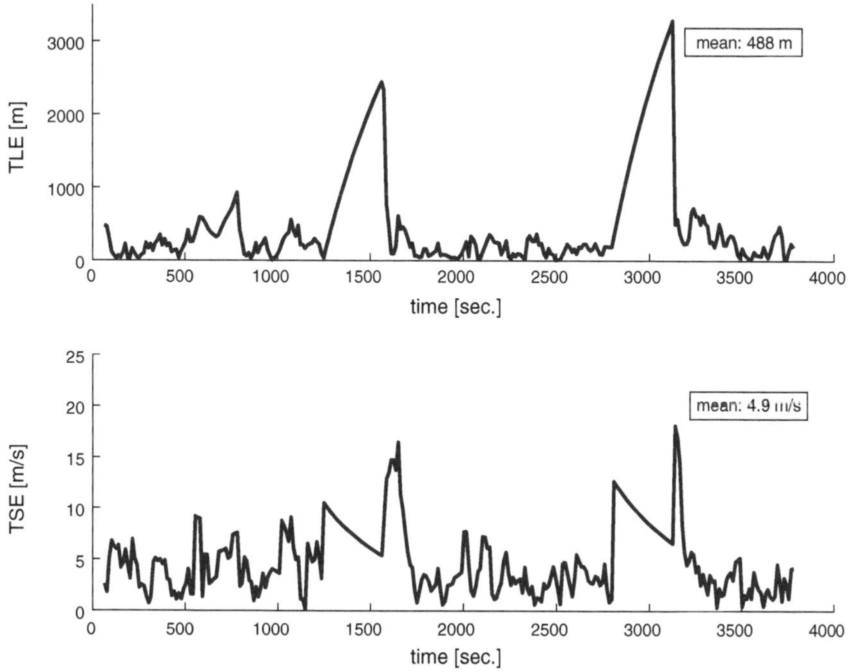

Fig. 9.11 Tracking error (localization, speed)

Simulation Results

Figure 9.10 shows the major and minor semi-axes of the error ellipses related to the position estimates of the vehicle as a function of the tracking time (solid and dashed lines, respectively). These quantities are simply obtained by applying Kalman filtering for tracking as there are no false returns or other vehicles in the vicinity. We observe four pronounced peaks which correspond to terrain obscurations, the vehicle stops, and the regions where the radar is Doppler-blind. The mean values of the semi-axes are 364 and 109 m, respectively.

In Fig. 9.11, the tracking error, i.e. the distance between the simulated true vehicle state and the corresponding estimates, are displayed for a single run as a function of the tracking time (TLE: Target Localization Error, TSE: Target Speed Error). The corresponding mean values are 498 m and 4.9 m/s, respectively. In the temporal evolution of the localization error, only two peaks are visible. The orientation of the road happens to be along the resulting predictions in the situations where the other peaks occurred in the previous figure.

Figure 9.12 shows the major and minor semi-axes of the error ellipses of the retrodicted position estimates of the vehicle as a function of the tracking time (solid and dashed lines, respectively). Obviously the error covariance matrices are much reduced in size (mean values 161 and 41 m, respectively) and are used for describ-

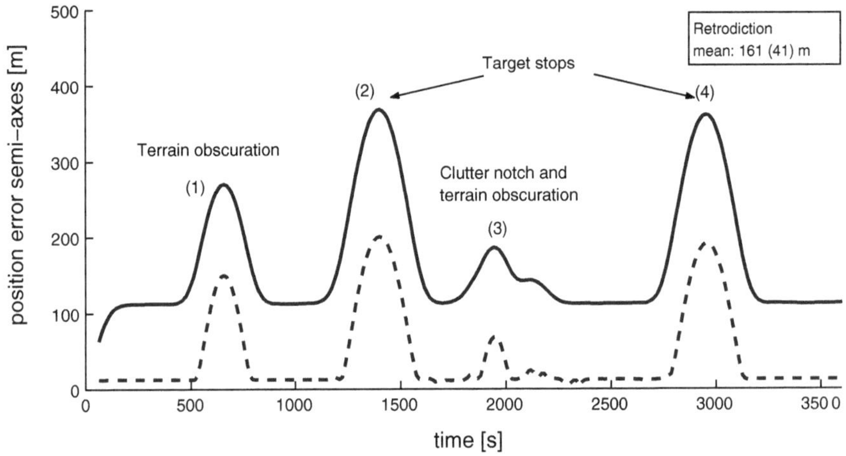

Fig. 9.12 Retrodiction covariances (major/minor eigenvalues)

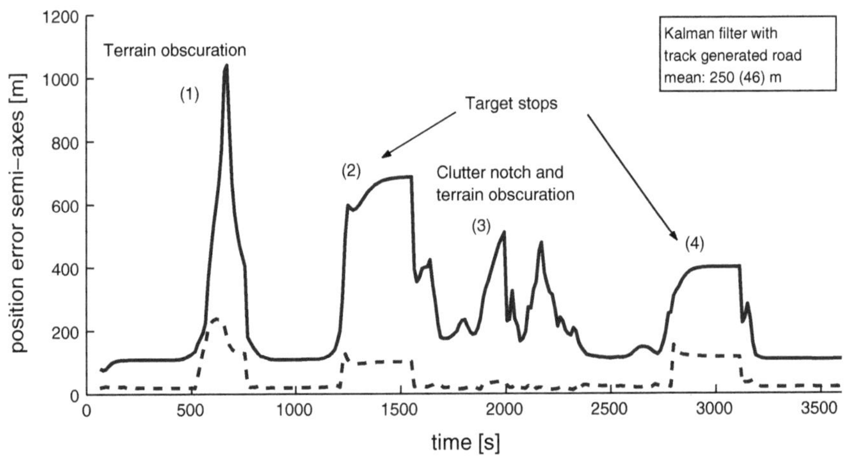

Fig. 9.13 Filtering covariances (major/minor eigenvalues)

ing the road-map errors as discussed in the previous sections. The retrodicted state estimates of the road-moving vehicle are used for approximating the road.

This first approximation of the road-map, which was reconstructed by the track of a first road-moving vehicle, is now used for 'road-map-assisted tracking' of a second vehicle using the same road. In Fig. 9.13 the resulting major and minor semi-axes of position error ellipses are shown. The mean values are 260 and 46 m, respectively. Evidently, these quantities are much smaller than the corresponding quantities obtained in the previous case. In particular, the pronounced peaks in the time periods when the target stops are significantly smaller. The same tendencies can be observed in Fig. 9.14, which shows the tracking errors for localization and speed (mean val-

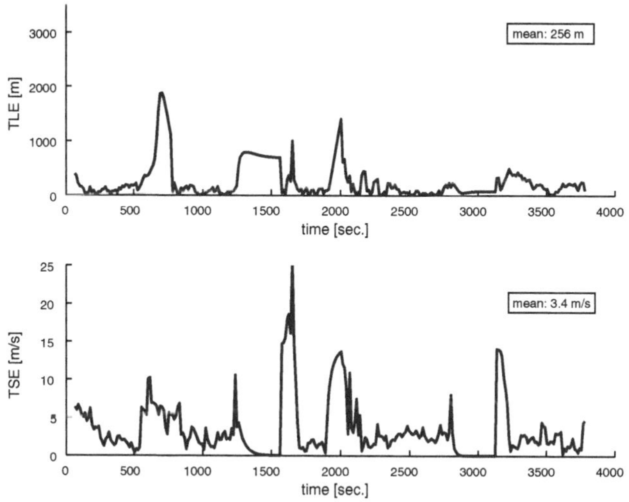

Fig. 9.14 Tracking error (localization, speed)

Table 9.1 Relevant accuracies (mean values)

Tracker	P [m]	P [m]	TLE [m]	TSE [m/s]
No road	364	109	489	4.9
Road	250	46	256	3.4

ues: 256 and 3.4 m/s, respectively). As expected, only a slight improvement in the velocity estimates is obtained by road-map information.

The next step for improving the underlying road-map to be extracted consists in applying retrodiction to this track, which is more accurate than the track used in the first step of the iteration.

In Table 9.1 the mean values of the accuracies previously discussed are summarized. P and p denote the major and minor eigenvalues of the corresponding covariance matrices. Even by this first iteration for reconstructing a road-map from GMTI tracks an error reduction in position of about 30–50 % can be obtained if this road is used for road-map-assisted tracking. We expect that by iterating this procedure under different sensor-to-target geometries and with a more refined continuous time retrodiction technique, such as described in Sect. 4, highly accurate roads can finally be obtained.

9.2.4 Summary of Observations

We discussed ground-moving vehicle tracking as a means for extracting road-map information from GMTI radar data. The resulting tracking-generated road-maps are highly up-to-date. By iteratively applying the described procedures, the produced maps can be highly precise as well. Moreover, their accuracy in each node is quantitatively described. The proposed approach to road-maps extraction is essentially based on a temporal integration of the received sensor data and by this differs in nature from methods based on pattern recognition in a single image.

We summarize some aspects, which might be of particular interest in view of sensing applications:

- Tracking-driven road extraction can be beneficial in situations or scenarios where reliable road-maps are not or not yet available, where the road-maps provided by geographical information systems are not up-to-date, or where the accuracy of existing road-maps is insufficient.
- In certain applications, roads or road-like 'lines of communication' exist only temporarily or may change with time. As practical evidence shows, even in typical off-road scenarios, structures similar to roads quickly evolve, as a 'second' vehicle usually moves in the 'tracks' of its precursor.
- As usually many targets use the same road, a significant advantage results from fusing several 'road tracks'. For airborne GMTI radar the sensor-to-target geometry is continuously changing. Therefore the fusion of 'road tracks' produced at different times improves the achievable accuracy even more.
- Sensor registration or misalignment errors usually cause serious problems in multiple sensor data fusion. For mitigating these phenomena, precisely extracted road-maps can be matched with each other, thus compensating relative bias errors. For removing absolute bias errors, matching with geo-referenced road-maps can be used.

Further Reading

A detailed discussion of this approach has been published in:

- W. Koch, J. Koller and M. Ulmke.
  Ground Target Tracking and Road-Map Extraction.
  ISPRS Journal of Photogrammetry & Remote Sensing, 61 (2006), 197–208, Elsevier.

  Abstract
  For analyzing dynamic scenarios with many ground moving vehicles, airborne Ground Moving Target Indicator (GMTI) radar is well-suited due to its wide-area, all-weather, day/night, and real time capabilities. The generation of GMTI tracks from these data is the backbone for ground surveillance and traffic flow analysis. In case of dense target situations, missing detections and false alarms, Multi-Hypotheses Tracking (MHT) is the method at choice to solve the inherent ambiguities in the data-targets assignment problem. The resulting MHT-tracks are suited to extract road-map information which is highly up-to-date and fairly precise. Moreover, their accuracy is quantitatively described. The precision of the

extracted road segments can be improved significantly using smoothed or retrodicted tracks. In turn, the extracted road information is exploited for the precise tracking of succeeding road targets. The proposed, fully Bayesian approach is illustrated by a simulated example including Doppler and terrain obscuration, providing hints to the achievable road-map accuracies.

Keywords: Tracking; Road extraction; Ground Moving Target Indicator (GMTI); Retrodiction; MHT (Multi-Hypotheses Tracking); Sequential track extraction

9.3 Integration of Ray Tracers

Complex sensing environments, such as urban terrain, come more and more into the focus of modern surveillance applications. As an example we may wish to localize an emitter in a city street using a receiver dislocated from it. The measured Direction of Arrival (DoA) of the emitted signal will generally not be related to the direct line of sight due to multipath propagation. Even if this is actually the case, we do not know it. Instead, there will be incident signals from several directions related to alternative propagation paths. To each measured DoA belongs a relative Time Difference of Arrival (rTDoA) with respect to first incoming signal.

9.3.1 Multipath Propagation Prediction

Evidently, this scenario is dominated by propagation phenomena. The key to localizing and tracking emitters in urban terrain thus lies in dealing with multipath phenomena appropriately. In urban scenarios, however, context information on the underlying road network and the location of the buildings is available. This context knowledge can serve as input for a ray-tracing algorithm to be used for evaluating the most likely propagation paths. In other words, for each randomly chosen candidate of the emitter position we can *algorithmically* calculate the likelihood function related to the measurements produced by the receiver. This pointwisely defined likelihood function can be directly exploited for emitter localization and tracking within the well established framework of particle filtering.

As an illustration, let us consider the situation shown in Fig. 9.3. OS denotes the passive receiver location, while MS indicates the position of the emitter to be localized and tracked. The buildings reflecting and refracting the emitted radiation are depicted in dark gray. The five major propagation paths are shown. In any realistic urban environment, however, propagation paths found by the ray tracer may actually be missing. This phenomenon must be modeled by the underlying likelihood function. The individual paths contribute differently to the localization accuracy achieved. The regions A, B, and C in Fig. 9.15 contain the emitter position with a probability of 95 %, where region A refers to the case where all paths are detected, in B path 5 is not detected, while in C path 3 is missing. False paths must also be taken into account and be modeled appropriately.

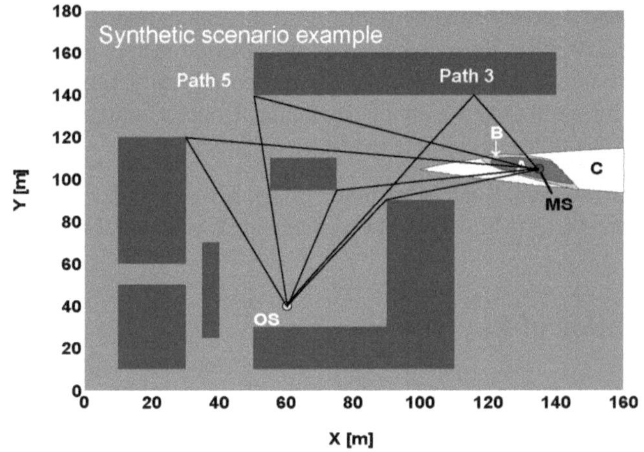

Fig. 9.15 Multipath channels in an urban environment: impact on localization

9.3.2 Particle Filter Approach

Figure 9.15 illustrates the principle of context-aided tracking, which may find application in many other fields where sophisticated models of the propagation channel exist ("ray tracers") and can be exploited for localization and tracking (ionospheric propagation such as in communications or over-the-horizon radar (OTR), shallow-water sonar, indoor navigation).

9.3.3 Conclusion

The key to blind mobile emitter tracking in urban scenarios is a likelihood function, which determines the proximity of the measured and predicted multipath components with respect to all possible association hypotheses between them. The measurements of the multipath components are provided by a receiver station equipped with an antenna array. The predicted temporal and spatial structure of the multipath components is generated by means of the 2D-ray tracking analysis using a priori information about the location of the scattering objects. In order to mitigate the impact of missing and false propagation paths on the positioning result, the measured path parameters have to be preprocessed. The likelihood function is algorithmically defined for a randomly distributed set of potential emitter MS positions and can be applied within a particle filtering realization of the Bayesian tracking paradigm. The simulation results in a synthetic environment show that the tracking technique provides a robust and accurate state initiation, which is essential for the subsequent track maintenance part of the emitter tracking algorithm.

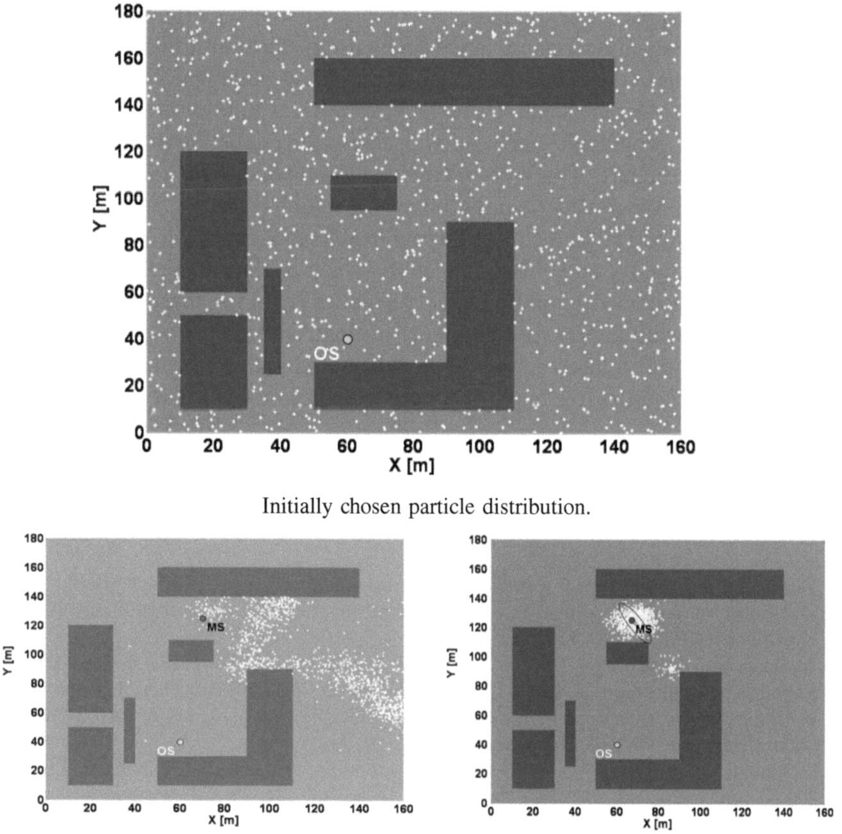

Initially chosen particle distribution.

Distribution: 1st filtering cycle

5th filtering cycle

Fig. 9.16 Emitter localization and tracking exploiting multipath propagation within a particle filter framework based on pointwisely defined likelihood functions calculated by using ray-tracing algorithms

Further Reading

A detailed discussion of this approach has been published in:

- V. Algeier, B. Demissie, W. Koch, and R. Thomä
  State Space Initiation for Blind Mobile Terminal Position Tracking
  EURASIP Journal on Advances in Signal Processing, Special Issue on Track-before-Detect Algorithms, Volume 2008 (2008), Article ID 394219, 14 pages

  Abstract
  Blind localization and tracking of mobile terminals in urban scenarios is an important requirement for offering new location-based services, handling emergency cases of unsubscribed users, public safety, countering IEDs, and so forth. In this

context, we propose a track-before-detect scheme that takes explicit advantage of multipath propagation in an urban terrain by using a priori information about the known locations of the main scattering objects such as buildings. This information is made available for localization and tracking by a real-time ray-tracing technique based on a 2D geographic database. This allows the prediction of the directional and temporal structure of the received multipath components for an arbitrary transmitter position. We consider a single observing station where the direction and the relative time of arrival of the received multipath components can be estimated by an antenna array. By a likelihood function, which is algorithmically defined for a randomly distributed set of potential transmitter positions, these measurements are compared with those expected by ray tracing. This likelihood function is the key component of a track-before-detect scheme providing initial state estimates for mobile transmitter tracking using a particle filtering technique.

References

1. M. Wunder, J. Grosche (eds.), *Verteilte Führungsinformationssysteme*. (Springer, Berlin, 2009)
2. M. Ulmke, W. Koch, Road-map assisted ground moving target tracking. IEEE Trans. Aerosp. Electron. Syst. AES-**42**(4), 1264–1274 (2006)
3. W. Koch, in *Ground Target Tracking with STAP Radar: Selected Tracking Aspects*, ed. by R. Klemm. The Applications of Space-Time Adaptive Processing, (IEE Publishers, London, 2004)
4. W. Ulmke, W. Koch, On Road-Map Assisted GMTI Tracking, in *Proceedings of: DGON German Radar Symposium GRS'02*, Bonn, Sept 2002, pp. 89–93
5. R. Klemm, *Principles of space-time adaptive processing*, 3rd edn. (IEE Publishers, London, 2006)

Chapter 10
Feed-Back to Acquisition: Sensor Management

Modern active phased-array radar [1] is an example of a multifunctional sensor system that requires sophisticated sensor management algorithms for its efficient operation. Such systems call for efficient exploitation of their degrees of freedom, which are variable over a wide range and may be chosen individually for each track. This is especially true in multiple object tracking tasks. Of special interest are air situations with agile objects significantly differing in their radar cross section. Unless properly handled, such situations can be highly allocation time- and energy-consuming. In this context, advanced sensor and dynamics models for combined tracking and sensor management are discussed, i.e. control of data innovation intervals, radar beam positioning, and transmitted energy management. By efficiently exploiting its limited resources, the total surveillance performance of such sensor systems can be significantly improved.

For track-while-scan sensor systems or operating modes, data acquisition and tracking are completely decoupled. For phased-array radar operated in an active tracking mode, however, the current signal-to-noise ratio of the object (i.e. the detection probability) strongly depends on the correct positioning of the pencil-beam, which is now taken into the responsibility of the tracking system. Sensor control and data processing are thus closely interrelated. This basically *local* character of the tracking process constitutes the principal difference between phased-array and track-while-scan applications from a tracking point of view. By using suitable sensor and object evolution models, however, this fact can be incorporated into the Bayesian formalism. The potential of this approach thus also applies to phased-array radar. The more difficult problem of global optimization, taking successive allocations into account, is not addressed here.

The discussion of a sensor management example seems to be well-suited as a concluding chapter of Part II of this thesis. Many methods discussed in Part I and specialized aspects of sensor and objects modeling must be combined appropriately. The methodologies discussed below were inspired and in early parts evaluated by the experimental system ELRA [1, Chap. 17]. Their applicability, however, includes modern on-board radar systems for interceptor aircraft and multifunctional radar systems for wide-area ground and sea surveillance. This application example thus

W. Koch, *Tracking and Sensor Data Fusion*,
Mathematical Engineering, DOI: 10.1007/978-3-642-39271-9_10,

illustrates in a particular way the inner structure and practical use of the underlying Bayesian formalism. The very success of the Bayesian paradigm may serve as retrospective justification of the approach as well as a motivation to apply this formalism to an even broader field of tracking, data fusion and sensor management applications. The material presented here is essentially taken from [2, Chap. 12].

10.1 Information Flow in Agile-Beam Radar

A simplified scheme illustrating the information flow in tracking-driven phased-array radar management is shown by Fig. 10.1. The starting point is the tracking system, which generates a request for new sensor information based on the current quality of an already established individual object track or on the requirement of initiating new tracks. We thus distinguish between track update and search requests, which enter into the priority management unit where its rank is evaluated based on the current threat or overload situation, for example, thus enabling graceful system degradation when necessary.

For each preparation of a radar system allocation, track-specific radar parameters must be set, such as the calculated radar revisit time and the corresponding radar

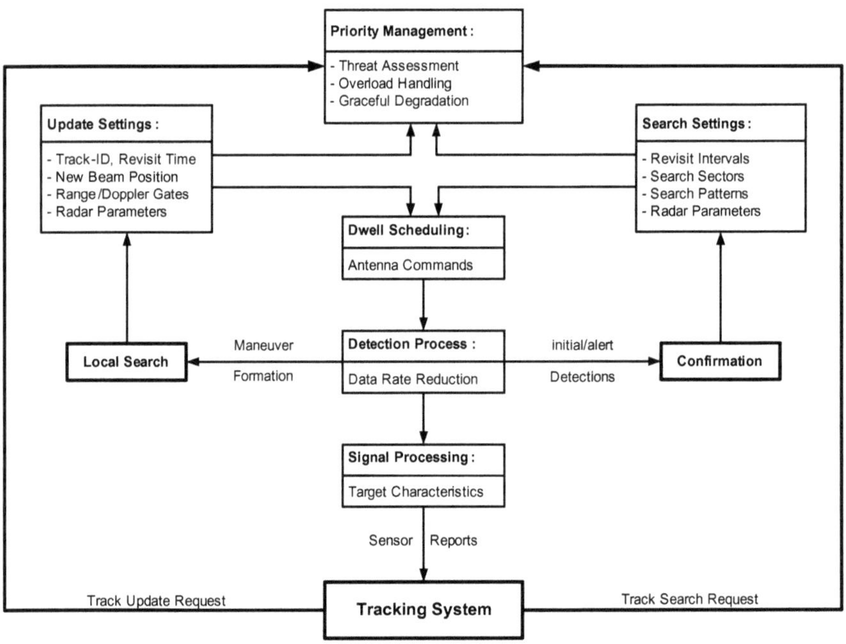

Fig. 10.1 Simplified scheme of the information flow in tracking-driven phased-array radar management

beam position, range- and Doppler-gates, or the type of the radar waveforms to be transmitted. Track search requests require the setting of appropriate revisit intervals, search sectors and patterns, and other radar parameters. In the dwell-scheduling unit, these preparations are transformed into antenna commands, by which the radar sensor is allocated and radar energy transmitted. The received echo signals pass a detection unit. If no detection occurs in the track maintenance mode, a local search procedure is initiated, new radar parameters are set, and a subsequent radar sensor allocation is started with as small a time delay as possible. This local search loop is repeated until either a valid detection is produced or the track is canceled. While a new beam position according to a global or sector search pattern is calculated if no detection occurs in the track search mode, a tentative detection has to be confirmed before a new track is finally established. After a successful detection, the received signal passes the signal processing unit, where characteristic object parameters, such as object range, azimuth angle, radial velocity, and the object strength, are estimated. These are the inputs for the tracking system. This closes the data processing and sensor management loop.

In certain applications, distinct maneuvering phases often exist, since even agile objects do not maneuver permanently. Nevertheless, abrupt transitions to high-g turns can easily occur. Allocation time and energy savings are thus to be expected if adaptive dynamics models of the object dynamics are used. Besides their kinematic characteristics, the mean radar cross section (RCS) of the objects to be tracked is usually unknown and variable over a wide range. By processing signal amplitude information, however, the energy spent for track maintenance can be adapted to the actual object strength. In this way, the total sensor load can also be significantly reduced.

Due to the locally confined object illumination by the pencil-beam of a phased-array radar, abrupt transitions into maneuvering flight phases are critical since, in contrast to more conventional track-while-scan radar, a periodic object illumination is no longer guaranteed. Any track reinitiation is thus highly allocation time- and energy-consuming. Track reinitiation is also in conflict with other sensor tasks, such as weapon guidance or providing communications links. This calls for intelligent algorithms for beam positioning and local search [3–5] that are crucial to phased-array radar tracking.

Resource management for a multi-functional radar depends on the particular application considered. We here discuss track maintenance for ground-based air surveillance while minimizing the allocation time and energy required. Track accuracy is important only insofar as stable tracks are guaranteed. Track initiation or implementation issues are not addressed here. To make the benefits of IMM modeling and amplitude information clearly visible, false detections (clutter, electronic counter measures), data association conflicts, or possibly unresolved measurements were excluded. Nevertheless, their impact might well be incorporated into the general Bayesian framework [6].

10.2 Sensor Modeling for Phased-Array Radar

In phased-array radar tracking, additional sensor information can be acquired when needed. Before each "radar resource allocation" [7], certain radar parameters must be selected by the tracking system depending on the current lack of information. We here consider the *object revisit time* t_k, the current *beam position* $\mathbf{b}_k$, i.e. a unit vector pointing into the direction where radar energy is to be transmitted, and the transmitted *energy per dwell* e_k. Other radar parameters (detection threshold λ_D, radar beam width B) are assumed to be constant for the sake of simplicity, i.e. we neglect the dependence of the radar beam width on the beam position, for example. After processing the skin echo produced by the illuminated object, the resource allocation R_k at time t_k results in measurements of direction cosines of the object and the object range, $\mathbf{z}_k = (\bar{u}_k, \bar{v}_k, \bar{r}_k)$, along with the signal amplitude a_k. A single dwell may be insufficient for object detection and subsequent fine localization. Let n_{B_k} denote the number of dwells needed for a successful detection and $B_k = \{\mathbf{b}_k^i\}_{i=1}^{n_{B_k}}$ the set of the corresponding beam positions. Each radar allocation is thus characterized by the tuple $R_k = (t_k, B_k, n_{B_k}, e_k, \mathbf{z}_k, a_k)$. The sequence of successive allocations is denoted by $\mathcal{R}^k = \{R_k, \mathcal{R}^{k-1}\}$.

10.2.1 Radar Cross Section Fluctuations

The instantaneous radar cross section σ_k of realistic objects strongly depends on the radar frequency used and the current aspect angle. For this reason, statistical models are used to describe the backscattering properties of the objects. In many practical cases, σ_k is described by gamma-densities,

$$p(\sigma_k|\bar{\sigma}, m) = \mathcal{G}_m\big(\sigma_k;\ \bar{\sigma},\ m\big) \tag{10.1}$$

$$= \frac{(m/\bar{\sigma})^m}{\Gamma(m)}\ \sigma_k^{m-1}\ \mathrm{e}^{-\sigma_k m/\bar{\sigma}}. \tag{10.2}$$

In this equation $\bar{\sigma}$ denotes the mean RCS of the object that is usually unknown, but constant in time and characteristic of a certain class of objects, while the parameter m denotes the number of "degrees of freedom". The individual samples σ_k are assumed to be statistically independent for subsequent dwells (guaranteed by frequency decorrelation, e.g.). The cases $m = 1, 2$ are referred to as Swerling-I and -III fluctuations [8].

Let the instantaneous object signal $\mathbf{v}_k = (v_1, v_2)$ with the two orthogonal signal components v_1 and v_2 be additively corrupted by Gaussian noise with variance σ_n^2 according to the standard modeling assumptions [8]. Since the signal components are assumed to be statistically independent, the pdf of the resulting sensor signal $\mathbf{s}_k = (s_1, s_2)$ is

$$p(\mathbf{s}_k|\mathbf{v}_k) = \mathcal{N}\big(s_1;\ v_1,\ \sigma_n^2\big)\ \mathcal{N}\big(s_2;\ v_2,\ \sigma_n^2\big). \tag{10.3}$$

The normalized scalar quantity $a_k^2 = (s_1^2 + s_2^2)/2\sigma_n^2$, derived from $\mathbf{s}_k$, is thus Rice-distributed [8]: $p(\mathcal{A}^k|\mathrm{sn}_k) = \mathrm{e}^{-\mathcal{A}^k - \mathrm{sn}_k} I_0\big(2a_k\sqrt{\mathrm{sn}_k}\big)$ with $\mathrm{sn}_k = (v_1^2+v_2^2)/2\sigma_n^2$. Hence, sn_k denotes the instantaneous signal-to-noise ratio of the object being proportional to the instantaneous radar cross section σ_k. The expectation value of a_k^2 with respect to $p(\mathcal{A}^k|\mathrm{sn}_k)$ is given by $\mathbb{E}[\mathcal{A}^k] = 1 + \mathrm{sn}_k$. According to the normalization chosen, pure noise ($\mathrm{sn}_k = 0$) has thus unit power. Due to the RCS model previously discussed, sn_k is gamma-distributed with the mean SNR: $p(\mathrm{sn}_k|\mathrm{SNR}) = \mathcal{G}_m\big(\mathrm{sn}_k;\ \mathrm{SNR},\ m\big)$. The conditional density of $\mathcal{A}^k$ given SNR is thus obtained by calculating:

$$p(\mathcal{A}^k|\mathrm{SNR}) = \int_0^\infty d\mathrm{sn}_k\ p(\mathcal{A}^k|\mathrm{sn}_k)\ p(\mathrm{sn}_k|\mathrm{SNR}). \tag{10.4}$$

The integration can be carried out (see [9], e.g.) yielding:

$$p(\mathcal{A}^k|\mathrm{SNR}) = \big(\tfrac{m+\mathrm{SNR}}{m}\big)^{-m}\ \mathrm{e}^{-m\mathcal{A}^k/(m+\mathrm{SNR})}\ L_{m-1}\big(\tfrac{-\mathcal{A}^k\mathrm{SNR}}{m+\mathrm{SNR}}\big) \tag{10.5}$$

where L_{m-1} denotes the Laguerre polynomials. For Swerling-I/III these polynomials are given by: $L_0(-x) = 1$, $L_1(-x) = 1 + x$. Obviously, $p(\mathcal{A}^k|\mathrm{SNR})$ can be interpreted as a gamma mixture with the expectation value $\mathbb{E}[\mathcal{A}^k] = 1 + \mathrm{SNR}$.

10.2.2 Mean Received Signal-to-Noise Ratio

Any sensor model for phased-array radar tracking has to provide a functional relationship linking the expected signal-to-noise ratio SNR_k at the revisit time t_k, the sensor parameters considered (here: transmitted energy, beam position) and the relevant object parameters (mean RCS, object position). With a Gaussian beam form model [3], well proven in applications, the radar range equation (see [8], e.g.), we assume:

$$\mathrm{SNR}_k = \mathrm{SN}_0\ \big(\tfrac{\bar\sigma}{\bar\sigma_0}\big)\ \big(\tfrac{e_k}{e_0}\big)\ \big(\tfrac{r_k}{r_0}\big)^{-4}\ \mathrm{e}^{-\log 2\Delta b_k} \tag{10.6}$$

$$\text{with}\quad \Delta b_k = |\mathbf{d}_k - \mathbf{b}_k|^2/B^2. \tag{10.7}$$

r_k is the actual object range at time t_k, while $\mathbf{d}_k = (u_k, v_k)^\top$ denotes the related direction cosines. With the beam position $\mathbf{b}_k = (b_k^u, b_k^v)^\top$ and the (one-sided) beam width B, Δb_k is a measure of relative beam positioning error. For $\Delta b_k = 1$, the signal-to-noise ratio is reduced by a factor of 2. The radar parameter SN_0 is the expected mean signal-to-noise ratio of a object with a standard mean cross section $\bar\sigma_0$ at a reference range r_0 that is directly ($\Delta b_k = 0$) illuminated by the beam with the energy e_0. Due to the functional relationship stated in Eq. 10.5, the signal strength $\mathcal{A}^k$ can be interpreted as a measurement of $\bar\sigma$.

10.2.3 Detection and Measurement Process

A detection is assumed if the received signal strength exceeds a certain detection threshold: $\mathcal{A}^k > \lambda_D$. For a given m in the fluctuation model (Eq. 10.2), the detection probability P_D is a function of SNR and λ_D:

$$P_D(\text{SNR}, \lambda_D, m) = \int_{\lambda_D}^{\infty} d\mathcal{A}^k \; p(\mathcal{A}^k|\text{SNR}). \tag{10.8}$$

The false alarm probability P_F is analogously obtained:

$$P_F(\lambda_D) = P_D(0, \lambda_D, m) = \mathrm{e}^{-\lambda_D}. \tag{10.9}$$

Integration results in explicit expressions for P_D [8]. For Swerling-I/III fluctuations, we obtain:

$$P_D^I(\text{SNR}, \lambda_D) = \mathrm{e}^{-\frac{\lambda_D}{1+\text{SNR}}} = P_F^{\frac{1}{1+\text{SNR}}} \tag{10.10}$$

$$P_D^{III}(\text{SNR}, \lambda_D) = \mathrm{e}^{-\frac{\lambda_D}{1+\text{SNR}/2}} \left(1 + \tfrac{(\text{SNR}/2)\lambda_D}{(1+\text{SNR}/2)^2}\right). \tag{10.11}$$

For object tracking $\mathcal{A}^k$ is available after a detection, i.e. $\mathcal{A}^k > \lambda_D$. We thus need the conditional density:

$$p(\mathcal{A}^k|\mathcal{A}^k > \lambda_D, \text{SNR}, m) = \begin{cases} \frac{p(\mathcal{A}^k|\text{SNR})}{P_D(\text{SNR},\lambda_D,m)} & \text{for} \quad \mathcal{A}^k > \lambda_D \\ 0 & \text{else} \end{cases}. \tag{10.12}$$

For strong objects, we can assume SNR $\approx 1 + \text{SNR} \approx \ldots \approx m + \text{SNR}$ and thus approximately obtain: $p(\mathcal{A}^k|\text{SNR}) \approx \left(\frac{\text{SNR}}{m}\right)^{-m} \mathrm{e}^{-m\mathcal{A}^k/\text{SNR}} \, L_{m-1}\left(-\mathcal{A}^k\right)$, which is similar to the expression in Eq. 10.5. On the other hand, let the detection probability for $m \neq 1$ be approximately given by: $P_D(\text{SNR}, \lambda_D, m) \approx P_D^I(\text{SNR}, \lambda_D)$ (i.e. by a Swerling-I-model). We can therefore write: $p(a|\mathcal{A}^k > \lambda_D, \text{SNR}, m) \approx \mathcal{S}_m\left(\mathcal{A}^k;\, \text{SNR},\, m\right)$ with:

$$\mathcal{S}_m\left(\mathcal{A}^k;\, \text{SNR},\, m\right) = \begin{cases} \left(\frac{\text{SNR}}{m}\right)^{-m} \mathrm{e}^{-(m\mathcal{A}^k+\lambda_D)/\text{SNR}} \, L_{m-1}\left(-\mathcal{A}^k\right) & \text{for} \mathcal{A}^k > \lambda_D \\ 0 & \text{else.} \end{cases} \tag{10.13}$$

Let us furthermore assume that monopulse localization after detection results in bias-free measurements u_k', v_k' of the direction cosines and range with Gaussian measurement errors. According to [8], the standard deviations $\sigma_k^{u,v}$ depend on the beam width B and the instantaneous sn_k in the following manner: $\sigma_k^{u,v} \propto B/\sqrt{\text{sn}_k} \approx B/\sqrt{\mathcal{A}^k - 1}$. Since sn_k is unknown, in the last approximation $\mathcal{A}^k$ is used as a bias-free estimate of sn_k ($\mathbb{E}[\mathcal{A}^k] = 1 + \text{sn}_k$). The range error is assumed to be Gaussian with a constant standard deviation σ^r, which can be related to the size r_{res} of discrete range resolution cells via: $\sigma^r = r_{\text{res}}/\sqrt{12}$. Evidently, this model of the measurement process does not depend on the RCS fluctuation model.

10.3 Bayesian Tracking Algorithms Revisited

According to the previous discussion, object tracking is an iterative updating scheme for conditional probability densities $p(\mathbf{x}_k|\mathcal{R}^k)$ that describe the current object state $\mathbf{x}_k$ given all available resource allocations $\mathcal{R}^k$ and the underlying a priori information in terms of statistical models. The processing of each new measurement $\mathbf{z}_k$ via Bayes' Theorem establishes a recursive relation between the densities at two consecutive revisit times (a prediction step followed by filtering).

10.3.1 Predictions: Basis for Allocation Decisions

The knowledge of the object state at time t_k before a new radar allocation has taken place is thus given by $p(\mathbf{x}_k|\mathcal{R}^{k-1})$. Allocation decisions for a certain time t_k must thus be based on this prediction, which essentially depends on the underlying dynamics model. For IMM dynamics model, $p(\mathbf{x}_k|\mathcal{R}^{k-1})$ is given by a finite mixture density:

$$p(\mathbf{x}_k|\mathcal{R}^{k-1}) = \sum_{\mathbf{j}_k} p^{\mathbf{j}_k}_{k|k-1} \mathcal{N}\big(\mathbf{x}_k;\ \mathbf{x}^{\mathbf{j}_k}_{k|k-1},\ \mathbf{P}^{\mathbf{j}_k}_{k|k-1}\big) \tag{10.14}$$

with $\mathbf{j}_k = (j_k, \ldots, j_{k-n+1})$ denoting a particular *model history*, i.e. a sequence of possible hypotheses regarding the object dynamics model from a certain observation at time t_{k-n+1} up to the most recent measurement at time t_k ("n scans back"). In the case of a single dynamics model ($r = 1$), the prediction densities $p(\mathbf{x}_k|\mathcal{R}^{k-1})$ are strictly given by Gaussians (standard Kalman prediction). For $n = 1$, $p(\mathbf{x}_k|\mathcal{R}^{k-1})$ is approximated by a mixture with r components according to the r dynamics models used. GPB2 and standard IMM algorithms are possible realizations of this scheme [10]. For standard IMM, the approximations are made after the prediction, but before the filtering step, while for GPB2 they are applied after the filtering step. Hence, GPB2 requires more computational effort. For details see [10].

10.3.2 Processing of Signal Strength Information

Let us treat the normalized mean RCS of the object, $s_k = \bar{\sigma}_k/\bar{\sigma}_0$, as an additional component of the state vector. Since the signal strength after a detection occurred may be viewed as a measurement of s_k, let us consider the augmented conditional density

$$p(\mathbf{x}_k, s_k|\mathcal{R}^k) = p(s_k|\mathbf{x}_k, \mathcal{R}^k)\, p(\mathbf{x}_k|\mathcal{R}^k). \tag{10.15}$$

The calculation of $p(\mathbf{x}_k|\mathcal{R}^k)$ was discussed in Sect. 10.2. For the remaining density $p(s_k|\mathbf{x}_k, \mathcal{R}^k)$, an application of Bayes' Theorem yields up to a normalizing constant:

$$p(s_k|\mathbf{x}_k, \mathcal{A}^k, \mathcal{R}^{k-1}) \propto \mathcal{S}_m\big(\mathcal{A}^k;\ \text{SNR},\ m\big)\ p(s_k|x_k, R^{k-1}). \tag{10.16}$$

Let us furthermore assume that $p(s_k|\mathbf{x}_k, \mathcal{R}^{k-1})$ are given by inverse gamma densities,

$$p(s_k|\mathbf{x}_k, \mathcal{R}^{k-1}) = \mathcal{I}_{\mu_{k|k-1}}\big(s_k;\ \hat{s}_{k|k-1},\ \mu_{k|k-1}\big), \tag{10.17}$$

which are defined by:

$$\mathcal{I}_\mu\big(s;\ \hat{s},\ \mu\big) = \big[((\mu-1)\hat{s})^\mu/\Gamma(\mu)\big]\, s^{-\mu-1}\, \mathrm{e}^{-\frac{(\mu-1)\hat{s}}{s}}, \tag{10.18}$$

where $\hat{s}$ is the expectation of this density, $\hat{s} = \mathbb{E}[s] > 0$, μ a parameter $\mu > 1$. For $\mu > 2$, the related variance exists: $\mathbb{V}[s] = \hat{s}^2/(\mu-2)$. This class of densities is invariant under the successive application of Bayes' Theorem according to Eq. 10.16, since up to normalization we obtain:

$$\mathcal{S}_m\big(\mathcal{A}^k;\ \text{SNR},\ m\big)\, \mathcal{I}_{\mu_{k|k-1}}\big(s_k;\ \hat{s}_{k|k-1},\ \mu_{k|k-1}\big) \tag{10.19}$$

$$\propto \alpha_k^{-m}\ s_k^{-\mu_{k|k-1}-m-1}\ \exp\Big(-\frac{(\mu_{k|k-1}-1)\hat{s}_{k|k-1}+\frac{m\mathcal{A}^k+\lambda_D}{\alpha_k}}{s_k}\Big) \tag{10.20}$$

$$\propto \mathcal{I}_{\mu_k}\big(s_k;\ \hat{s}_k,\ \mu_k\big), \tag{10.21}$$

where the parameters α_k, $\hat{s}_k$, and μ_k are given by:

$$\alpha_k = \text{SN}_0\, \Big(\frac{e_k}{e_0}\Big)\, \Big(\frac{r_k}{r_0}\Big)^{-4}\, \mathrm{e}^{-2\Delta b_k}, \tag{10.22}$$

$$\hat{s}_k = \frac{\mu_{k|k-1}-1}{\mu_{k|k-1}+m-1}\hat{s}_{k|k-1} + \frac{(m\mathcal{A}^k+\lambda_D)/\alpha_k}{\mu_{k|k-1}+m-1} \tag{10.23}$$

$$\mu_k = \mu_{k|k-1} + m. \tag{10.24}$$

With reference to s_k the density $\mathcal{I}_{\mu_k}\big(s_k;\ \hat{s}_k,\ \mu_k\big)$ is correctly normalized. Evidently, α_k depends on the object position ($\alpha_k = \alpha_k(r_k, u_k, v_k)$). In order to preserve the factorization of $p(\mathbf{x}_k, s_k|\mathcal{R}^k)$ in a normal mixture related to the kinematic properties of the object $\mathbf{x}_k$ and an inverse gamma density related to its RCS s_k, we use the approximation:

$$\alpha_k \approx \Big(\frac{e_k}{e_0}\Big)\, \Big(\frac{\hat{\mathbf{r}}_k}{r_0}\Big)^{-4}\, \mathrm{e}^{-2\{(\hat{\mathbf{u}}_k-b_k^u)^2+(\hat{\mathbf{v}}_k-b_k^v)^2\}/B^2}, \tag{10.25}$$

where $\hat{\mathbf{r}}_k$, $\hat{\mathbf{u}}_k$, $\hat{\mathbf{v}}_k$ are the MMSE estimates for r_k, u_k and v_k derived from $p(\mathbf{x}_k|\mathcal{R}^k)$. Hence, α_k compensates both the estimated positioning error of the radar beam and the propagation loss due to the radar equation. Assuming s_k to be constant, we have

$\mathcal{I}_{\mu_{k|k}}(s_k;\ \hat{s}_{k|k-1},\ \mu_{k|k}) = \mathcal{I}_{\mu_{k|k-1}}(s_k;\ \hat{s}_{k-1},\ \mu_{k|k-1})$. In principle, a dynamics model describing temporal changes of the radar cross section could be introduced.

10.4 Adaptive Bayesian Sensor Management

The predicted tracking performance is essential for adaptive radar revisit time control, the selection of the transmitted radar energy, and the design of intelligent algorithms for local search.

10.4.1 Adaptive Radar Revisit Time Control

The time t_k when a radar allocation R_k should take place is determined by the current lack of information conveniently described [3] by the error covariance matrix $\mathbf{P}_{k|k-1}$ of the predicted state estimate $\mathbf{x}_{k|k-1}$. Since $p(\mathbf{x}_k|\mathcal{R}_{k-1})$ is a normal mixture, $\mathbf{x}_{k|k-1}$ and $\mathbf{P}_{k|k-1}$ are given by:

$$\mathbf{x}_{k|k-1} = \sum_{\mathbf{j}_k} p^{\mathbf{j}_k}_{k|k-1}\, \mathbf{x}^{\mathbf{j}_k}_{k|k-1} \tag{10.26}$$

$$\mathbf{P}_{k|k-1} = \sum_{\mathbf{j}_k} p^{\mathbf{j}_k}_{k|k-1}\, \Big(\mathbf{P}^{\mathbf{j}_k}_{k|k-1} \tag{10.27}$$

$$+ (\mathbf{x}^{\mathbf{j}_k}_{k|k-1} - \mathbf{x}_{k|k-1})\, (\mathbf{x}^{\mathbf{j}_k}_{k|k-1} - \mathbf{x}_{k|k-1})^{\top}\Big). \tag{10.28}$$

The covariance matrix $\mathbf{P}^{\mathbf{j}_k}_{k|k-1}$ of the individual mixture components grow the faster in time the more often maneuvers are assumed in the corresponding model histories. This has an impact on the total covariance matrix $\mathbf{P}_{k|k-1}$ according to the corresponding weighting factors $p^{\mathbf{j}_k}_{k|k-1}$. In addition, $\mathbf{P}_{k|k-1}$ is "broadened" by the positively definite spread terms $(\mathbf{x}^{\mathbf{j}_k}_{k|k-1} - \mathbf{x}_{k|k-1})\, (\mathbf{x}^{\mathbf{j}_k}_{k|k-1} - \mathbf{x}_{k|k-1})^{\top}$. Obviously, the adaptive IMM modeling affects $\mathbf{P}_{k|k-1}$ in a rather complicated way.

A scalar measure of the information deficit is provided, e.g., by the largest eigenvalue of the covariance matrix of the predicted object direction (in terms of u, v). Let it be denoted by $G_{k|k-1}$. A track update is allocated when the $G_{k|k-1}$ exceeds a predetermined proportion of the squared radar beam width B:

$$G_{k|k-1} > (v_0 B)^2. \tag{10.29}$$

The *relative track accuracy* v_0 introduced by this criterion is a measure of the minimum track quality required and a parameter to be optimized. In many practical applications, $v_0 = 0.3$ is a reasonable choice [3].

10.4.2 Transmitted Radar Energy Selection

In view of the tracking system, the sensor performance is mainly characterized by the signal-to-noise ratio that determines both, the detection probability and the measurement error. By suitably choosing the transmitted energy per dwell e_k, the expected signal-to-noise ratio $\mathrm{SN}_{k|k-1}$ can be kept constant during tracking. Besides v_0, $\mathrm{SN}_{k|k-1}$ is an additional parameter subject to optimization. Since v_0 may be viewed as a measure of the beam positioning error, the energy e_k at time t_k is defined by this condition (Eq. 10.6):

$$\mathrm{SN}_{k|k-1} \overset{!}{=} \text{const.} \tag{10.30}$$

$$\rightarrow \qquad \tfrac{e_k}{e_0} = \left(\tfrac{\mathrm{SN}_0}{\mathrm{SN}_{k|k-1}}\right) \left(\tfrac{\bar{\sigma}_0}{\bar{\sigma}}\right) \left(\tfrac{\hat{\mathbf{r}}_{k|k-1}}{r_0}\right)^4 \mathrm{e}^{2v_0^2}. \tag{10.31}$$

By this particular choice, the influence of the radar range equation is compensated (at least for a certain range interval). For the mean radar cross section $\bar{\sigma}$ either a worst-case assumption or estimates from object amplitude information can be used. The track quality v_0 also affects the transmitted energy. As a side effect of this choice, the standard deviations $\sigma_k^{u,v}$ of the u, v-measurements are kept constant on average.

10.4.3 Bayesian Local Search Procedures

Intelligent algorithms for beam positioning and local search are crucial for IMM-type phased-array tracking. Overly simple strategies may easily destroy the benefits of the adaptive dynamics model, because track loss immediately after a model switch can easily occur. To avoid this phenomenon, we adapt the optimal approach based on the predicted densities $p(\mathbf{x}_k|\mathcal{R}^{k-1})$ proposed in [3] to IMM tracking [4].

1. The beam position $\mathbf{b}_k^1$ of the first dwell at time t_k is simply given by the predicted direction $\mathbf{d}_{k|k-1}$ to be derived from the predicted density function $p(\mathbf{x}_k|\mathcal{R}^{k-1})$.
2. If no detection occurs in the first dwell, even this non-detection provides useful information on the target. We thus have to calculate the conditional density of the target state given the event $\neg \mathrm{D}_k^1$: 'no detection at time t_k in the direction $\mathbf{b}_k^1$.'
3. An application of Bayes' Theorem directly yields:

$$p(\mathbf{d}_k|\neg \mathrm{D}_k^1, \mathcal{R}^{k-1}) \propto \left(1 - P_D(\mathbf{d}_k; \mathbf{b}_k^1)\right) p(\mathbf{d}_k|\mathcal{R}^{k-1}) \tag{10.32}$$

   up to a normalizing factor. In this expression, the detection probability P_D depends on the expected SNR (Eq. 10.6) and thus on the current beam and target position $\mathbf{b}_k$, $\mathbf{d}_k$.
4. The two dimensional density $p(\mathbf{d}_k|\neg \mathrm{D}_k^1, \mathcal{R}^{k-1})$ can easily be calculated on a grid. The beam position for the next dwell is then simply provided by its maximum.

5. This computational scheme for Bayesian local search is repeated until a detection occurs. Since the maximum of $p(\mathbf{d}_k|\neg\mathrm{D}_k^1, \neg\mathrm{D}_k^2, \dots, \mathcal{R}^{k-1})$ is searched, the computation of the normalization integral is not required. Numerically efficient realizations are possible.

Alternatively, $p(\mathbf{d}_k|\neg\mathrm{D}_k^1, \mathcal{R}^{k-1})$ might be used for calculating the expected SNR in a certain direction $\mathbf{b}_k$:

$$\mathrm{SNR}(\mathbf{b}_k) = \int d\mathbf{d}_k \; \mathrm{SNR}(\mathbf{b}_k, \mathbf{d}_k) \; p(\mathbf{d}_k|\neg\mathrm{D}_k^1, \mathcal{R}^{k-1}).$$

Searching the maximum of $\mathrm{SNR}(\mathbf{b}_k)$ results in a different local search strategy. In the examples considered below, however, no significant performance improvements were observed. Nevertheless, there might be applications where the maximization of $\mathrm{SNR}(\mathbf{b}_k)$ is advantageous (e.g. for track recovery in case of intermittent operating modes).

This local search scheme exploits *'negative' evidence*, as also here the lack of an expected measurement carries information on the current target position. In particular, we here observe a direct impact on adaptive sensor management. Again, the prerequisite for dealing with negative evidence is an adequate sensor performance model. As in the case of resolution phenomena (Sect. 10.2), the processing of negative sensor evidence implies mixture densities with potentially *negative* mixture coefficients, i.e. not each mixture component has a direct probabilistic interpretation. Since the mixture coefficients sum up to one, the overall density still has a well-defined probabilistic meaning.

Figure 10.2 illustrates this scheme of Bayesian local search for a particular example. In Fig. 10.2a, the predicted pdf $p(\mathbf{d}_k|\mathcal{R}^{k-1})$, a mixture density, is shown for some time t_k. The target is expected to be in the bright region with high probability; the true target position is indicated by a green dot. The blue dot denotes the beam position of the next dwell. The related detection probability is 26 %. However, no detection occurred during the first dwell. We thus calculate the conditional pdf $p(\mathbf{d}_k|\neg\mathrm{D}_k^1, \mathcal{R}^{k-1})$ given that event. As visible in Fig. 10.2b, it differs significantly from $p(\mathbf{d}_k|\mathcal{R}^{k-1})$. The previous maximum decreased in height, while the global maximum is at a different location. Again no detection occurred; the resulting density $p(\mathbf{d}_k|\neg\mathrm{D}_k^1, \neg\mathrm{D}_k^2, \mathcal{R}^{k-1})$ reflecting the two pieces of 'negative' evidence $\neg\mathrm{D}_k^1$ and $\neg\mathrm{D}_k^2$ is shown in Fig. 10.2c. Now the search algorithm decides to look again near the position at dwell 1. Although wrong in this case, this does not seem to be unreasonable. In addition, two smaller local maxima appear that increase in size as in the next dwell no detection occurs either. According to Fig. 10.2d, the next decision is ambiguous. We finally obtain a decision which leads to success. The last picture shows the updated pdf (Fig. 10.2f).

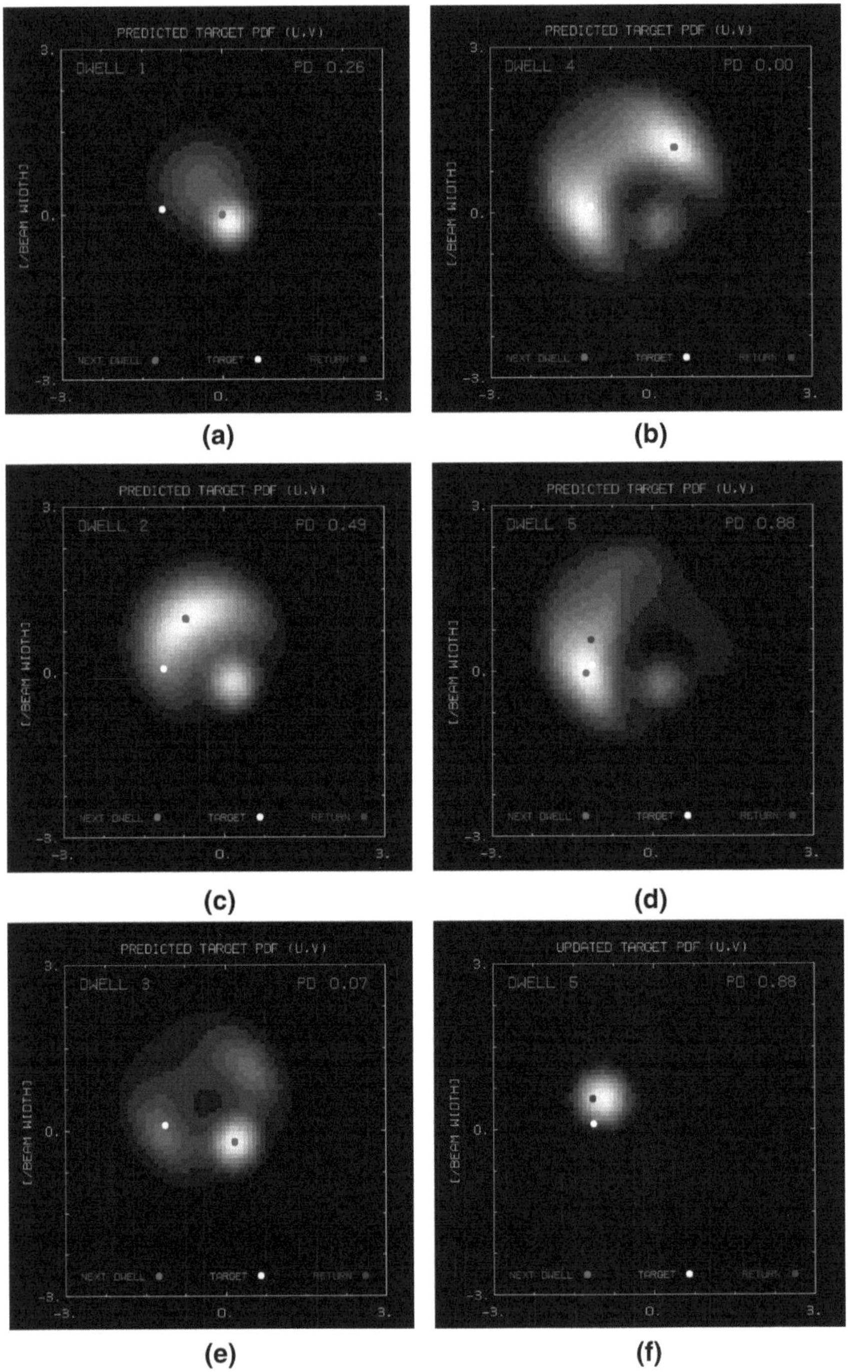

Fig. 10.2 Bayesian local search: five consecutive dwells

10.5 Discussion of Numerical Simulation Results

Simulation results provide hints as to what extent the total performance of multiple-object air surveillance by phased-array can be improved by using adaptive techniques for combined tracking and sensor control. The following four questions are addressed:

1. What resource savings (allocation time, energy) can be expected by using adaptive dynamics models?
2. How should the IMM dynamics modeling be designed (e.g. number of models, transition matrix)?
3. What energy savings can be expected if object amplitude information for sensor control is exploited?
4. Why is Bayesian local search important when adaptive dynamics models are used for revisit time control?

10.5.1 Discussion of Simulation Scenarios

In general we follow the parameter and threshold settings recommended in [3]. To exclude false alarms due to receiver noise, the false alarm probability is $P_F = 10^{-4}$. False returns due to clutter or ECM are not considered. The standard deviation of the measurement errors in object range is $\sigma_r = 100\,\text{m}$, while the the radar beam width is $B = 1°$. We assume a minimum time interval of 20 ms between consecutive dwells on a particular object and statistically independent signal amplitudes (achievable by frequency agility, e.g.). The reference range is set to $r_0 = 80\,\text{km}$.

IMM Modeling Parameters

Antenna coordinates (direction cosines, range) are used also for tracking; non-linearities introduced by these non-Cartesian coordinates are taken into account [6]. In each component u_k, v_k, r_k the state vector is given by position, speed, and acceleration. For the sake of simplicity, we consider a block diagonal system matrix defined by

$$\mathbf{F}_{k|k-1} = \begin{pmatrix} \mathbf{1} & \Delta t_k \mathbf{1} & \frac{1}{2}\Delta t_k^2 \mathbf{1} \\ \mathbf{O} & \mathbf{1} & \Delta t_k \mathbf{1} \\ \mathbf{O} & \mathbf{O} & \mathrm{e}^{-\Delta t_k/\theta} \end{pmatrix} \tag{10.33}$$

$$\mathbf{D}_{k|k-1} = \Sigma^2(1 - \mathrm{e}^{-2\Delta t_k/\theta}) \begin{pmatrix} \mathbf{O}\,\mathbf{O}\,\mathbf{O} \\ \mathbf{O}\,\mathbf{O}\,\mathbf{O} \\ \mathbf{O}\,\mathbf{O}\,\mathbf{1} \end{pmatrix}, \tag{10.34}$$

with $\Delta t_k = t_k - t_{k-1}$. The maneuvering capability of the objects is thus characterized by two parameters: *maneuver correlation time* θ and *acceleration width* Σ. For $r = 2, 3$ we consider the parameter sets:

- M_1 (worst-case model): $\Sigma_1 = 60\ \mathrm{m/s^2}$, $\theta_1 = 30\,\mathrm{s}$
- M_2 (best-case model): $\Sigma_2 = 1\ \mathrm{m/s^2}$, $\theta_2 = 10\,\mathrm{s}$
- M_3 (medium-case model): $\Sigma_3 = 30\ \mathrm{m/s^2}$, $\theta_3 = 30\,\mathrm{s}$

The matrices of the model transition probabilities are given by:

$$(p_{ij})_{i=1,j=1}^{2} = \begin{pmatrix} 0.8 & 0.1 \\ 0.2 & 0.9 \end{pmatrix}, \qquad (p_{ij})_{i=1,j=1}^{3} = \begin{pmatrix} 0.8 & 0.1 & 0 \\ 0 & 0.9 & 0.2 \\ 0.2 & 0 & 0.8 \end{pmatrix}. \tag{10.35}$$

We observed that the performance does not critically depend on the particular switching probabilities p_{ij} chosen. A detailed mismatch analysis, however, has not been performed. A track is considered to be lost if more than 50 dwells occur in the local search or if the beam positioning error Δb_k is greater than $3B$. We thus permit even a rather extensive local search that correspondingly burdens the total energy budget. In all simulations considered below (1000 runs) the relative frequency of track loss is less than 2 %.

Selected Benchmark Trajectories

The horizontal projection of four standard benchmark trajectories (cargo aircraft, medium bomber, interceptor aircraft, and anti-ship missile) is shown in Fig. 10.3 along with representative kinematical characteristics such as acceleration (solid line), range (dashed), height (dotted), and speed (solid). They have been proposed in [11, 12] and cover a rather wide range of militarily relevant objects. The missile trajectory might serve to explore the performance limits of the algorithms. In principle, missiles can execute even stronger maneuvers. It is questionable, however, if for those objects and their individual missions the dynamics models discussed above remain applicable. All objects are tracked over a period of 180 s. The RCS fluctuations are described by a Swerling-III model. The mean cross sections significantly vary from object to object (4.0, 2.0, 1.2, 0.5 $\mathrm{m^2}$).

Measures of Performance Considered

The discussion is confined to a few intuitively clear and simple performance measures obtained by Monte-Carlo simulation (1000 runs). In general a single performance measure is not sufficient as there may exist applications where the transmitted energy is the limiting factor, while in a different scenario the number of radar allocations must be kept low.

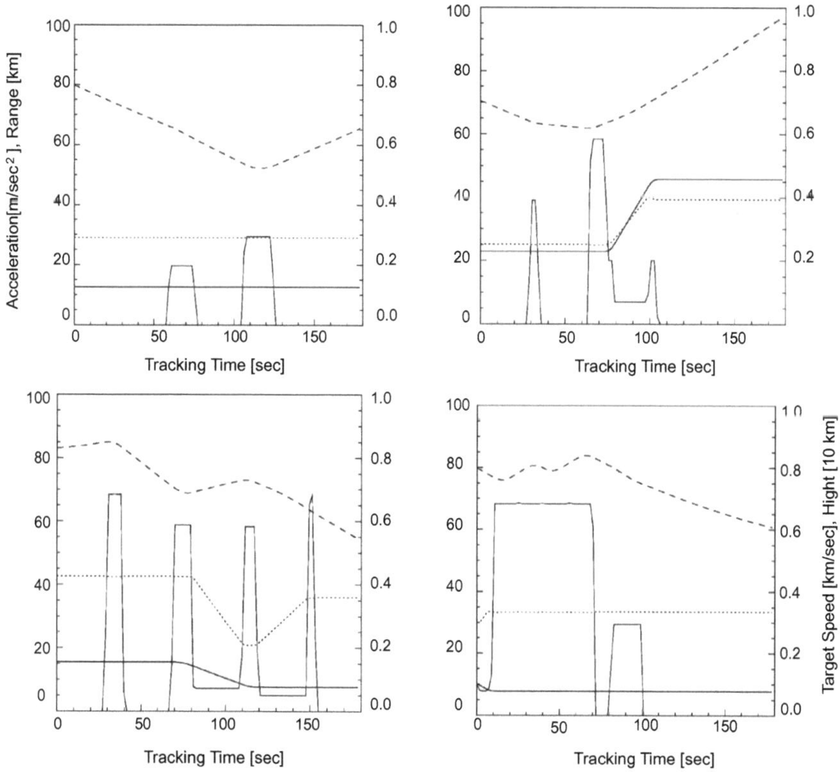

Fig. 10.3 Horizontal projections and kinematical quantities (cargo aircraft, medium bomber, interceptor aircraft, and anti-ship missile)

The adaptivity becomes visible if the performance is evaluated as a function of the tracking time that can be compared with the kinematics of the individual trajectories (Fig. 10.3). Here, histograms with 100 cells were used. In particular, it was considered: the mean revisit intervals, the mean number of dwells for a successful update, the mean number of sensor allocations in total required for track maintenance, the mean energy spent for a successful allocation, the mean energy totally spent for track maintenance, and the mean RCS of the objects estimated during tracking.

Four tracking filters were compared: worst-case Kalman filter (KF), standard IMM filter with two or three models, respectively (S-IMM$_{2,3}$), and IMM-MHT filtering with model histories of length $n = 4$. For IMM-MHT with $n > 4$, the performance characteristics change only slightly. We thus conclude that $n = 4$ already provides a good approximation to optimal filtering (at least for the scenarios considered here). With reference to object amplitude information we considered three cases: (1) the object RCS $\bar{\sigma}$ is known and used for energy management. (2) The mean RCS $\bar{\sigma}$ is unknown and to be estimated during tracking. (3) A worst-case assumption is used for all objects ($\bar{\sigma} = 0.5\,\mathrm{m}^2$).

10.5.2 Remarks on IMM Modeling Design

Practically, the question of how many models should be used in the IMM approach arises. In addition, it must be clarified whether each trajectory needs an individual modeling or if the same IMM modeling can be used without significant loss of performance. For the interceptor scenario, a worst/best-case model should be appropriate at first sight. Trajectory 1 (Cargo Aircraft), however, shows that objects can occur for which medium-case models are sufficient. To answer these questions, we used IMM with two ($r = 2, M_1, M_2$) and three models ($r = 3, M_1, M_2, M_3$), respectively, with $v_0 = 0.3$, $\mathrm{SN}_{k|k-1} = 50$. How these parameters affect the performance is discussed further below. Figure 10.4 shows the resulting mean revisit intervals for all trajectories. The kinematic object characteristics are clearly mirrored. We observed:

1. As expected, Kalman filtering ($r = 1, M_1$) leads to constant revisit intervals that are comparable for all trajectories. This is no longer true for S-IMM. The resultant curves related to $r = 2$ (solid) and $r = 3$ (dashed) differ from each other significantly. The onset of maneuvers (Fig. 10.3) strongly affects the mean update intervals and thus illustrates the adaptivity of the algorithm.
2. The difference between the cases $r = 2, 3$ vanishes however, if IMM-MHT is used. If model histories are permitted (here $n = 4$), it seems to be irrelevant if, besides worst/best-case assumptions, additional medium-case models are used. Even longer histories or further models ($r > 3$) do not significantly improve the performance obtained with $r = 2$ and $n = 4$. For a suitable (!) choice of the switching probabilities, the performance of S-IMM$_4$ approaches closes to B-IMM$_2$; for B-IMM$_4$ no improvement over B-IMM$_2$ was observed.
3. For the bomber and the interceptor, S-IMM$_3$ (M_1, M_2, M_3) outperforms S-IMM$_2$ (M_1, M_2), in spite of the fact that for these trajectories only worst-case maneuvers occur and the medium-case model appeared to be unnecessary at first sight. The difference between $r = 2$ and $r = 3$, however, is not as clear as for scenario 1 (cargo aircraft).
4. For moderately maneuvering cargo aircraft, the question arises whether the performance can be improved using a medium/best-case IMM modeling. We found that worst/best-case IMM-MHT and medium/best-case IMM-MHT differ, but not very much. This indicates that worst/best-case IMM-MHT has a more or less "universal" character, i.e. it does not critically depend of the scenario considered (at least within certain limits).

These observations indicate that the mixtures $p(\mathbf{x}_k|\mathcal{R}^k)$ for $n = 4, r = 2$ have enough internal degrees of freedom to provide an adequate representation of the actual object behavior. Refined approximations using even more mixture components seem to be irrelevant for the trajectories considered. A rule of thumb: A worst/best-case analysis of the problem along with IMM-MHT seems to be sufficient to achieve a nearly optimal tracking performance. Obviously, for two dynamics models, reasonable and intuitive assumptions for the switching probabilities are easily obtained. IMM-MHT

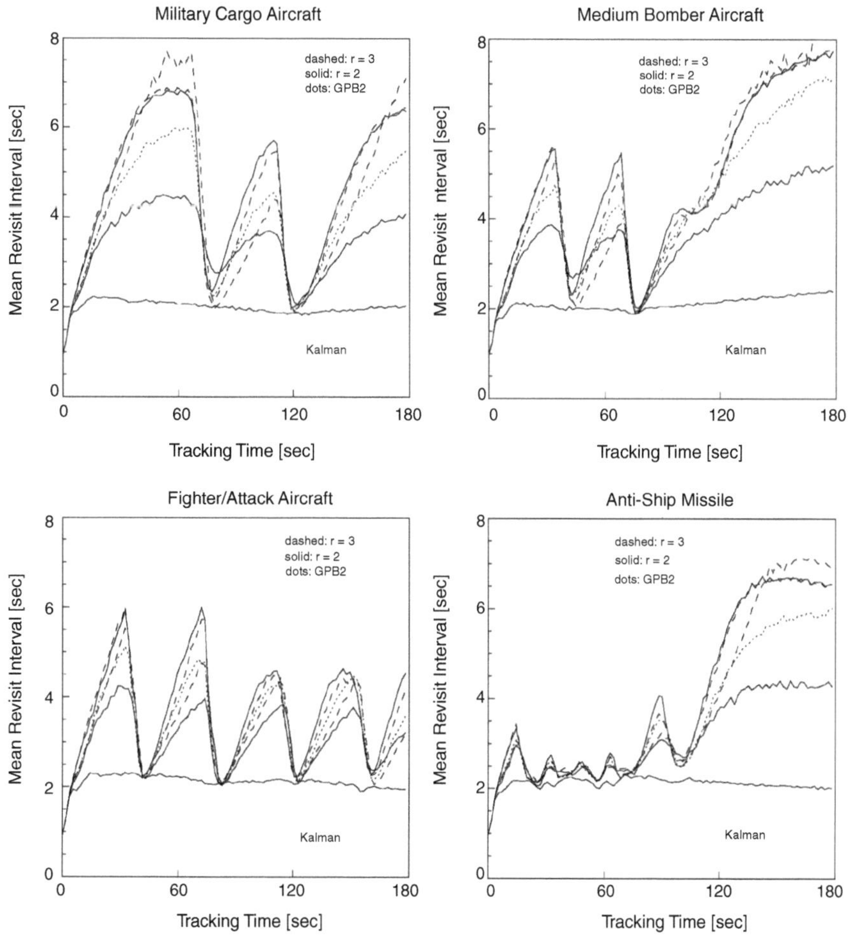

Fig. 10.4 Revisit intervals for different filters: cargo aircraft, medium bomber, interceptor aircraft, and anti-ship missile

thus enables a more simplified dynamics modelling without significant loss of performance.

10.5.3 Gain by IMM Modelling

To investigate the gain by adaptive dynamics models, let us for the present assume that the mean RCS of the object is known and used for energy management. Figure 10.5 shows the mean number of allocations required for track maintenance (KF, S-IMM_2, B-IMM_2). As expected, for KF the mean number of revisits increases linearly with

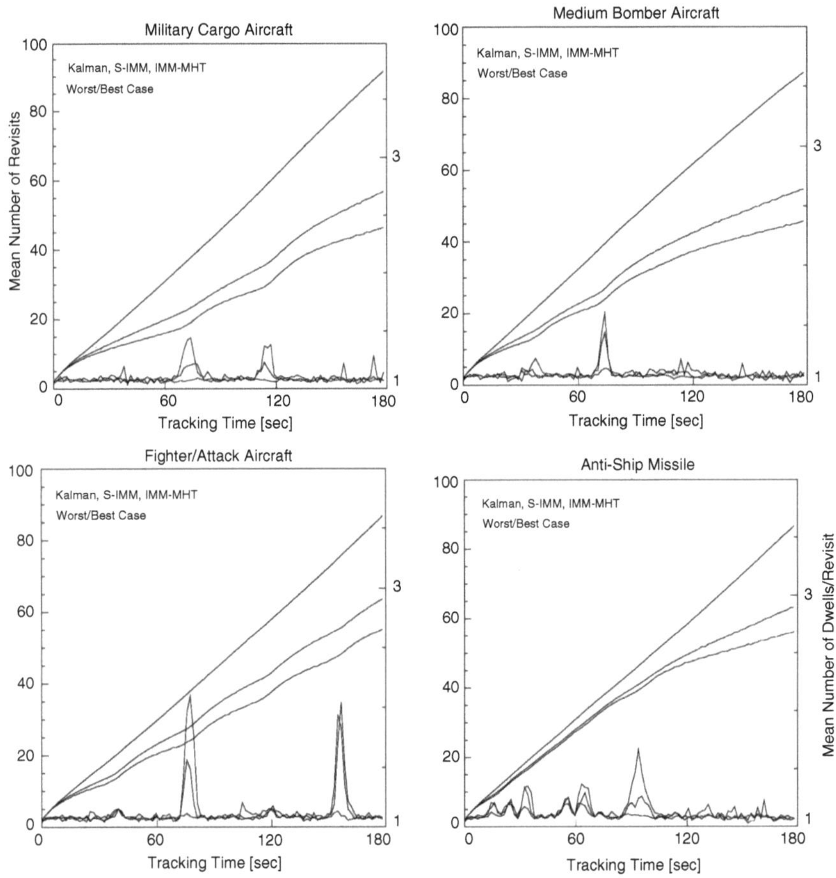

Fig. 10.5 Radar allocations for different filters: cargo aircraft, medium bomber, interceptor aircraft, and anti-ship missile

increasing tracking time and is nearly the same for all trajectories. By adaptive dynamics modeling, however, the number of sensor allocations is reduced.

1. Compared with KF, IMM results in significant resource savings. There is an improvement by IMM-MHT over S-IMM; the difference, however, is less significant than between S-IMM and KF. Besides simplified modeling assumptions, the practical use of IMM-MHT therefore consists in the exploration of the limiting bounds for performance improvements.
2. The largest gain is observed for the cargo aircraft and the bomber. In the case of the interceptor aircraft, the allocations required are reduced by about 50 % compared with worst-case Kalman filtering. Even during the 7 g weaving of the missile, some advantages of the IMM modeling can be observed.

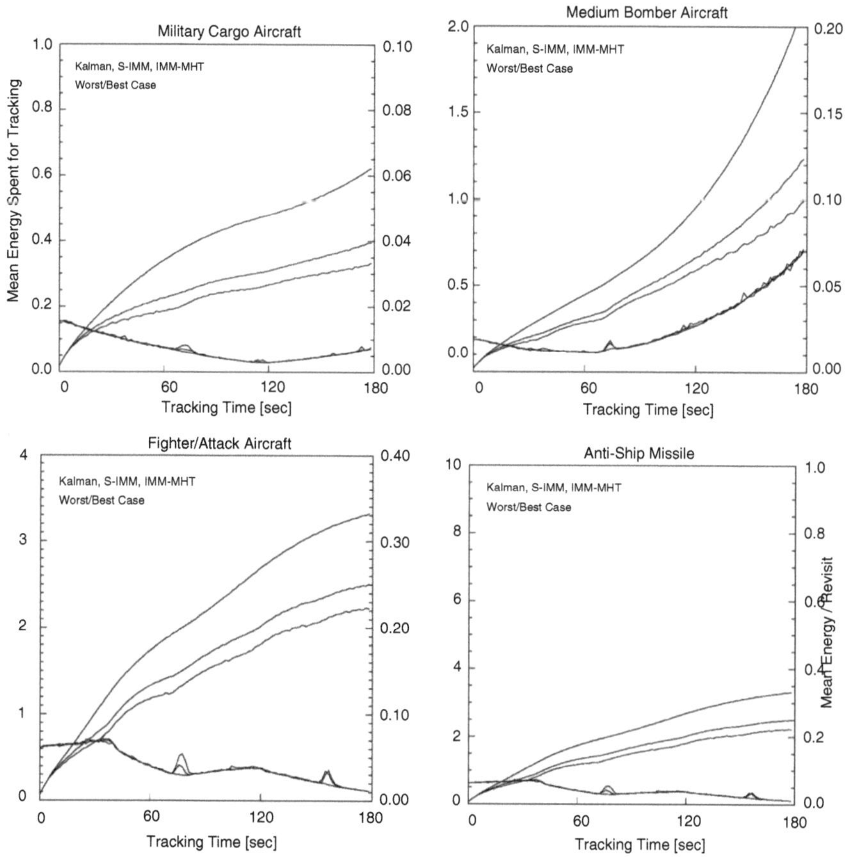

Fig. 10.6 Radar energy spent for different filters: cargo aircraft, medium bomber, interceptor aircraft, and anti-ship missile

Figure 10.5 shows the mean number of dwells per revisit. Up to peaks corresponding with the onset of maneuvers, it is constant and roughly equal for all filters and trajectories. The more adaptive the filter is, the higher the peaks are, i.e. the larger the revisit intervals can be during inertial flight. The peaks thus indicate that, for abrupt maneuvers, a local search might be required. This is the cost of increased adaptivity. Evidently, intelligent algorithms for beam positioning and local search are essential for IMM phased-array tracking.

These observations are consistent with Fig. 10.6, which shows the mean energy spent for track maintenance (relative units). Besides the object maneuvers, these curves are influenced by the current object range (Fig. 10.3, dotted line). In addition, the mean energy spent per revisit is displayed. Up to characteristic peaks, the energy per revisit is roughly the same for all tracking filters.

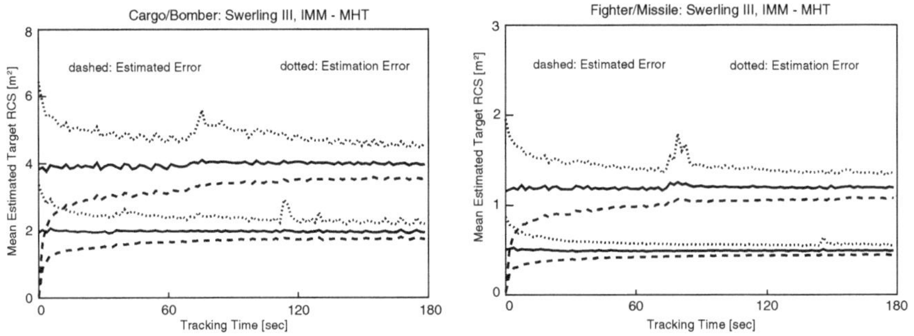

Fig. 10.7 On the quality of RCS estimates

10.5.4 On the Quality of RCS Estimates

In a practical application, the mean RCS of the objects to be tracked is unknown and might be estimated from object amplitude information. In general, the estimators used should be at least approximately bias-free, the estimated error and the empirical error should be roughly identical, and the estimators should show a certain robustness against model mismatch. As indicated by Fig. 10.7, the estimator previously proposed provides rather satisfying results for all trajectories. Using IMM-MHT for tracking, the recursion was initiated with $\bar{\sigma} = 0.5\,\mathrm{m}^2$ (worst-case assumption) and $m_0 = 1.01$.

The solid lines show the mean RCS estimates as a function of the tracking time. For all scenarios it is roughly constant and corresponds with the actual values (4, 2, 1.2, $0.5\,\mathrm{m}^2$). The dotted lines indicate the mean estimation error (available in the simulation). The curves show peaks that are related to the onset of maneuvers and the corresponding lack of track accuracy. The dashed lines denote the mean standard deviation calculated by the estimator itself. Tracking and RCS estimation are closely interrelated: Only when tracked over a certain period of time, the estimates are reliable enough to distinguish between the object classes. A satisfying RCS estimation by signal processing only, i.e. without a temporal integration along the estimated trajectory does not seem to be possible. In this context, IMM retrodiction techniques [13] might be considered that can provide more accurate estimates of the trajectory and thus more accurate RCS estimates.

10.5.5 RCS Model Mismatch

The backscattering properties of real objects are highly complex. A practicable method for estimating the RCS must thus show some robustness against model mismatch. To get a first hint, we simulated amplitude information according to both Swerling I and III being processed according to both modelling assumptions. The results for the four possible combinations are summarized in Table 10.1. Besides

Table 10.1 Mismatch regarding the fluctuation model

Object type	Processed	Simulated	RCS [m^2]	Estimated error	Estimation error	Energy
Bomber	III	III	1.96	0.22	0.28	0.37
	III	I	2.34	0.25	0.57	0.42
	I	III	1.77	0.28	0.33	0.39
	I	I	2.03	0.32	0.34	0.44
Fighter	III	III	1.19	0.13	0.16	0.82
	III	I	1.41	0.15	0.34	1.
	I	III	1.07	0.16	0.19	0.86
	I	I	1.22	0.18	0.19	0.99

the quantities already shown in Fig. 10.7, we also listed the total energy spent for tracking (relative units).

1. For matching models, the RCS estimates are nearly bias-free and more or less roughly consistent.
2. For Swerling III fluctuations, the estimates are more accurate than in case of Swerling I.
3. For Swerling I (no mismatch), more energy is spent than for Swerling III (keeping $\mathrm{SN}_{k|k-1}$ constant).
4. If Swerling I amplitudes are processed according to Swerling III, the RCS is overestimated, consistency is lost.
5. It is underestimated if Swerling III amplitudes are processed according to Swerling I.
6. Mismatch does not greatly affect the performance (energy).

10.5.6 Adaptive Energy Management

Finally, we have to show to what degree the transmitted radar energy can be reduced by estimating the RCS in comparison to worst-case assumptions. In Fig. 10.7 the mean radar energy spent for track maintenance is displayed. The dotted lines refer to IMM-MHT tracking using the true RCS of the objects (as previously discussed). In a practical application, this cannot be realized; the resultant curves, however, may serve as a reference to discuss the performance of RCS-adaptive algorithms. The solid lines denote methods that exploit signal strength information for estimating the RCS (Worst-Case Kalman filter, IMM-MHT). Dashed lines indicate algorithms that use a worst-case assumption (here: $0.5\,\mathrm{m}^2$, missile) on the RCS (KF, IMM-MHT).

A comparison between sensor control by using the true RCS (not available in a real application) and methods exploiting recursive RCS estimates is of particular interest. The largest deviation is observed for scenario 1 ($\bar{\sigma} = 4\ \mathrm{m}^2$). This is to be expected, as the recursion was started with a worst-case assumption. The discrepancy

Table 10.2 Gain by RCS-adaptive energy control

Object type	RCS	Filter	ΔT [s]	Revisits	Energies	Rel.
Cargo	Worst case	Kalman	3.3	55.1	2828	8.5
		IMM-MHT	6.5	32.6	1664	5.0
	Estimated	Kalman	2.0	90.3	750	2.2
		IMM-MHT	4.9	45.0	453	1.4
	Known		4.9	46.4	334	1
Bomber	Worst case	Kalman	3.1	60.1	5488	5.5
		IMM-MHT	6.4	34.0	2868	2.9
	Estimated	Kalman	2.2	85.5	2257	2.3
		IMM-MHT	5.1	44.2	1095	1.1
	Known		5.0	45.7	993	1
Fighter	Worst case	Kalman	2.8	67.3	5786	2.6
		IMM-MHT	4.9	43.3	3882	1.7
	Estimated	Kalman	2.2	85.6	3563	1.6
		IMM-MHT	3.9	54.0	2420	1.1
	Known		3.8	55.1	2226	1
Missile	Worst case	Kalman	2.1	86.6	8657	1.3
		IMM-MHT	4.1	56.1	6593	1
	Estimated	Kalman	2.2	85.4	9036	1.4
		IMM-MHT	4.2	55.4	7042	1.1
	Known		4.1	56.1	6593	1

between both curves, however, is not very significant in all four cases. Compared with IMM-MHT (Worst-Case RCS) it can be neglected. The difference between sensor control with known and estimated RCS is roughly constant during tracking. We thus conclude that it is caused primarily in the initiation phase, where not much signal strength information is yet available. As soon as reliable RCS estimates have been produced, the performance is practically identical. Figure 10.7 also shows how the resource savings due to adaptive dynamics models and RCS-adaptive energy management are related to each other.

In Table 10.2, scalar performance measures are summarized for all scenarios and processing methods: object revisit intervals (ΔT), sensor allocations required, energy spent for track maintenance (time averages taken over the tracking time). The last column shows the energy spent by the various methods relative to IMM-MHT with known RCS. Compared with IMM-MHT (Worst-Case RCS), the gain is: 3.8 (cargo aircraft), 2.6 (Bomber), 1.5 (Fighter), 0.9 (anti-ship missile). Hence, in the missile-scenario, where the worst-case assumption is correct, a small loss of performance must be taken into account.

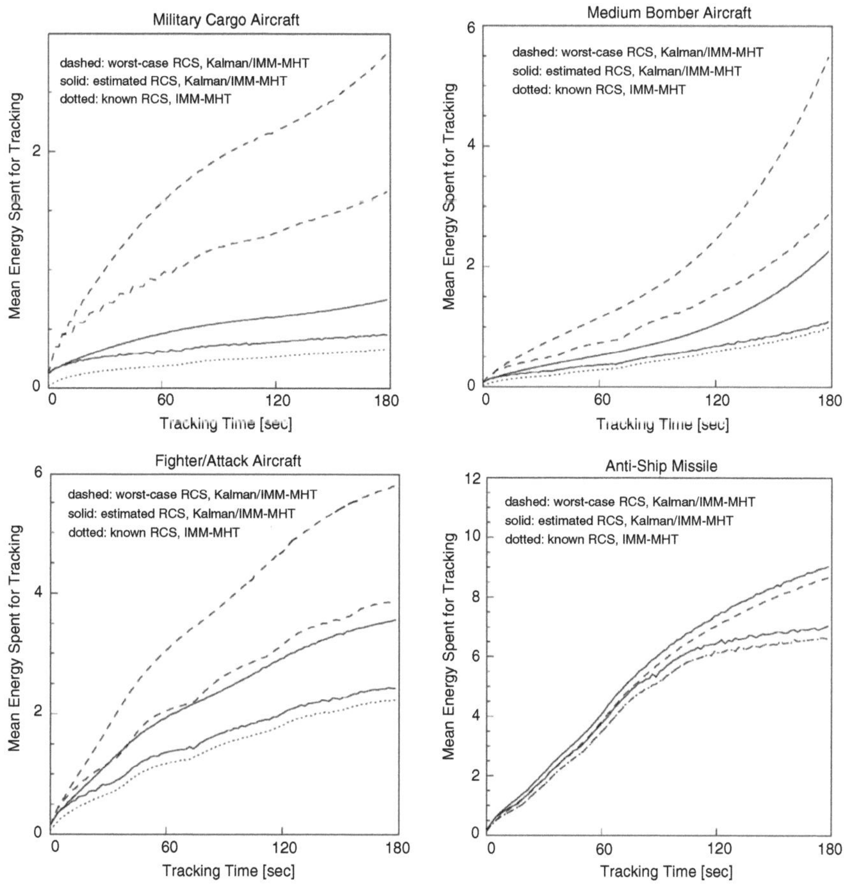

Fig. 10.8 Mean number of allocations for different filters and trajectories

10.6 Adaptive Sensor Management: Summary of Results

The gain by exploiting adaptive dynamics models and signal amplitude information is demonstrated by simulations with standard benchmark trajectories representative of typical objects (cargo aircraft, medium bomber, interceptor, and anti-ship missile) [11, 12]. Preliminary results were published in [14].

1. In the case of IMM-MHT, simple worst/best-case considerations seem to be sufficient for modelling the object dynamics. Medium-case models implying additional, a priori unknown parameters (e.g. transitions matrices) result in significant performance improvements only for standard IMM algorithms. IMM-MHT thus permits simplified, more *qualitative* models without significant loss of performance.

2. Compared with worst-case Kalman filtering, IMM results in considerable resource savings. The reduction with respect to the number of allocations required and the energy spent for track maintenance is roughly comparable and varies between 50 and 100 % depending on the scenario considered. Essentially, the savings are due to longer revisit intervals on average.
3. IMM-MHT improves on standard IMM algorithms. The difference, however, is less significant than between standard IMM and worst-case Kalman filtering. Besides simplified modelling assumptions, the practical use of IMM-MHT primarily consists in the exploration of the theoretical boundaries that limit the performance improvements achievable by adaptive dynamics models.
4. Due to abrupt maneuvers after a longer inertial flight, IMM-type tracking must necessarily be complemented by efficient Bayesian algorithms for adaptive beam positioning and local search. If used, however, the tracking process remains highly stable, because all information on the possible dynamical behavior of the objects is taken into account.
5. By processing object amplitude information along the estimated trajectory, the a priori unknown RCS of the objects can (roughly) be estimated. The estimate is approximately bias-free; its variance corresponds with the empirical variance. It is closely related to the tracking process and might provide a contribution to object classification. Within certain limits, the method seems to be rather robust against model mismatch.
6. Compared with worst-case assumptions on the object RCS, significant energy savings can be obtained by exploiting amplitude information. Depending on the scenario considered, the gain is larger than the improvement achievable by adaptive dynamics models. The difference compared to algorithms that use of the correct object RCS (available in a simulation) is comparatively small and arises mainly in the initiation phase.

References

1. W.-D. Wirth, *Radar Techniques Using Array Antennas.* (IEE Press, London, 2nd Edition, 2013)
2. M. Wunder, J. Grosche (eds.), Verteilte Führungsinformationssysteme. Springer (2009)
3. G. van Keuk, S. Blackman, On phased-array tracking and parameter control. IEEE Trans. Aerosp. Electron. Syst. **AES-29**(1) (1993)
4. W. Koch, On adaptive parameter control for phased-array tracking. *Proceedings of SPIE Signal and Data Processing of Small Targets*, vol. 3809 (Denver, USA, 1999), pp. 444–455
5. W. Koch, On exploiting 'negative' sensor evidence for object tracking and sensor data fusion. Inf. Fusion J. **8**(1), 28–39 (2007)
6. G. van Keuk, Multihypothesis tracking with electronically scanned radar. IEEE Trans. Aerosp. Electron. Syst. **31**(3), 916 (1995)
7. R.J. Dempster, S. Blackman, S.H. Roszkowski, and D.M. Sasaki, IMM/MHT solution to radar and multisensor benchmark tracking problems. SPIE Signal and Data Processing of Small, Targets, vol. 3373 (1998)
8. P.L. Bogler, *Radar Principles with Applications to Tracking Systems* (John Wiley & Sons, New York, 1990)

9. M. Abramowitz, I.A. Segun, *Handbook of Mathematical Functions*. Dover, New York (1965)
10. Y. Bar-Shalom, X.-R. Li, *Estimation and Tracking: Principles Techniques, Software* (Artech House, Boston, MA, 1993)
11. W.D. Blair, G.A. Watson, T. Kirubarajan, Y. Bar-Shalom, Benchmark for radar resource allocation and tracking objects in the presence of ECM. IEEE Trans. Aerosp. Electron. Syst. Mag. **35**(4) (1998)
12. W.D. Blair, G.A. Watson, K.W. Kolb, Benchmark problem for beam pointing control of phased-array radar in the presence of false alarms and ECM. *Proceedings 1995 American Control Conference*, Seattle WA (1994)
13. W. Koch, Fixed-interval retrodiction approach to bayesian IMM-MHT for maneuvering multiple targets. IEEE Aerosp. Electron. Syst. **AES-36**(1) (2000)
14. W. Koch, G. van Keuk, On Bayesian IMM-tracking for phased-array radar. *Proceedings of the DGON/ITG International Radar Symposium, IRS '98, Conference Proceedings*, pp. 715–723, München (1998)

Appendix A

A.1 List of Acronyms Used

C^4ISTAR	Command, control, communications, computers, intelligence, surveillance, target acquisition, and reconnaissance
E/O	Electro-optical sensor
ESM	Electronic support measures
FDoA	Frequency difference of arrival
GMTI	Ground moving target indicator radar
HMI	Human-machine interaction (HMI)
IMM	Interacting multiple models
IR	Infrared sensor
JDL	Joint directors of laboratories
SAR	Synthetic aperture radar
SDI	Stratecic defence initiative
SPD	Symmetric and positively definite
TDoA	Time difference of arrival
ToA	Time of arrival

A.2 List of Symbols Used

$\mathbb{N}$	Set of natural numbers
k, l, m, n	Integers
$\mathbb{R}$	Set of real numbers
$a, b, \ldots, x, y, z$	Scalars
$\mathbf{a}, \mathbf{b}, \ldots, \mathbf{x}, \mathbf{y}, \mathbf{z}$	Vectors

W. Koch, *Tracking and Sensor Data Fusion*,
Mathematical Engineering, DOI: 10.1007/978-3-642-39271-9,

$\mathbf{A}, \mathbf{B}, \ldots, \mathbf{X}, \mathbf{Y}, \mathbf{Z}$	Matrices
t	Time
$\lvert\cdots\rvert$	Determinant of a matrix, norm of a vector
$(\cdots)^\top$	Transpose
$(\cdots)^{-1}$	Inverse
$\mathrm{trace}(\cdots)$	Trace
$\mathrm{etr}(\cdots)$	$\exp(\mathrm{trace}(\cdots))$
$\mathcal{N}(\mathbf{x};\ \mathbb{E}[\mathbf{x}],\ \mathbb{C}[\mathbf{x}])$	Gaußian with expectation $\mathbb{E}[\mathbf{x}]$, covariance matrix $\mathbb{C}[\mathbf{x}]$
$\mathbb{E}[\cdots]$	Expectation
$\mathbb{C}[\cdots]$	Covariance matrix
$\mathcal{LW}(\mathbf{Z};\ m,\ \mathbf{X})$	$\lvert\mathbf{X}\rvert^{-\frac{m-1}{2}}\ \mathrm{etr}\big(-\frac{1}{2}(\mathbf{Z}\mathbf{X}^{-1})\big)$
$\mathbf{r}_k, \dot{\mathbf{r}}_k, \ddot{\mathbf{r}}_k$	Position, velocity, acceleration at time t_k
$\mathbf{x}_k$	Kinematic state vector at time t_k
Z_k	A set of measurements at time t_k
Z^k	Time series of measurements up to time t_k
m_k	Number of measurements at time t_k
$\mathbf{z}_k$	Measurement vector at time t_k
X_k	A set of state quantities at time t_k
i_k	Characteristic object feature (e.g. dynamics mode)
$\mathbf{X}_k$	SDP matrix modeling object extension at time t_k
$\mathbf{F}_{k\lvert k-1}$	Evolution matrix
$\mathbf{D}_{k\lvert k-1}$	Evolution covariance matrix
q_t	Acceleration bandwidth
θ_t	Maneuver correlation time
p_{ij}	Transition probabilities
$\ell(\cdots;\cdots)$	Likelihood function
$p(\cdots\lvert\cdots)$	Conditional probability density function
$\mathbf{R}_k$	Measurement error covariance matrix
$\mathbf{h}_k$	Measurement function
$\mathbf{H}_k$	Measurement matrix
$r_k, \dot{r}_k, \phi_k$	Range, range-rate, azimuth
$\mathbf{t}, \mathbf{T}$	Transform, corresponding Jacobian
$\mathbf{S}_{k\lvert k-1}$	Innovation covariance matrix
$\chi^2(P)$	χ^2-value at P
P_c	Correlation probability
P_D	Detection probability
P_u	Probability of being irresolved
P_F	False alarm probability
λ	Detection threshold
j_k	Index of measurements/data interpretations at time t_k
ΔT	Data innovation interval
ρ_F	Spatial density of false measurements/clutter

$p_F(m)$	Poisson distribution (number of false measurements)
D, ¬D	Object has/has not been detected
$\bar{m}_F$	Mean number of false measurements
$\lvert\text{FoV}\rvert$	Field of view of one ore more sensors
$\mathbf{d}_k$	Distance between two objects at time t_k
z_k	Strength of received object signals at time t_k
c_k	Strength of clutter objects at time t_k
x_k	Strength of an object at time t_k

A.3 Elementary Facts on Probability Densities

In the following several facts from elementary probability theory are collected:

1. Information on a vector variate random variable $\mathbf{x}$ is gained by *integrating* the corresponding probability density function $p(\mathbf{x})$ properly. As an example, integration over a volume V yields the probability that the event '$\mathbf{x} \in V$' occurs:

$$P(\mathbf{x} \in V) = \int_V \mathrm{d}\mathbf{x}\; p(\mathbf{x}). \tag{A.1}$$

   According to this interpretation, a probability density function must be *non-negative*, $p(\mathbf{x}) \geq 0$, and *normalized*, i.e. the probability of $\mathbf{x}$ 'being somewhere in the domain of $\mathbf{x}$' is given by $\int \mathrm{d}\mathbf{x}\; p(\mathbf{x}) = 1$.
2. An *observable* of $\mathbf{x}$ is a scalar-, vector- or matrix-valued function $g : \mathbf{x} \mapsto g(\mathbf{x})$ of the random variable $\mathbf{x}$. The *expectation* of an observable is defined by the integral

$$\mathbb{E}[g(\mathbf{x})] = \int \mathrm{d}\mathbf{x}\; g(\mathbf{x})\; p(\mathbf{x}). \tag{A.2}$$

   An important example is the expectation of a random variable $\mathbf{x}$, $\bar{\mathbf{x}} = \mathbb{E}[\mathbf{x}]$, defined by the 'centroid' of the corresponding probability density $p(\mathbf{x})$. Another important expectation is the 'expected error of the expectation of $\mathbf{x}$', i.e. a quality measure for $\bar{\mathbf{x}} = \mathbb{E}[\mathbf{x}]$. Using a matrix-valued quadratic observable $g : \mathbf{x} \mapsto (\mathbf{x} - \bar{\mathbf{x}})(\mathbf{x} - \bar{\mathbf{x}})^\top$, it is called *covariance matrix* and defined by the integral:

$$\mathbb{E}[(\mathbf{x} - \bar{\mathbf{x}})(\mathbf{x} - \bar{\mathbf{x}})^\top] = \int \mathrm{d}\mathbf{x}\; (\mathbf{x} - \bar{\mathbf{x}})(\mathbf{x} - \bar{\mathbf{x}})^\top\; p(\mathbf{x}). \tag{A.3}$$

3. Forming an expectation is a a linear operation, i.e. with two observables g, f and scalars α_1, α_2 the following identity holds:

$$\mathbb{E}[\alpha_1 g(\mathbf{x}) + \alpha_2 f(\mathbf{x})] = \alpha_1 \mathbb{E}[g(\mathbf{x})] + \alpha_2 \mathbb{E}[f(\mathbf{x})]. \tag{A.4}$$

4. A *conditional probability density* $p(\mathbf{x}|\mathbf{y})$ of a random variable $\mathbf{x}$ describes how available knowledge about another random variable $\mathbf{y}$ affects knowledge avail-

able on $\mathbf{x}$. It is defined by:

$$p(\mathbf{x}|\mathbf{y}) = \frac{p(\mathbf{x},\mathbf{y})}{p(\mathbf{y})} \tag{A.5}$$

with $p(\mathbf{x},\mathbf{y})$ denoting the *joint pdf* of both random variables $\mathbf{x}$ and $\mathbf{y}$. In this notation we do not distinguish between a random variable and a realization of it in order to keep the formulae lean.

5. By writing the probability density function $p(\mathbf{x})$ of a random variable $\mathbf{x}$ as a *marginal probability density*,

$$p(\mathbf{x}) = \int \mathrm{d}\mathbf{y}\; p(\mathbf{x},\mathbf{y}) = \int d\mathbf{y}\; p(\mathbf{x}|\mathbf{y})\; p(\mathbf{y}), \tag{A.6}$$

we are able to bring another random variable $\mathbf{y}$ into the play, which is related to $\mathbf{x}$.

6. By using *Bayes' formula* we can calculate how information on a random variable $\mathbf{y}$ affects our knowledge on $\mathbf{x}$, provided the probability density functions $p(\mathbf{y}|\mathbf{x})$ and $p(\mathbf{x})$ are known. According to $p(\mathbf{x}|\mathbf{y})\; p(\mathbf{y}) = p(\mathbf{x},\mathbf{y}) = p(\mathbf{y},\mathbf{x}) = p(\mathbf{y}|\mathbf{x})\; p(\mathbf{x})$, Bayes' rule it is a direct consequence of the last two statements and is given by:

$$p(\mathbf{x}|\mathbf{y}) = \frac{p(\mathbf{y}|\mathbf{x})p(\mathbf{x})}{\int \mathrm{d}\mathbf{x}\; p(\mathbf{y}|\mathbf{x})p(\mathbf{x})}. \tag{A.7}$$

7. *Precise knowledge* that a random variable $\mathbf{x}$ is equal to a certain value $\bar{\mathbf{x}}$ fits well into the description of uncertainty by using probability densities if Dirac's δ-distributions $p(\mathbf{x}) = \delta(\mathbf{x} - \bar{\mathbf{x}})$ are considered. In this case, we have for any reasonable observable $g : \mathbf{x} \mapsto g(\mathbf{x})$:

$$\mathbb{E}[g(\mathbf{x})] = \int \mathrm{d}\mathbf{x}\; g(\mathbf{x})\; \delta(\mathbf{x} - \bar{\mathbf{x}}) = g(\bar{\mathbf{x}}). \tag{A.8}$$

8. The probability density of a random variable $\mathbf{y} = \mathbf{t}(\mathbf{x})$, which results from $\mathbf{x}$ via an invertible transformation $\mathbf{t} : \mathbf{x} \mapsto \mathbf{t}[\mathbf{x}]$ is given by:

$$p(\mathbf{y}) = \int \mathrm{d}\mathbf{x}\; p(\mathbf{y}|\mathbf{x})\; p(\mathbf{x}) \tag{A.9}$$

$$= \int \mathrm{d}\mathbf{x}\; \delta(\mathbf{y} - \mathbf{t}[\mathbf{x}])\; p(\mathbf{x}) \tag{A.10}$$

$$= \int \mathrm{d}\mathbf{z}\; |\mathbf{T}^{-1}(\mathbf{z})|\; \delta(\mathbf{y} - \mathbf{z})\; p(\mathbf{t}^{-1}[\mathbf{z}]) \tag{A.11}$$

$$= |\mathbf{T}^{-1}(\mathbf{y})|\; p(\mathbf{t}^{-1}[\mathbf{y}]), \tag{A.12}$$

where the substitution $\mathbf{x} = \mathbf{t}^{-1}[\mathbf{z}]$ with the corresponding Jacobi determinant $|\mathbf{T}^{-1}(\mathbf{z})| = |\frac{\partial \mathbf{t}^{-1}[\mathbf{z}]}{\partial \mathbf{z}}|$ was used. The transformation of $p(\mathbf{x})$ induced by $\mathbf{t}$ can thus be described by a *Transfer Operator*:

$$\mathcal{T}[p](\mathbf{y}) = |\mathbf{T}^{-1}(\mathbf{y})|\; p(\mathbf{t}^{-1}[\mathbf{y}]). \tag{A.13}$$

Under certain conditions, a similar result can be obtained for piecewise invertible transforms $\mathbf{t}$.

9. An important special case is the *Gaußian probability density* characterized by a single maximum concentrated around $\bar{\mathbf{x}}$. Let the quadratic form $q(\mathbf{x}) = \frac{1}{2}(\mathbf{x} - \bar{\mathbf{x}})^\top \mathbf{C}^{-1}(\mathbf{x} - \bar{\mathbf{x}})$ be a measure of the distance between the random variable $\mathbf{x}$ and the 'center' $\bar{\mathbf{x}}$ weighted by a symmetric and positively definite matrix $\mathbf{C}$. Evidently, by $q(\mathbf{x}) = \text{const.}$ ellipsoids are defined that are centered around $\bar{\mathbf{x}}$ and whose volume and orientation are determined by the eigenvectors and the eigenvalues of the matrix $\mathbf{C}$. As a special density function that decays with an increasing distance of $\mathbf{x}$ from $\bar{\mathbf{x}}$, let us consider:

$$p(\mathbf{x}) = \frac{\mathrm{e}^{-q(\mathbf{x})}}{\int \mathrm{d}\mathbf{x}\, \mathrm{e}^{-q(\mathbf{x})}}. \tag{A.14}$$

Evidently, $p(\mathbf{x})$ is positive and correctly normalized. After integration we obtain:

$$p(\mathbf{x}) = |2\pi\mathbf{C}|^{-\frac{1}{2}} \exp\big\{ -\tfrac{1}{2}(\mathbf{x} - \bar{\mathbf{x}})^\top \mathbf{C}^{-1}(\mathbf{x} - \bar{\mathbf{x}})\big\} \tag{A.15}$$

$$=: \mathcal{N}\big(\mathbf{x};\, \bar{\mathbf{x}},\, \mathbf{C}\big) \tag{A.16}$$

with $|2\pi\mathbf{C}|$ denoting the determinant of the matrix $2\pi\mathbf{C}$ and an expectation vector and covariance matrix given by:

$$\mathbb{E}[\mathbf{x}] = \bar{\mathbf{x}} \tag{A.17}$$

$$\mathbb{E}[(\mathbf{x} - \bar{\mathbf{x}})(\mathbf{x} - \bar{\mathbf{x}})^\top] = \mathbf{C}, \tag{A.18}$$

respectively. By this, the covariance matrix $\mathbf{C}$ has a simple and intuitive geometrical interpretation, since it defines ellipsoidal contours.

10. Let $\mathbf{x}$ be a Gaußian random variable. The probability function of an affine transform of $\mathbf{x}$,

$$\mathbf{t} : \mathbf{x} \mapsto \mathbf{y} = \mathbf{a} + \mathbf{A}\mathbf{x}, \tag{A.19}$$

with a fixed vector $\mathbf{a}$ and matrix $\mathbf{A}$ is given by

$$\mathcal{N}\big(\mathbf{x};\, \bar{\mathbf{x}},\, \mathbf{X}\big) \xrightarrow{\mathbf{y}=\mathbf{a}+\mathbf{A}\mathbf{x}} \mathcal{N}\big(\mathbf{y};\, \mathbf{a} + \mathbf{A}\bar{\mathbf{x}},\, \mathbf{A}\mathbf{X}\mathbf{A}^\top\big). \tag{A.20}$$

This directly follows from: $p(\mathbf{y}) = |\frac{\partial \mathbf{t}^{-1}[\mathbf{y}]}{\partial \mathbf{y}}|\, p\big(\mathbf{t}^{-1}[\mathbf{y}]\big)$ with $\mathbf{t}^{-1} : \mathbf{y} \mapsto \mathbf{t}^{-1}[\mathbf{y}] = \mathbf{A}^{-1}(\mathbf{y} - \mathbf{a})$ and $\frac{\partial \mathbf{t}^{-1}[\mathbf{y}]}{\partial \mathbf{y}} = \mathbf{A}^{-1}$.

11. A d-dimension random vector $\mathbf{x}$ is *Student-t-distributed* with n degrees of freedom and a parameter matrix $\mathbf{A}$, if its density is given by [1, p. 133]:

$$\mathcal{T}\big(\mathbf{x};\ n,\ \mathbf{a}, \mathbf{A}\big) = \tfrac{1}{Z}\left(1 + (\mathbf{x}-\mathbf{a})^\top \mathbf{A}^{-1}(\mathbf{x}-\mathbf{a})\right)^{-\frac{n+d}{2}}. \tag{A.21}$$

with a normalizing constant Z. It has the following expectation and covariance:

$$\mathbb{E}[\mathbf{x}] = \mathbf{a}, \quad \mathbb{E}[(\mathbf{x}-\mathbf{a})(\mathbf{x}-\mathbf{a})^\top] = \tfrac{1}{n-2}\,\mathbf{A}. \tag{A.22}$$

A.4 Facts on Inverse Block Matrices

In several calculations involving multivariate Gaußian probability density functions, matrices with a block structure have to be inverted. In this case, the following results can be useful (see [2], e.g.), which provide the inverse of a suitably partitioned symmetric matrix:

$$\begin{pmatrix} \mathbf{A} & \mathbf{C} \\ \mathbf{C}^\top & \mathbf{B} \end{pmatrix}^{-1} = \begin{pmatrix} \mathbf{A}^{-1} + \mathbf{A}^{-1}\mathbf{C}\mathbf{S}^{-1}\mathbf{C}^\top\mathbf{A}^{-1} & -\mathbf{A}^{-1}\mathbf{C}\mathbf{S}^{-1} \\ -\mathbf{S}^{-1}\mathbf{C}^\top\mathbf{A}^{-1} & \mathbf{S}^{-1} \end{pmatrix} \tag{A.23}$$

$$= \begin{pmatrix} \mathbf{T}^{-1} & -\mathbf{T}^{-1}\mathbf{C}\mathbf{B}^{-1} \\ -\mathbf{B}^{-1}\mathbf{C}^\top\mathbf{T}^{-1} & \mathbf{B}^{-1} + \mathbf{B}^{-1}\mathbf{C}^\top\mathbf{T}^{-1}\mathbf{C}\mathbf{B}^{-1} \end{pmatrix} \tag{A.24}$$

where the auxilliary matrices:

$$\mathbf{S} = \mathbf{B} - \mathbf{C}^\top\mathbf{A}^{-1}\mathbf{C} \tag{A.25}$$

$$\mathbf{T} = \mathbf{A} - \mathbf{C}\mathbf{B}^{-1}\mathbf{C}^\top \tag{A.26}$$

are called the *Schur Complements* of the matrix $\mathbf{A}$ and $\mathbf{B}$, respectively. The useful *Matrix Inversion Lemma* directly results from comparing corresponding diagonal blocks:

$$\big(\mathbf{A} - \mathbf{C}\mathbf{B}^{-1}\mathbf{C}^\top\big)^{-1} = \mathbf{A}^{-1} + \mathbf{A}^{-1}\mathbf{C}(\mathbf{B} - \mathbf{C}^\top\mathbf{A}^{-1}\mathbf{C})^{-1}\mathbf{C}^\top\mathbf{A}^{-1}. \tag{A.27}$$

A.5 A Product Formula for Gaußians

For matrices and vectors of compatible dimensions, the following formula for products of Gaussians holds:

$$\mathcal{N}\big(\mathbf{z};\ \mathbf{H}\mathbf{x},\ \mathbf{R}\big)\,\mathcal{N}\big(\mathbf{x};\ \bar{\mathbf{x}},\ \mathbf{P}\big) = \mathcal{N}\big(\mathbf{z};\ \mathbf{H}\bar{\mathbf{x}},\ \mathbf{S}\big)\ \mathcal{N}\big(\mathbf{x};\ \mathbf{q},\ \mathbf{Q}\big), \tag{A.28}$$

where for $\mathbf{S}$, $\mathbf{q}$, and $\mathbf{Q}$ several equivalent representations exist:

$$\mathbf{S} = \begin{cases} \mathbf{HPH}^\top + \mathbf{R} \\ (\mathbf{R}^{-1} - \mathbf{R}^{-1}\mathbf{HQH}^\top\mathbf{R}^{-1})^{-1} \end{cases} \tag{A.29}$$

$$\mathbf{q} = \begin{cases} \bar{\mathbf{x}} + \mathbf{W}(\mathbf{z} - \mathbf{H}\bar{\mathbf{x}}) \\ \bar{\mathbf{x}} + \mathbf{W}\boldsymbol{\nu} \\ \mathbf{Q}(\mathbf{P}^{-1}\bar{\mathbf{x}} + \mathbf{H}^\top\mathbf{R}^{-1}\mathbf{z}) \end{cases} \tag{A.30}$$

$$\mathbf{Q} = \begin{cases} \mathbf{P} - \mathbf{WSW}^\top \\ (\mathbf{1} - \mathbf{WH})\mathbf{P} \\ (\mathbf{P}^{-1} + \mathbf{H}^\top\mathbf{R}^{-1}\mathbf{H})^{-1} \end{cases} \tag{A.31}$$

with the following abbreviations:

$$\boldsymbol{\nu} = \mathbf{z} - \mathbf{H}\bar{\mathbf{x}}, \qquad \mathbf{W} = \mathbf{PH}^\top\mathbf{S}^{-1}. \tag{A.32}$$

Note that the vector $\mathbf{x}$ appearing in both Gaußians of the left side of Eq. A.28 only exists in one of the factors on the right side.

Proof

The product $\mathcal{N}\big(\mathbf{z};\ \mathbf{Hx},\ \mathbf{R}\big)\ \mathcal{N}\big(\mathbf{x};\ \bar{\mathbf{x}},\ \mathbf{P}\big)$ can be interpreted as a joint probability density $p(\mathbf{z}, \mathbf{x}) = p(\mathbf{z}|\mathbf{x})\ p(\mathbf{x})$ with $p(\mathbf{z}|\mathbf{x}) = \mathcal{N}\big(\mathbf{z};\ \mathbf{Hx},\ \mathbf{R}\big)$ and $p(\mathbf{x}) = \mathcal{N}\big(\mathbf{x};\ \bar{\mathbf{x}},\ \mathbf{P}\big)$. It can be written as:

$$p(\mathbf{z}, \mathbf{x}) = |2\pi\mathbf{R}|^{-\frac{1}{2}}|2\pi\mathbf{P}|^{-\frac{1}{2}}\ \exp\{-\tfrac{1}{2}q(\mathbf{z}, \mathbf{x})\}, \tag{A.33}$$

where $q(\mathbf{z}, \mathbf{x})$ is a sum of two quadratic forms:

$$q(\mathbf{z}, \mathbf{x}) = (\mathbf{z} - \mathbf{Hx})^\top\mathbf{R}^{-1}(\mathbf{z} - \mathbf{Hx}) + (\mathbf{x} - \bar{\mathbf{x}})^\top\mathbf{P}^{-1}(\mathbf{x} - \bar{\mathbf{x}}). \tag{A.34}$$

By a completion of the square, $q(\mathbf{z}, \mathbf{x})$ can be transformed in such a way that it contains the joint vector $\mathbf{u} = (\mathbf{z}^\top, \mathbf{x}^\top)^\top$ in a quadratic form $(\mathbf{u} - \bar{\mathbf{u}})^\top\mathbf{U}^{-1}(\mathbf{u} - \bar{\mathbf{u}})$ with a suitably chosen matrix $\mathbf{U}$ and vector $\bar{\mathbf{u}}$. Moreover, $\mathbf{u}$, $\mathbf{z}$, or $\mathbf{x}$ appear nowhere else. For this reason, $\mathbf{u}$ must be a Gaußian random variable with $p(\mathbf{u}) = \mathcal{N}\big(\mathbf{u};\ \bar{\mathbf{u}},\ \mathbf{U}\big)$, since it is already known that $p(\mathbf{u}) = \mathcal{N}\big(\mathbf{z};\ \mathbf{Hx},\ \mathbf{R}\big)\ \mathcal{N}\big(\mathbf{x};\ \bar{\mathbf{x}},\ \mathbf{P}\big)$ is a probability density, i.e. normalized.

With $\mathbf{A} = (\mathbf{1}, -\mathbf{H})$, $\mathbf{B} = (\mathbf{O}, \mathbf{1})$, $\mathbf{b} = (\mathbf{O}, \bar{\mathbf{x}})^\top$, $q(\mathbf{u})$ can be written as:

$$\begin{aligned} q(\mathbf{u}) &= (\mathbf{Au})^\top\mathbf{R}^{-1}\mathbf{Au} + \big(\mathbf{B}(\mathbf{u} - \mathbf{b})\big)^\top\mathbf{P}^{-1}(\mathbf{B}(\mathbf{u} - \mathbf{b})) && \text{(A.35)} \\ &= \mathbf{u}^\top(\mathbf{A}^\top\mathbf{R}^{-1}\mathbf{A})\mathbf{u} + (\mathbf{u} - \mathbf{b})^\top(\mathbf{B}^\top\mathbf{P}^{-1}\mathbf{B})(\mathbf{u} - \mathbf{b}) && \text{(A.36)} \\ &= \mathbf{u}^\top(\mathbf{A}^\top\mathbf{R}^{-1}\mathbf{A} + \mathbf{B}^\top\mathbf{P}^{-1}\mathbf{B})\mathbf{u} - 2\mathbf{u}^\top(\mathbf{B}^\top\mathbf{P}^{-1}\mathbf{B})\mathbf{b} + \text{const.} && \text{(A.37)} \\ &= \mathbf{u}^\top\mathbf{U}^{-1}\mathbf{u} - 2\mathbf{u}^\top\mathbf{U}^{-1}\bar{\mathbf{u}} + \text{const.} && \text{(A.38)} \\ &= (\mathbf{u} - \bar{\mathbf{u}})^\top\mathbf{U}^{-1}(\mathbf{u} - \bar{\mathbf{u}}) + \text{const.}' && \text{(A.39)} \end{aligned}$$

where $\mathbf{U}^{-1}$ and $\bar{\mathbf{u}}$ is given by:

$$\mathbf{U}^{-1} = \mathbf{A}^\top\mathbf{R}^{-1}\mathbf{A} + \mathbf{B}^\top\mathbf{P}^{-1}\mathbf{B} \tag{A.40}$$

$$= \begin{pmatrix} \mathbf{R}^{-1} & -\mathbf{R}^{-1}\mathbf{H} \\ & -\mathbf{H}^\top\mathbf{R}^{-1}\mathbf{P}^{-1} + \mathbf{H}^\top\mathbf{R}^{-1}\mathbf{H} \end{pmatrix} \tag{A.41}$$

$$= \begin{pmatrix} \mathbf{S} & \mathbf{HP} \\ \mathbf{PH}^\top & \mathbf{P} \end{pmatrix}^{-1} \tag{A.42}$$

$$\bar{\mathbf{u}} = \mathbf{U}(\mathbf{B}^\top\mathbf{P}^{-1}\mathbf{B})\mathbf{b} \tag{A.43}$$

$$= \begin{pmatrix} \mathbf{H}\bar{\mathbf{x}} \\ \bar{\mathbf{x}} \end{pmatrix} \quad . \tag{A.44}$$

These calculations make use of the inversion formulae for partitioned matrices discussed in the previous section.

Since the joint density $p(\mathbf{z}, \mathbf{x}) = p(\mathbf{x}, \mathbf{z})$ is available now, it can be used to calculate the marginal density $p(\mathbf{z}) = \int \mathrm{d}\mathbf{x}\; p(\mathbf{z}, \mathbf{x})$ and the conditional density $p(\mathbf{x}|\mathbf{z}) = p(\mathbf{z}, \mathbf{x})/p(\mathbf{z})$. To this end, we rewrite via another completion of the square the quadratic form $(\mathbf{u} - \bar{\mathbf{u}})\mathbf{U}^{-1}(\mathbf{u} - \bar{\mathbf{u}})$ as a sum of two quadratic forms in such a way that the integration variable $\mathbf{x}$ in $\int \mathrm{d}\mathbf{x}\; p(\mathbf{z}, \mathbf{x})$ appears in one of them only:

$$(\mathbf{u} - \bar{\mathbf{u}})^\top\mathbf{U}^{-1}(\mathbf{u} - \bar{\mathbf{u}})$$

$$= \left((\mathbf{z} - \mathbf{H}\bar{\mathbf{x}})^\top \;\; (\mathbf{x} - \bar{\mathbf{x}})^\top\right) \begin{pmatrix} \mathbf{R}^{-1} & -\mathbf{R}^{-1}\mathbf{H} \\ -\mathbf{H}^\top\mathbf{R}^{-1} & \mathbf{Q}^{-1} \end{pmatrix} \begin{pmatrix} \mathbf{z} - \mathbf{H}\bar{\mathbf{x}} \\ \mathbf{x} - \bar{\mathbf{x}} \end{pmatrix} \tag{A.45}$$

$$= \left((\mathbf{z} - \mathbf{H}\bar{\mathbf{x}})^\top \;\; (\mathbf{x} - \bar{\mathbf{x}})^\top\right) \begin{pmatrix} \mathbf{R}^{-1}(\mathbf{z} - \mathbf{H}\bar{\mathbf{x}}) - \mathbf{R}^{-1}\mathbf{H}(\mathbf{x} - \bar{\mathbf{x}}) \\ -\mathbf{H}^\top\mathbf{R}^{-1}(\mathbf{z} - \mathbf{H}\bar{\mathbf{x}}) + \mathbf{Q}^{-1}(\mathbf{x} - \bar{\mathbf{x}}) \end{pmatrix} \tag{A.46}$$

$$= (\mathbf{z} - \mathbf{H}\bar{\mathbf{x}})^\top\mathbf{R}^{-1}(\mathbf{z} - \mathbf{H}\bar{\mathbf{x}}) - 2(\mathbf{x} - \bar{\mathbf{x}})^\top\mathbf{H}^\top\mathbf{R}^{-1}(\mathbf{z} - \mathbf{H}\bar{\mathbf{x}})$$
$$+ (\mathbf{x} - \bar{\mathbf{x}})^\top\mathbf{Q}^{-1}(\mathbf{x} - \bar{\mathbf{x}}) \tag{A.47}$$

$$= (\mathbf{z} - \mathbf{H}\bar{\mathbf{x}})^\top\mathbf{S}^{-1}(\mathbf{z} - \mathbf{H}\bar{\mathbf{x}}) + (\mathbf{x} - \mathbf{q})\mathbf{Q}^{-1}(\mathbf{x} - \mathbf{q})^\top + \text{const.}. \tag{A.48}$$

The various versions for $\mathbf{S}$, $\mathbf{q}$, and $\mathbf{Q}$ result from the matrix inversion lemma discussed in the previous section. We thus have $p(\mathbf{x}, \mathbf{z}) = p(\mathbf{x}|\mathbf{z})\; p(\mathbf{z})$ with $p(\mathbf{x}|\mathbf{z}) = \mathcal{N}(\mathbf{x};\; \mathbf{q},\; \mathbf{Q})$ and $p(\mathbf{z}) = \mathcal{N}(\mathbf{z};\; \mathbf{H}\bar{\mathbf{x}},\; \mathbf{S})$. This completes the proof.

A.6 Approximation by Moment Matching

Moment matching is an important approximation method, by which a probability density function $p(x)$ with expectation $\mathbb{E}_p[x] = \mathbf{x}$ and a covariance matrix $\mathbb{E}_p[(x - \bar{\mathbf{x}})(x - \bar{\mathbf{x}})^\top] = \mathbf{P}$ is approximated by $p(x) \approx \mathcal{N}(x;\; \mathbf{x},\; \mathbf{P})$. In the context of this thesis, moment matching is applied to mixture densities of the form $p(x) = \sum_i p_i\, \mathcal{N}(x; \mathbf{x}_i, \mathbf{P}_i)$, i.e. to normal mixtures. In this case, $\mathbf{x}$ and $\mathbf{P}$ are given by:

$$\mathbf{x} = \sum_i p_i \mathbf{x}_i \tag{A.49}$$

$$\mathbf{P} = \sum_i p_i \big(\mathbf{P}_i + (\mathbf{x}_i - \mathbf{x})(\mathbf{x}_i - \mathbf{x})^\top\big). \tag{A.50}$$

The matrix $\sum_i p_i \, (\mathbf{x}_i - \mathbf{x})(\mathbf{x}_i - \mathbf{x})^\top$ is called *spread matrix*. These formulae are a consequence of the following calculations:

$$\begin{aligned}
\mathbb{E}_p[x] &= \int \mathrm{d}x \; x \;\, p(x) && \text{(A.51)}\\
&= \sum_i p_i \int \mathrm{d}x \; x \, \mathcal{N}(x; \mathbf{x}_i, \mathbf{P}_i) = \sum_i p_i \; \mathbf{x}_i =: \mathbf{x} && \text{(A.52)}\\
\mathbb{C}_p[x] &= \int \mathrm{d}x \; p(x) \; (x - \mathbb{E}_p[x])(x - \mathbb{E}_p[x])^\top && \text{(A.53)}\\
&= \sum_i p_i \int \mathrm{d}x \; (x - \mathbf{x})(x - \mathbf{x})^\top \, \mathcal{N}(x; \mathbf{x}_i, \mathbf{P}_i) && \text{(A.54)}\\
&= \sum_i p_i \int \mathrm{d}x \; \big\{(x - \mathbf{x})(x - \mathbf{x})^\top - 2(x - \mathbf{x}_i)(\mathbf{x}_i - \mathbf{x})^\top\big\} \, \mathcal{N}(x; \mathbf{x}_i, \mathbf{P}_i) &&\\
&= \sum_i p_i \int \mathrm{d}x \; \big\{(x - \mathbf{x}_i)(x - \mathbf{x}_i)^\top + (\mathbf{x}_i - \mathbf{x})(\mathbf{x}_i - \mathbf{x})^\top\big\} \, \mathcal{N}(x; \mathbf{x}_i, \mathbf{P}_i) &&\\
&= \sum_i p_i \; \big\{\mathbf{P}_i + (\mathbf{x}_i - \mathbf{x})(\mathbf{x}_i - \mathbf{x})^\top\big\} = \mathbf{P}. && \text{(A.55)}
\end{aligned}$$

Figure A.1 provides a schematic illustration of moment matching. A particular mixture density $p(x) = c_1 p_1(x) + c_2 p_2(x)$ is displayed along with the related mixture components $c_1 p_1(x), c_1 p_1(x)$ (Fig. A.1a). In Fig. A.1b the mixture $p(x)$ is compared with the Gaussian density $\mathcal{N}(x;\; \mathbf{x},\; \mathbf{P})$ with $\mathbf{x} = \mathbb{E}_p[x]$, $\mathbf{P} = \mathbb{E}_p[(x - \mathbf{x})^2]$. The bars at the bottom line indicate the relative size of the mixture coefficients c_1, c_2 in this example. Evidently, moment matching can provide a satisfactory approximation to a mixture as long as it is unimodal.

A.7 Retrodiction: Dependency Analysis

We wish to show: $p(\mathbf{x}_l|\mathbf{x}_{l+1}, Z^k) = p(\mathbf{x}_l|\mathbf{x}_{l+1}, Z^l)$. With a Markovian evolution model and under the assumptions $p(Z_k, n_k|\mathbf{x}_k, \ldots) = p(Z_k, n_k|\mathbf{x}_k)$, this follows from an application of Bayes rule:

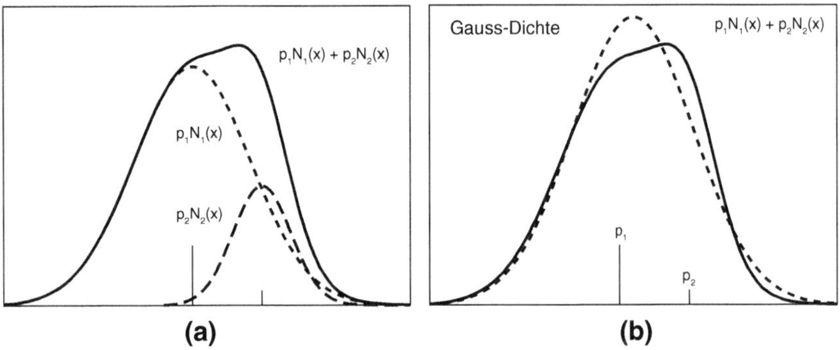

Fig. A.1 Scheme of moment matching

$$p(\mathbf{x}_l|\mathbf{x}_{l+1}, \mathcal{Z}^k) = \frac{p(Z_k, n_k, \ldots, Z_{l+1}, n_{l+1}|\mathbf{x}_{l+1}, \cancel{\mathbf{x}_l}, \mathcal{Z}^l)\ p(\mathbf{x}_l|\mathbf{x}_{l+1}, \mathcal{Z}^l)}{\int \mathrm{d}\mathbf{x}_l\ p(Z_k, n_k, \ldots, Z_{l+1}, n_{l+1}|\mathbf{x}_{l+1}, \cancel{\mathbf{x}_l}, \mathcal{Z}^l)\ p(\mathbf{x}_l|\mathbf{x}_{l+1}, \mathcal{Z}^l)} \tag{A.56}$$

and a closer look at the likelihood function $p(Z_k, n_k, \ldots, Z_{l+1}, n_{l+1}|\mathbf{x}_{l+1}, \mathbf{x}_l, \mathcal{Z}^l)$ herein:

$$p(Z_k, n_k, \ldots, Z_{l+1}, n_{l+1}|\mathbf{x}_{l+1}, \mathbf{x}_l, \mathcal{Z}^l)$$

$$= \int \mathrm{d}\mathbf{x}_k \ldots \mathrm{d}\mathbf{x}_{l+2} \times\ p(Z_k, n_k, \mathbf{x}_k, \ldots, \mathbf{x}_{l+2}, Z_{l+1}, n_{l+1}|\mathbf{x}_{l+1}, \mathbf{x}_l, \mathcal{Z}^l) \tag{A.57}$$

$$= \int \mathrm{d}\mathbf{x}_k \ldots \mathrm{d}\mathbf{x}_{l+2}\ p(Z_k, n_k|\mathbf{x}_k, \ldots) \times\ p(\mathbf{x}_k, Z_{k-1}, n_{k-1}, \mathbf{x}_{k-1}, \ldots, \mathbf{x}_{l+2}, Z_{l+1}, n_{l+1}|\mathbf{x}_{l+1}, \mathbf{x}_l, \mathcal{Z}^l) \tag{A.58}$$

$$= \int \mathrm{d}\mathbf{x}_k \ldots \mathrm{d}\mathbf{x}_{l+2}\ p(Z_k, n_k|\mathbf{x}_k)\ p(\mathbf{x}_k|\mathbf{x}_{k-1}, \ldots) \times\ p(Z_{k-1}, n_{k-1}, \mathbf{x}_{k-1}, \ldots, Z_{l+1}, n_{l+1}|\mathbf{x}_{l+1}, \mathbf{x}_l, \mathcal{Z}^l) \tag{A.59}$$

$$= p(Z_{l+1}, n_{l+1}|\mathbf{x}_{l+1}, \mathbf{x}_l, \mathcal{Z}^l) \int \mathrm{d}\mathbf{x}_k \ldots d\mathbf{x}_{l+2} \prod_{i=l+2}^{k} p(Z_i, n_i|\mathbf{x}_i)\ p(\mathbf{x}_i|\mathbf{x}_{i-1})$$

$$= p(Z_{l+1}, n_{l+1}|\mathbf{x}_{l+1}) \int \mathrm{d}\mathbf{x}_k \ldots d\mathbf{x}_{l+2} \prod_{i=l+2}^{k} p(Z_i, n_i|\mathbf{x}_i)\ p(\mathbf{x}_i|\mathbf{x}_{i-1}). \tag{A.60}$$

A.8 Gaußian Accumulated States Densities

For the accumulated object state $\mathbf{x}_{k:n} = (\mathbf{x}_k^\top, \mathbf{x}_{k-1}^\top, \ldots, \mathbf{x}_{n+1}^\top, \mathbf{x}_n^\top)^\top$, the accumulated state density $p(\mathbf{x}_{k:n}|Z^k)$ is a Gaußian given by:

$$p(\mathbf{x}_{k:n}|Z^k) = \mathcal{N}\big(\mathbf{x}_{k:n};\ \mathbf{x}_{k:n}^k,\ \mathbf{P}_{k:n}^k\big), \tag{A.61}$$

where the accumulated expectation vector $\mathbf{x}_{k:n}^k$ and the corresponding covariance matrix $\mathbf{P}_{k:n}^k$ are completely determined by the underlying evolution model, i.e. by $\mathbf{F}_{l+1|k}$ and $\mathbf{D}_{l+1|l}$, $n \leq l \leq k-1$, by the covariance matrices $\mathbf{P}_{l|l}$, $n \leq l \leq k$ provided by filtering, and by the expectation vectors $\mathbf{x}_{l|k}$ and the corresponding covariance matrices $\mathbf{P}_{l|k}$, $n \leq l \leq k$ available after retrodiction. While the expectation vector $\mathbf{x}_{k:n}^k$ of the accumulated object states $\mathbf{x}_{k:n}$ is defined by:

$$\mathbf{x}_{k:n}^k = (\mathbf{x}_{k|k}^\top, \mathbf{x}_{k-1|k}^\top, \ldots, \mathbf{x}_{n+1|k}^\top, \mathbf{x}_{n|k}^\top)^\top, \tag{A.62}$$

the corresponding covariance matrix $\mathbf{P}_{k:n}^k$ can recursively be written as:

$$\mathbf{P}_{l:n}^k = \begin{pmatrix} \mathbf{P}_{l|k} & \mathbf{P}_{l|k}\mathbf{W}_{l-1:n}^\top \\ \mathbf{W}_{l-1:n}\mathbf{P}_{l|k} & \mathbf{P}_{l-1:n}^k \end{pmatrix}, \quad n+1 \leq l \leq k \tag{A.63}$$

with $\mathbf{P}_{n:n}^k = \mathbf{P}_{n|k}$ and $\mathbf{W}_{l:n}$ given by:

$$\mathbf{W}_{l:n} = \begin{pmatrix} \mathbf{W}_{l|l+1} \\ \mathbf{W}_{l-1:n}\mathbf{W}_{l|l+1} \end{pmatrix} \tag{A.64}$$

with $\mathbf{W}_{n:n} = \mathbf{W}_{n|n+1}$ and the retrodiction gain matrices $\mathbf{W}_{l|l+1}$:

$$\mathbf{W}_{l|l+1} = \mathbf{P}_{l|l}\mathbf{F}_{l+1|l}^\top\mathbf{P}_{l+1|l}^{-1} \tag{A.65}$$

$$= \mathbf{P}_{l|l}\mathbf{F}_{l+1|l}^\top(\mathbf{F}_{l+1|l}\mathbf{P}_{l|l}\mathbf{F}_{l+1|l}^\top + \mathbf{D}_{l+1|l})^{-1}. \tag{A.66}$$

Proof

This statement directly follows from a straightforward induction argument, though the necessary calculations are perhaps somewhat tedious. Evidently, the proposition holds for $k = n$. Let us assume that it is true at time t_k. Due to the standard conditional independence and Markov assumptions regarding the measurement and evolution process, the accumulated state density at t_{k+1} can be represented by:

$$p(\mathbf{x}_{k+1:n}|Z^{k+1}) = p(\mathbf{x}_{k+1}, \mathbf{x}_{k:n}|\mathbf{z}_{k+1}, Z^k) \tag{A.67}$$

$$= \frac{p(\mathbf{z}_{k+1}|\mathbf{x}_{k+1})\; p(\mathbf{x}_{k+1}, \mathbf{x}_{k:n}|Z^k)}{\int d\mathbf{x}_{k+1:n}\; p(\mathbf{z}_{k+1}|\mathbf{x}_{k+1})\; p(\mathbf{x}_{k+1}, \mathbf{x}_{k:n}|Z^k)} \tag{A.68}$$

$$= \frac{p(\mathbf{z}_{k+1}|\mathbf{x}_{k+1})\; p(\mathbf{x}_{k+1}|\mathbf{x}_k)\; p(\mathbf{x}_{k:n}|Z^k)}{\int d\mathbf{x}_{k+1:n}\; p(\mathbf{z}_{k+1}|\mathbf{x}_{k+1})\; p(\mathbf{x}_{k+1}|\mathbf{x}_k)\; p(\mathbf{x}_{k:n}|Z^k)}. \tag{A.69}$$

With modeling assumptions compatible with Kalman filtering and using the projection matrices $\boldsymbol{\Pi}_k = (\mathbf{1}, \mathbf{O}, \ldots, \mathbf{O})$ defined by $\boldsymbol{\Pi}_k \mathbf{x}_{k:n} = \mathbf{x}_k$ and $\boldsymbol{\Pi}_{k:n} = (-\mathbf{W}_{k:n}, \mathbf{1})$ defined by $\boldsymbol{\Pi}_{k:n}(\mathbf{x}_{k+1}^\top, \mathbf{x}_{k:n}^\top)^\top = -\mathbf{W}_{k:n}\mathbf{x}_{k+1} + \mathbf{x}_{k:n}$, a repeated use of the product formula A.28 yields:

$$p(\mathbf{z}_{k+1}|\mathbf{x}_{k+1})\; p(\mathbf{x}_{k+1}|\mathbf{x}_k)\; p(\mathbf{x}_{k:n}|Z^k) \tag{A.70}$$

$$= \mathcal{N}\big(\mathbf{z}_k;\, \mathbf{H}_k\mathbf{x}_{k+1},\, \mathbf{R}_{k+1}\big)\; \mathcal{N}\big(\mathbf{x}_{k+1};\, \mathbf{F}_{k+1|k}\boldsymbol{\Pi}_k\mathbf{x}_{k:n},\, \mathbf{D}_{k+1|k}\big)$$
$$\times\; \mathcal{N}\big(\mathbf{x}_{k:n};\, \mathbf{x}_{k:n}^k,\, \mathbf{P}_{k:n}^k\big) \tag{A.71}$$

$$= \mathcal{N}\big(\mathbf{z}_k;\, \mathbf{H}_k\mathbf{x}_{k+1},\, \mathbf{R}_{k+1}\big)\; \mathcal{N}\big(\mathbf{x}_{k+1};\, \mathbf{x}_{k+1|k},\, \mathbf{P}_{k+1|k}\big) \times\; \mathcal{N}\big(\mathbf{x}_{k:n};\, \mathbf{x}_{k:n}^k$$
$$+\, \mathbf{W}_{k:n}(\mathbf{x}_{k+1} - \mathbf{x}_{k+1|k}),\, \mathbf{R}_{k:n}\big) \tag{A.72}$$

$$= \mathcal{N}\big(\mathbf{z}_k;\, \mathbf{H}_k\mathbf{x}_{k+1|k},\, \mathbf{S}_{k+1|k}\big)\; \mathcal{N}\big(\mathbf{x}_{k+1};\, \mathbf{x}_{k+1|k+1},\, \mathbf{P}_{k+1|k+1}\big)$$
$$\times\; \mathcal{N}\big(\boldsymbol{\Pi}_{k:n}\mathbf{x}_{k+1:n};\, \boldsymbol{\Pi}_{k:n}(\mathbf{x}_{k+1|k}^\top, \mathbf{x}_{k:n}^{k})^\top,\, \mathbf{R}_{k:n}\big) \tag{A.73}$$

with $\mathbf{W}_{k:n}$ and $\mathbf{R}_{k:n}$ given by:

$$\mathbf{W}_{k:n} = \mathbf{P}_{k:n}^k \boldsymbol{\Pi}_k^\top \mathbf{F}_{k+1|k}^\top \mathbf{P}_{k+1|k}^{-1} \tag{A.74}$$

$$= \begin{pmatrix} \mathbf{P}_{k|k} & \mathbf{P}_{k|k}\mathbf{W}_{k-1:n}^\top \\ \mathbf{W}_{k-1:n}\mathbf{P}_{k|k} & \mathbf{P}_{k-1:n}^k \end{pmatrix} \boldsymbol{\Pi}_k^\top \mathbf{F}_{k+1|k}^\top \mathbf{P}_{k+1|k}^{-1} \tag{A.75}$$

$$= \begin{pmatrix} \mathbf{W}_{k|k+1} \\ \mathbf{W}_{k-1:n}\mathbf{W}_{k|k+1} \end{pmatrix} \tag{A.76}$$

$$\mathbf{R}_{k:n} = \mathbf{P}_{k:n}^k - \mathbf{W}_{k:n}\mathbf{P}_{k+1|k}\mathbf{W}_{k:n}^\top \tag{A.77}$$

$$\boldsymbol{\Pi}_{k:n}^\top \mathbf{R}_{k:n}^{-1} \boldsymbol{\Pi}_{k:n} = \begin{pmatrix} -\mathbf{W}_{k:n}^\top \mathbf{R}_{k:n}^{-1} \\ \mathbf{R}_{k:n}^{-1} \end{pmatrix} \big(-\mathbf{W}_{k:n}, \mathbf{1}\big) \tag{A.78}$$

$$= \begin{pmatrix} \mathbf{W}_{k:n}^\top \mathbf{R}_{k:n}^{-1}\mathbf{W}_{k:n} & -\mathbf{W}_{k:n}^\top \mathbf{R}_{k:n}^{-1} \\ -\mathbf{R}_{k:n}^{-1}\mathbf{W}_{k:n} & \mathbf{R}_{k:n}^{-1} \end{pmatrix}. \tag{A.79}$$

By a second use of the product formula for Gaußians, we thus obtain up to a constant independent of the state vectors:

$$p(\mathbf{z}_{k+1}|\mathbf{x}_{k+1})\; p(\mathbf{x}_{k+1}|\mathbf{x}_k)\; p(\mathbf{x}_{k:n}|Z^k)$$
$$\propto \mathcal{N}\big(\mathbf{z}_k;\, \mathbf{H}_k\mathbf{x}_{k+1|k},\, \mathbf{S}_{k+1|k}\big)\; \mathcal{N}\big(\mathbf{x}_{k+1:n};\, \mathbf{x}_{k+1:}^{k+1},\, \mathbf{P}_{k+1:n}^{k+1}\big), \tag{A.80}$$

where the covariance matrix $\mathbf{P}_{k+1:n}^{k+1}$ it is given by:

$$\mathbf{P}_{k+1:n}^{k+1} = \left(\boldsymbol{\Pi}_{k+1}^{\top}\mathbf{P}_{k+1|k+1}^{-1}\boldsymbol{\Pi}_{k+1} + \boldsymbol{\Pi}_{k:n}^{\top}\mathbf{R}_{k:n}^{-1}\boldsymbol{\Pi}_{k:n}\right)^{-1} \tag{A.81}$$

$$= \begin{pmatrix} \mathbf{P}_{k+1|k+1}^{-1} + \mathbf{W}_{k:n}^{\top}\mathbf{R}_{k:n}^{-1}\mathbf{W}_{k:n} & -\mathbf{W}_{k:n}^{\top}\mathbf{R}_{k:n}^{-1} \\ -\mathbf{R}_{k:n}^{-1}\mathbf{W}_{k:n} & \mathbf{R}_{k:n}^{-1} \end{pmatrix}^{-1}. \tag{A.82}$$

This block matrix can directly be inverted by using Eq. A.24. The corresponding Schur Complement is particularly simple and given by (Eq. A.26):

$$\mathbf{T} = \mathbf{P}_{k+1|k+1}^{-1} + \mathbf{W}_{k:n}^{\top}\mathbf{R}_{k:n}^{-1}\mathbf{W}_{k:n} - \mathbf{W}_{k:n}^{\top}\mathbf{R}_{k:n}^{-1}\mathbf{R}_{k:n}\mathbf{R}_{k:n}^{-1}\mathbf{W}_{k:n} \tag{A.83}$$

$$= \mathbf{P}_{k+1|k+1}^{-1}. \tag{A.84}$$

We thus obtain for the covariance matrix of the ASD:

$$\mathbf{P}_{k+1:n}^{k+1} = \begin{pmatrix} \mathbf{P}_{k+1|k+1} & \mathbf{P}_{k+1|k+1}\mathbf{W}_{k:n}^{\top} \\ \mathbf{W}_{k:n}\mathbf{P}_{k+1|k+1} & \mathbf{R}_{k:n} + \mathbf{W}_{k:n}\mathbf{P}_{k+1|k+1}\mathbf{W}_{k:n}^{\top} \end{pmatrix} \tag{A.85}$$

$$= \begin{pmatrix} \mathbf{P}_{k+1|k+1} & \mathbf{P}_{k+1|k+1}\mathbf{W}_{k:n}^{\top} \\ \mathbf{W}_{k:n}\mathbf{P}_{k+1|k+1} & \mathbf{P}_{k:n}^{k} + \mathbf{W}_{k:n}(\mathbf{P}_{k+1|k+1} - \mathbf{P}_{k+1|k})\mathbf{W}_{k:n}^{\top} \end{pmatrix}. \tag{A.86}$$

Using the identity $\mathbf{W}_{k|k+1}(\mathbf{P}_{k+1|k+1} - \mathbf{P}_{k+1|k})\mathbf{W}_{k|k+1}^{\top} = \mathbf{P}_{k|k+1} - \mathbf{P}_{k|k}$ resulting from the Rauch-Tung-Striebel equations, the matrix $\mathbf{W}_{k:n}(\mathbf{P}_{k+1|k+1} - \mathbf{P}_{k+1|k})\mathbf{W}_{k:n}^{\top}$ can be transformed yielding:

$$\mathbf{W}_{k:n}(\mathbf{P}_{k+1|k+1} - \mathbf{P}_{k+1|k})\mathbf{W}_{k:n}^{\top} \tag{A.87}$$

$$= \begin{pmatrix} \mathbf{W}_{k|k+1}(\mathbf{P}_{k+1|k+1} - \mathbf{P}_{k+1|k}) \\ \mathbf{W}_{k-1:n}\mathbf{W}_{k|k+1}(\mathbf{P}_{k+1|k+1} - \mathbf{P}_{k+1|k}) \end{pmatrix} \begin{pmatrix} \mathbf{W}_{k|k+1}^{\top} & \mathbf{W}_{k|k+1}^{\top}\mathbf{W}_{k-1:n}^{\top} \end{pmatrix}$$

$$= \begin{pmatrix} \mathbf{P}_{k|k+1} - \mathbf{P}_{k|k} & (\mathbf{P}_{k|k+1} - \mathbf{P}_{k|k})\mathbf{W}_{k-1:n}^{\top} \\ \mathbf{W}_{k-1:n}(\mathbf{P}_{k|k+1} - \mathbf{P}_{k|k}) & \mathbf{W}_{k-1:n}(\mathbf{P}_{k|k+1} - \mathbf{P}_{k|k})\mathbf{W}_{k-1:n}^{\top} \end{pmatrix}. \tag{A.88}$$

With this result, the block matrix $\mathbf{P}_{k:n}^{k} + \mathbf{W}_{k:n}(\mathbf{P}_{k+1|k+1} - \mathbf{P}_{k+1|k})\mathbf{W}_{k:n}^{\top}$ on the right-lower corner on the right side of Eq. A.86 is given by:

$$\mathbf{P}_{k:n}^{k} + \mathbf{W}_{k:n}(\mathbf{P}_{k+1|k+1} - \mathbf{P}_{k+1|k})\mathbf{W}_{k:n}^{\top} \tag{A.89}$$

$$= \begin{pmatrix} \mathbf{P}_{k|k+1} & \mathbf{P}_{k|k+1}\mathbf{W}_{k-1:n}^{\top} \\ \mathbf{W}_{k-1:n}\mathbf{P}_{k|k+1} & \mathbf{P}_{k-1:n}^{k} + \mathbf{W}_{k-1:n}(\mathbf{P}_{k|k+1} - \mathbf{P}_{k|k})\mathbf{W}_{k-1:n}^{\top} \end{pmatrix}. \tag{A.90}$$

An induction argument for the block matrix on the right-lower corner directly yields:

$$\mathbf{P}_{k:n}^{k} + \mathbf{W}_{k:n}(\mathbf{P}_{k+1|k+1} - \mathbf{P}_{k+1|k})\mathbf{W}_{k:n}^{\top} = \mathbf{P}_{k:n}^{k+1}. \tag{A.91}$$

In the sequel, the expectation vector $\mathbf{x}_{k+1:n}^{k+1}$ will be calculated. According to the product formula A.28, $\mathbf{x}_{k+1:n}^{k+1}$ is the sum of the following vectors:

$$\mathbf{P}_{k+1:n}^{k+1}\boldsymbol{\Pi}_{k:n}^{\top}\mathbf{R}_{k:n}^{-1}\boldsymbol{\Pi}_{k:n}(\mathbf{x}_{k+1|k}^{\top},\mathbf{x}_{k:n}^{k\top})^{\top} \tag{A.92}$$

$$= \begin{pmatrix} \mathbf{P}_{k+1|k+1} & \mathbf{P}_{k+1|k+1}\mathbf{W}_{k:n}^{\top} \\ \mathbf{W}_{k:n}\mathbf{P}_{k+1|k+1} & \mathbf{R}_{k:n}+\mathbf{W}_{k:n}\mathbf{P}_{k+1|k+1}\mathbf{W}_{k:n}^{\top} \end{pmatrix} \begin{pmatrix} -\mathbf{W}_{k:n}^{\top} \\ \mathbf{1} \end{pmatrix}$$

$$\times \left(-\mathbf{R}_{k:n}^{-1}\mathbf{W}_{k:n} \;\; \mathbf{R}_{k:n}^{-1}\right) \begin{pmatrix} \mathbf{x}_{k+1|k} \\ \mathbf{x}_{k:n}^{k} \end{pmatrix} \tag{A.93}$$

$$= \begin{pmatrix} \mathbf{O} \\ \mathbf{R}_{k:n} \end{pmatrix} \left(-\mathbf{R}_{k:n}^{-1}\mathbf{W}_{k:n} \;\; \mathbf{R}_{k:n}^{-1}\right) \begin{pmatrix} \mathbf{x}_{k+1|k} \\ \mathbf{x}_{k:n}^{k} \end{pmatrix} \tag{A.94}$$

$$= \begin{pmatrix} \mathbf{O} & \mathbf{O} \\ -\mathbf{W}_{k:n} & \mathbf{1} \end{pmatrix} \begin{pmatrix} \mathbf{x}_{k+1|k} \\ \mathbf{x}_{k:n}^{k} \end{pmatrix} \tag{A.95}$$

$$= \begin{pmatrix} \mathbf{O} \\ -\mathbf{W}_{k:n}\mathbf{x}_{k+1|k}+\mathbf{x}_{k:n}^{k} \end{pmatrix} \tag{A.96}$$

$$\mathbf{P}_{k+1:n}^{k+1}\boldsymbol{\Pi}_{k+1}^{\top}\mathbf{P}_{k+1|k+1}^{-1}\boldsymbol{\Pi}_{k+1}\mathbf{x}_{k+1|k+1} \tag{A.97}$$

$$= \begin{pmatrix} \mathbf{P}_{k+1|k+1} \\ \mathbf{W}_{k:n}\mathbf{P}_{k+1|k+1} \end{pmatrix} \mathbf{P}_{k+1|k+1}^{-1}\mathbf{x}_{k+1|k+1} \tag{A.98}$$

$$= \begin{pmatrix} \mathbf{x}_{k+1|k+1} \\ \mathbf{W}_{k:n}\mathbf{x}_{k+1|k+1} \end{pmatrix}. \tag{A.99}$$

By using an induction argument, we thus obtain:

$$\mathbf{x}_{k+1:n}^{k+1} = \begin{pmatrix} \mathbf{x}_{k+1|k+1} \\ \mathbf{x}_{k:n}^{k}+\mathbf{W}_{k:n}(\mathbf{x}_{k+1|k+1}-\mathbf{x}_{k+1|k}) \end{pmatrix} \tag{A.100}$$

$$= \begin{pmatrix} \mathbf{x}_{k+1|k+1} \\ \mathbf{x}_{k|k}+\mathbf{W}_{k|k+1}(\mathbf{x}_{k+1|k+1}-\mathbf{x}_{k+1|k}) \\ \mathbf{x}_{k-1:n}^{k}+\mathbf{W}_{k-1:n}\mathbf{W}_{k|k+1}(\mathbf{x}_{k+1|k+1}-\mathbf{x}_{k+1|k}) \end{pmatrix} \tag{A.101}$$

$$= \begin{pmatrix} \mathbf{x}_{k+1|k+1} \\ \mathbf{x}_{k|k+1} \\ \mathbf{x}_{k-1:n}^{k}+\mathbf{W}_{k-1:n}(\mathbf{x}_{k|k+1}-\mathbf{x}_{k|k}) \end{pmatrix}. \tag{A.102}$$

An induction argument concludes the proof.

A.9 Some Facts on Kronecker Products

The Kronecker product $\mathbf{A}\otimes\mathbf{B}$ of two matrices $\mathbf{A}=(a_{ij})_{i=1,\,j=1}^{m,n}$, $\mathbf{B}$ is defined by:

$$\mathbf{A}\otimes\mathbf{B} = \begin{pmatrix} a_{11}\mathbf{B} & a_{12}\mathbf{B} & \cdots & a_{1n}\mathbf{B} \\ a_{21}\mathbf{B} & a_{22}\mathbf{B} & \cdots & a_{2n}\mathbf{B} \\ \vdots & \vdots & & \vdots \\ a_{m1}\mathbf{B} & a_{m2}\mathbf{B} & \cdots & a_{mn}\mathbf{B} \end{pmatrix}. \tag{A.103}$$

For matrices **A**, **B**, **C** and a scalar α (e.g. [2]):

$$(\mathbf{A} \otimes \mathbf{B}) \otimes \mathbf{C} = \mathbf{A} \otimes (\mathbf{B} \otimes \mathbf{C}) \tag{A.104}$$
$$\alpha \otimes \mathbf{A} = \alpha\mathbf{A} = \mathbf{A}\alpha = \mathbf{A} \otimes \alpha \tag{A.105}$$
$$(\mathbf{A} \otimes \mathbf{B})(\mathbf{C} \otimes \mathbf{D}) = \mathbf{AC} \otimes \mathbf{BD} \tag{A.106}$$
$$(\mathbf{A} \otimes \mathbf{B})^\top = \mathbf{A}^\top \otimes \mathbf{B}^\top \tag{A.107}$$
$$(\mathbf{A} \otimes \mathbf{B})^{-1} = \mathbf{A}^{-1} \otimes \mathbf{B}^{-1}. \tag{A.108}$$

For quadratic matrices **A**, **B** we obtain:

$$\mathrm{tr}[\mathbf{A} \otimes \mathbf{B}] = (\mathrm{tr}\mathbf{A})\,(\mathrm{tr}\mathbf{B}). \tag{A.109}$$

The determinant of $\mathbf{A} \otimes \mathbf{B}$ is given by the determinants of **A**, **B** with $m = \dim(\mathbf{A})$, $n = \dim(\mathbf{B})$:

$$|\mathbf{A} \otimes \mathbf{B}| = |\mathbf{A}|^n |\mathbf{B}|^m. \tag{A.110}$$

A.10 Extended Object Likelihood: Details

For column vectors **x**, **y** of equal dimension, the following identities are valid:

$$\mathbf{x}^\top \mathbf{y} = \mathrm{tr}\Big[\mathbf{x}\mathbf{y}^\top\Big] \tag{A.111}$$
$$\mathbf{x}^\top \mathbf{A}^{-1}\mathbf{x} = \mathrm{tr}\Big[\mathbf{x}\mathbf{x}^\top \mathbf{A}^{-1}\Big] = \mathrm{tr}\Big[\mathbf{A}^{-1}\mathbf{x}\mathbf{x}^\top\Big] \tag{A.112}$$
$$\exp\Big[-\tfrac{1}{2}\mathbf{x}^\top \mathbf{A}^{-1}\mathbf{x}\Big] = \mathrm{etr}\Big[-\tfrac{1}{2}\mathbf{x}\mathbf{x}^\top \mathbf{A}^{-1}\Big] \tag{A.113}$$
$$\Big|\mathbf{1} + \mathbf{x}\mathbf{y}^\top\Big| = 1 + \mathbf{x}^\top \mathbf{y}. \tag{A.114}$$

etr[**A**] is an abbreviation for exp[tr**A**]. For proofs see e.g. [2].

By applying the product formula for Gaussians (Eq. A.28) repeatedly we obtain:

$$p(Z_k|n_k, \mathbf{x}_k, \mathbf{X}_k)$$
$$= \mathcal{N}\big(\mathbf{z}_k;\, \mathbf{H}_k\mathbf{x}_k,\, \tfrac{1}{n_k}\mathbf{X}_k\big) \prod_{i=1}^{n_k-1} \mathcal{N}\big(\mathbf{z}_k^{i+1};\, \bar{\mathbf{z}}_k^i,\, \tfrac{i+1}{i}\mathbf{X}_k\big) \tag{A.115}$$

with the following quantities:

$$\bar{\mathbf{z}}_k^i = \tfrac{1}{i}\textstyle\sum_{j=1}^i \mathbf{z}_k^j, \quad \mathbf{z}_k = \bar{\mathbf{z}}_k^{n_k} = \tfrac{1}{n_k}\textstyle\sum_{j=1}^i \mathbf{z}_k^j. \tag{A.116}$$

Only the first factor of the right side of this equation depends on the kinematic state variable $\mathbf{x}_k$. The remaining $n_k - 1$ factors are functions of the extension $\mathbf{X}_k$ alone. An induction argument yields:

$$\prod_{i=1}^{n_k-1} \mathcal{N}\big(\mathbf{z}_k^{i+1};\ \bar{\mathbf{z}}_k^i,\ \tfrac{i+1}{i}\mathbf{X}_k\big) \propto |\mathbf{X}_k|^{-\frac{n_k-1}{2}} \operatorname{etr}\Big[-\tfrac{1}{2}\mathbf{Z}_k\mathbf{X}_k^{-1}\Big] \tag{A.117}$$

$$\propto \mathcal{LW}\big(\mathbf{Z}_k;\ n_k - 1,\ \mathbf{X}_k\big). \tag{A.118}$$

A.11 Facts on Matrix-variate Densities

Wishart Density

A $d \times d$ SPD random matrix $\mathbf{X}$ is "Wishart-distributed", if its density is given by [1, p. 87]:

$$\mathcal{W}\big(\mathbf{X};\ a,\ \mathbf{A}\big) = \tfrac{1}{Z}\ |\mathbf{A}|^{-\frac{1}{2}a}\ |\mathbf{X}|^{\frac{1}{2}(a-d-1)} \operatorname{etr}\Big[-\tfrac{1}{2}\mathbf{A}^{-1}\mathbf{X}\Big],\quad a \geq d \tag{A.119}$$

with a scalar parameter a, a $d \times d$ SPD matrix $\mathbf{A}$, and a normalizing constant Z. Its first and second moments are given by [1, p. 98]:

$$\mathbb{E}[\mathbf{X}] = a\,\mathbf{A} \tag{A.120}$$

$$\mathbb{C}\big[x_{ij}, x_{kl}\big] = a\,(a_{ik}a_{jl} + a_{il}a_{jk}) \tag{A.121}$$

with $\mathbf{A} = \begin{pmatrix} a_{11} & a_{12} \\ a_{12} & a_{22} \end{pmatrix}$, $\mathbf{X} = \begin{pmatrix} x_{11} & x_{12} \\ x_{12} & x_{22} \end{pmatrix}$. As a scalar measure for the second moments, we consider

$$\mathbb{E}\Big[(x_{11})^2\Big] + \mathbb{E}\Big[(x_{22})^2\Big] = 2a\big((a_{11})^2 + (a_{22})^2\big). \tag{A.122}$$

Inverted Wishart Density

A $d \times d$ SPD random matrix $\mathbf{X}$ is 'inverted-Wishart-distributed', if its density is given by [1, p. 111]:

$$\mathcal{IW}\big(\mathbf{X};\ a,\ \mathbf{A}\big) = \tfrac{1}{Z}\ |\mathbf{A}|^{\frac{1}{2}(a-d-1)}\ |\mathbf{X}|^{-\frac{1}{2}a} \operatorname{etr}\Big[-\tfrac{1}{2}\mathbf{A}\mathbf{X}^{-1}\Big],\quad a > 2d. \tag{A.123}$$

with a scalar parameter a, a $d \times d$ SPD parameter matrix $\mathbf{A}$, and a normalizing constant Z. Its first and second moments are given by [1, p. 113]:

$$\mathbb{E}[\mathbf{X}] = \mathbf{A}/(a - 2d - 2),\quad a - 2d - 2 > 0 \tag{A.124}$$

$$\mathbb{C}\big[x_{ij}, x_{kl}\big] = \frac{2(a - 2d - 2)^{-1}a_{ij}a_{kl} + a_{ik}a_{jl} + a_{il}a_{kj}}{(a - 2d - 1)(a - 2d - 2)(a - 2d - 4)}, \tag{A.125}$$

$a - 2d - 4 > 0$. As a scalar measure for the second moments, we consider

$$\mathbb{C}\left[(x_{11})^2\right] + \mathbb{C}\left[(x_{22})^2\right] = \frac{2(a-2d-2)^{-1}+2}{(a-2d-1)(a-2d-2)(a-2d-4)}\left((a_{11})^2 + (a_{22})^2\right). \quad \text{(A.126)}$$

The following is obviously true: if $\mathbf{X}$ is an Inverted Wishart distributed random matrix with the density $p(\mathbf{X}) \propto |\mathbf{X}|^{-a/2}\ \mathrm{etr}[-\frac{1}{2}\mathbf{X}^{-1}\mathbf{A}]$, then the inverse matrix $\mathbf{Y} = \mathbf{X}^{-1}$ is Wishart-distributed with $p(\mathbf{Y}) \propto |\mathbf{Y}|^{(a-d-1)/2}\ \mathrm{etr}[-\frac{1}{2}\mathbf{YA}]$. This is due to the fact that the Jacobi matrix of the inverse transformation is given by $J(\mathbf{X} \to \mathbf{X}^{-1}) = |\mathbf{X}|^{d+1}$ [1, p. 14]. This, however, is a Wishart density given by $p(\mathbf{Y}) = \mathcal{W}(\mathbf{Y};\ b,\ \mathbf{B})$ with parameters $\mathbf{b} = a - d - 1$, $\mathbf{B} = \mathbf{A}^{-1}$.

Beta Density
A $d \times d$ SPD random matrix $\mathbf{X}$ is 'generalized-beta-type-II-distributed', if its density is given by [1, p. 167]:

$$\mathcal{GB}_d^{\mathrm{II}}(\mathbf{X}; a, b, \mathbf{A}, \mathbf{B} = \mathbf{O}) := \mathcal{B}(\mathbf{X};\ a,\ b,\ \mathbf{A}) \quad \text{(A.127)}$$

$$:= \tfrac{1}{Z}\ |\mathbf{A}|^{b}\ |\mathbf{X}|^{a-\frac{d+1}{2}}\ |\mathbf{A}+\mathbf{X}|^{-(a+b)}. \quad \text{(A.128)}$$

with scalar parameters a, b, a $d \times d$ SPD parameter matrix $\mathbf{A}$, and a normalizing constant Z. Its first and second moments are given by [1, p. 179]:

$$\mathbb{E}[\mathbf{X}] = \tfrac{2a}{2b-d-1}\,\mathbf{A}, \quad 2b-d-1>0 \quad \text{(A.129)}$$

$$\mathbb{C}\left[x_{ij}x_{kl}\right] = 2a\,\frac{(2a(2b-d-2)+2)a_{ij}a_{kl}+(2(a+b)-d-1)(a_{jl}a_{ik}+a_{il}a_{kj})}{(2b-d)(2b-d-1)(2b-d-3)} \quad \text{(A.130)}$$

with $2b - d - 3 > 0$. As a scalar measure for the second moments we consider

$$\mathbb{E}\left[(x_{11})^2\right] + \mathbb{E}\left[(x_{22})^2\right] = \tfrac{4a(2b-d)(a+1)}{(2b-d)(2b-d-1)(2b-d-3)}\left((a_{11})^2 + (a_{22})^2\right). \quad \text{(A.131)}$$

References

1. A.K. Gupta, D.K. Nagar, *Matrix Variate Distributions* (Chapman & Hall/CRC, New York, 1999)
2. D.A. Harville, *Matrix Algebra from a Statistician's Perspective* (Springer, New York, 1997)